Handbook of Quality Control

Handbook of Quality Control

Edited by **Theresa Heen**

CLANRYE INTERNATIONAL

New Jersey

Published by Clanrye International,
55 Van Reypen Street,
Jersey City, NJ 07306, USA
www.clanryeinternational.com

Handbook of Quality Control
Edited by Theresa Heen

© 2015 Clanrye International

International Standard Book Number: 978-1-63240-285-1 (Hardback)

Printed in the United States of America.

Contents

Permissions

List of Contributors

Preface

This book was inspired by the evolution of our times; to answer the curiosity of inquisitive minds. Many developments have occurred across the globe in the recent past which has transformed the progress in the field.

Quality control is a procedure or set of procedures formulated for the purpose of maintaining standards in manufactured products. Rapid development has been made in the past decade in the area of quality control methods and techniques. The objective of this book is to illustrate quality control procedures in several fields, like pharmaceutics and medicine, construction engineering and data quality. A wide range of techniques and methods have been presented in this book.

This book was developed from a mere concept to drafts to chapters and finally compiled together as a complete text to benefit the readers across all nations. To ensure the quality of the content we instilled two significant steps in our procedure. The first was to appoint an editorial team that would verify the data and statistics provided in the book and also select the most appropriate and valuable contributions from the plentiful contributions we received from authors worldwide. The next step was to appoint an expert of the topic as the Editor-in-Chief, who would head the project and finally make the necessary amendments and modifications to make the text reader-friendly. I was then commissioned to examine all the material to present the topics in the most comprehensible and productive format.

I would like to take this opportunity to thank all the contributing authors who were supportive enough to contribute their time and knowledge to this project. I also wish to convey my regards to my family who have been extremely supportive during the entire project.

Editor

Part 1

Quality Control
for Medical Research and Process

Quality Control and Quality Assurance in Human Experimentation

Stahl, Edmundo
LatAmScience, LLC
U.S.A.

1. Introduction

During the 20th, century the awareness of the need for the ethical treatment of human subjects participating in experimentation has evolved. Various incidents over the years have sparked the creation of government entities dedicated to the regulation of human experimentation. This has brought about the creation of regulations whose objective is the protection human subjects throughout the experimentation process. These regulations call for many checks and balances with the objective of protecting the individual under experimentation through quality control procedures in the monitoring process of the experiment. The quality is assured through auditing the process by independent professionals. This chapter will describe the history of the development of Good Clinical Practices (GCP) and an analysis of some applicable documents and practices developed by the Food and Drug Administration of the USA, and the International Conference on Harmonization of Technical Requirements for Registration of Pharmaceuticals for Human Use (ICH). FDA (USA), EMA (EU) and Pharmaceutics and Medicines Safety Bureau (Japan) as well as pharmaceutical industry representatives of the USA, EU and Japan form the ICH. ICH guidelines provide a unified standard for designing, conducting, recording and reporting clinical trials involving the participation of human subjects and other necessary activities related to human experimentation. ICH is especially concerned with harmonizing the regulatory requirements of its sponsor countries; USA, EU and Japan. It describes the necessary activities and documentation that would allow the evaluation of the ethical conduct of a clinical trial and assure the quality of the information derived from such a study. Many countries all over the world are now including these guidelines in their regulations and are effectively adhering to them.

A significant part of human experimentation is conducted in the development of new drugs for the treatment of human disease as well as devices and instruments used in medical practice. This chapter will also describe the development process, the logic behind it, the non-clinical testing that is necessary for the drug/device development process, the clinical phases of drug development, the role of the ethics committees and Institutional Review Boards in the approval process to conduct human experimentation as well as the role of the government agencies which regulate human experimentation.

2. Evolution of ethical conduct in human experimentation

Since the 5th century B.C. the most prevalent code of ethical conduct for the medical profession has been and still is the "Hippocratic Oath"[1]. It is widely believed to have been written by Hippocrates, often regarded as the father of western medicine. The original text of the Hippocratic Oath is usually interpreted as one of the first statements of a moral of conduct to be used by physicians, assuming the respect for all human life. It has been modified over time in many occasions but the spirit of the concept has been preserved.

It was not until after the end of World War II that the United States authorities conducted in their occupied zone several trials for war crimes committed by the Nazis in Nüremberg[2]. The trials were formally named the "Trials of War Criminals before the Nuremberg Military Tribunals". They were held before US military courts, not before the International Military Tribunal. The defendants were accused of unethical human experimentation and other atrocities. On August 19th, 1947 the tribunal delivered its verdict including their opinion on human experimentation. The Nüremberg Code that emerged from these trials consists of 10 points that represent a set of ethical research principles for human experimentation. The Nüremberg Code includes concepts like: voluntary consent of the research subject; experimentation with clear fruitful objectives; experimentation in humans should be preceded by animal experimentation; the conduct of research in humans should not produce physical or mental injury nor results in death of the study subjects; the experimentation should be conducted with a view of introducing the minimal possible risk to the individual during the experimentation and conducted by a qualified person. It also includes the concept that the subject should be at liberty to stop the experiment at any time for any reason. Likewise, the experimenter should be prepared to terminate the experiment if in their judgment there is any reasonable chance that it may harm the research subject. Subsequently in 1948 the World Medical Association introduced the Declaration of Geneva [3] as a modernization of the Hippocratic Oath. It was designed as a formulation of that oath's moral truths that could be comprehended and acknowledged modernly. The Declaration of Geneva has been amended in 1968, 1984, 1994, 2005 and 2006.

Another important historical document addressing human experimentation is the Declaration of Helsinki[4] adopted in 1964 by the World Medical Association in Helsinki, Finland. It is a set of ethical principles for the medical community specifically related to human experimentation and is widely regarded as a cornerstone document for human research. It has been revised six times since its adoption, the last revision being in 2008. The Declaration of Helsinki adopted the ten principles first stated in the Nüremberg Code and tied them to the Declaration of Geneva. It addresses clinical research reflecting the changes in medical practice. Its various revisions introduced the concept of independent review committees, now known as Institutional Review Board or Independent Ethics Committees; the management of the inclusion of minors in clinical research and the recognition of vulnerable groups; addressed the use of placebos; and the inclusion of human volunteers in clinical trials. This document was not meant to be legally binding but has influenced national and regional regulation and legislation around the world. It introduced the concept that ethical considerations must take precedence over laws and regulations.

In the USA, the Belmont Report[5] was created by the now named Department of Health and Human Services with the title "Ethical Principles and Guidelines for the Protection of Human Subjects of Research". The report was issued in April 1979 prompted in part by problems arising from the Tuskegee Syphilis Study (1932-1972). The Tuskegee Syphilis Study was designed to observe the clinical evolution of syphilis. The patients, 399 impoverished Black individuals from Macon county, Alabama, who thought they were receiving free health care from the government were never told they had syphilis nor were they treated for it. The Belmont Report incorporates the principles of the Nuremberg Code, the Declaration of Geneva and the Declaration of Helsinki. These documents influenced significantly the legislation and creation of regulations for the ethical conduct of human experimentation in the USA. Sections 45 (government sponsored studies, 45 CFR) and 21 (private and industry sponsored studies, 21 CFR) of the Code of Federal Regulations (CFR) base many of their regulations on these important ethical documents and have influenced in many important ways human experimentation in the US and around the world.

3. Regulatory environment

The U.S. Food and Drug Administration (FDA) [6] was created in 1906 by the Federal Food, Drug, and Cosmetic Act, the Wiley Act. The purpose was to prevent the manufacturing, sale, or transportation of adulterated or misbranded or poisonous or deleterious foods, drugs, medicines, and liquors. The FDA evolved over the years to require manufacturers to submit a New Drug Application (NDA) for each newly introduced drug and provide data that demonstrates the safety of the product (1938), and later (1962) to establish efficacy, in order to show that the products were effective for their claimed indication. Several amendments to the law have followed to reflect the evolution and emerging issues in the drug development and approval process, which remain today in the crossroads where science, medicine, politics and business intersect. Because a new drug approval is based largely on clinical data obtained by experiments in humans, the FDA has vested significant effort in ensuring the quality of the clinical data and the conditions under which they are obtained. The set of regulations and guidelines the FDA publishes constitute what is collectively known as good clinical practices or GCP. Through these FDA sets the minimum standards for the conduct of clinical trials, the collection of data and data management and reporting of clinical studies.

The European Medicine Agency, EMA[7] (formerly known as EMEA, European Agency for the Evaluation of Medicinal Products), was founded in 1995 and is a decentralized agency of the European Union responsible for the scientific evaluation of medicines developed by pharmaceutical companies for use in the European Union. Its main function is the promotion and protection of public human and animal health, through the evaluation and supervision of medicines for human and animal use. They are responsible for the evaluation of European Marketing Authorizations for human and veterinary use medicines. The agency monitors the safety of marketed products and provides scientific advice to companies on the development of new medicines. The agency constantly works to forge close ties with partner organizations around the world, including the World Health Organization, the FDA and the other regulatory authorities.

In Japan the Pharmaceuticals and Medical Devices Agency (PMDA) [8] working with the Ministry of Health implements measures for securing the efficacy and safety of drugs, cosmetics and medical devices. The PMDA also has forged close ties with other regulatory agencies, namely the EMA and FDA as partners in the formation of the International Conference on Harmonization (ICH).

The International Conference on Harmonization of Technical Requirements for Registration of Pharmaceuticals for Human Use (ICH) [9] was created in 1990. This organization brings together the regulatory authorities and pharmaceutical industry of Europe, Japan and the

US. The purpose is to discuss scientific and technical aspects of drug registration with the goal of harmonizing drug development and registration across the world. The Global Cooperation Group of this organization has been working to harmonize the increasingly global approach to drug development, so that the benefits of international harmonization for better global health can be realized worldwide. ICH's mission is to achieve greater harmonization to ensure that safe, effective, and high quality medicines are developed and registered in the most resource-efficient manner. ICH has developed a series of guidelines to help regulate the clinical drug development process. The instruments developed include a standardized medical terminology system (MedDRA) to use in the capturing, registering, documenting and reporting adverse events during human experimentation. ICH maintains, develops and distributes MedDRA. ICH also developed a standardized package for the submission of new drug applications, the Common Technical Document (CTD) which has been adopted around the world as the gold standard for new medical products submissions. The CTD assembles all the quality, safety and efficacy information required for regulatory submissions in a common format to facilitate the regulatory review process. The CTD has simplified the assembly of the regulatory packages since reformatting for submissions to different regulatory agencies is not necessary anymore. Another area where ICH has worked to improve is international electronic communication by evaluating and recommending Electronic Standards for the Transfer of Regulatory Information (ESTRI). ESTRI has developed recommendations for electronic individual case safety reports and electronic CTDs. Furthermore, ICH has developed guidelines to standardize and harmonize the areas involved in the drug development process. These include guidelines on quality of the product being developed; on safety focusing on nonclinical studies to uncover potential risks for humans; on efficacy, which is concerned with the design, conduct and safety clinical trials; and other guideless like the ones developed by ESTRI. The clinical guidelines include clinical safety (E1-2), clinical study reports (E3), dose response studies (E4), ethnic factors (E5), Good Clinical Practice (E6), clinical trials (E7-11), clinical evaluation by therapeutic category (12), clinical evaluation (E14), and pharmacogenomics (E15-16). The GCP (E6) document describes the responsibilities and expectations of all participants in the conduct of clinical trials, including investigators, monitors, sponsors and IRBs. It is one of several GCP guidelines published by various organizations, like the World Health Organization (WHO), Pan American Health Organization (PAHO), FDA and EMA. ICH's E6 is the GCP guideline most commonly accepted worldwide. It has been adopted by many countries in their regulations and accepted as the gold standard of GCP for clinical drug development. We will discuss E6 in more detail.

4. Nonclinical drug testing

Animal testing is an imperfect predictor of drug activity in humans. It constitutes the best practical experimental models for identifying and measuring the pharmacologic activity of the drug and predicting its effects in humans. *In vivo* and *in vitro* animal testing is the first major activity in the drug development process. The purpose is to characterize the toxicology, pharmacokinetic activity, and pharmacological activity of the candidate compounds prior to administration to human beings. FDA (21CFR58) as well as ICH (S guidelines) have developed standards for such testing. Initially, short-term effects are evaluated to decide if the drug is sufficiently safe for administration to humans and at what dose should the human testing start. As the drug development in human beings progresses, additional animal studies are conducted. These animal studies include long-term drug administration, and specialized animal tests are conducted to support longer administration in humans. These experiments allow the observation of drug effects that would be impractical or unethical to study in humans. Researchers can observe the effects of the compound over the lifespan of an animal, test dose responses and maximum doses; assess the effects on reproduction, pregnancy and the embryos; effects on genes; assess potential for carcinogenicity; evaluate mechanisms of action of the drug; and characterize the site, degree and duration of action of the compound. Regulatory agencies are involved in determining the amount and type of animal testing required to initiate drug development in humans as well as the requirements to support the whole clinical development program.

The regulatory agencies, specifically FDA, set the minimum standards for laboratories conducting these nonclinical tests through the publication and enforcement of Good Laboratory Practice (GLP)[10]. To ensure the quality and integrity of the data derived, nonclinical laboratories are required to implement quality systems to conduct their experiments and to abide by the animal welfare laws of the country. GLPs establish basic standards for the conduct and reporting of non clinical safety testing, including the organization of the laboratory, personnel qualifications, physical structure of the facility, equipment, maintenance procedures, and operating procedures. It requires the use of a written protocol and its structure, including its purpose, who is sponsoring the study, procedure for identification and evaluation of the test animals or test system. GLP details how to report nonclinical studies, the storage and retrieval of records and data, and the retention of records. FDA conducts inspections to monitor compliance with GLP requirements. Nonclinical laboratories may be disqualified if the laboratory facility fails to comply with the regulations, and the noncompliance affects adversely the results of the study.

In addition, FDA may provide advice to sponsors on the adequacy of the nonclinical testing plans before animal testing has begun, and evaluate independently the results and conclusions of the nonclinical testing. FDA has developed guidances for nonclinical testing also. Other regulatory authorities, namely European Community and Japan, have also developed their testing standards. ICH has stepped in in an effort to harmonize these standards with the Safety guidelines (S).

The basic toxicology studies undertaken to identify and measure a drug's adverse effects in the short- and long-term may include any or all of the studies shown in table 1 depending on the drug, intended use and duration of exposure in clinical trials (Table 1).

The responsibility of the conduct of these animal experiments falls on the sponsor, the animal laboratory and the regulatory authorities. Quality systems are required to guarantee the quality of the data generated. The Sponsor monitors the study and conducts audits, the laboratory needs to have proper standard operating procedures and guidelines in accordance with the regulations and prevailing laws as well as a quality group to ensure compliance with said regulations and laws, and the regulatory authorities perform inspections to make sure the regulations and laws are being complied with.

5. The clinical phases of drug development

In the FDA regulations and regulations by health authorities around the world accept 3 phases of drug development[11]. A fourth phase is frequently included during the post

Acute toxicity studies	Measure the short-term adverse effects of one of more doses administered over no more than 24 hours. Provide information on appropriate dosage for multiple dose studies, potential target organs, timeline of drug induced effects, species specific toxicity, potential acute toxicity in humans and estimate the safe acute dose for humans.
Subacute or subchronic toxicity studies	Evaluate toxic potential over 14 to 90 days depending on the proposed clinical indication and duration of exposure. They are designed to assess the progression and regression of drug induced lesions.
Chronic toxicity studies	Determine the risk in relation to the anticipated dose and duration of treatment, potential target organs, reversibility of observed toxicity and the no observed effect level. These studies last 180 days to 1 year of exposure.
Carcinogenicity studies	To observe the generation of malignant tumors in animals. Generally they are required for drugs which are intended to be used for chronic conditions for 6 months or more, or to be intermittently used over the years for chronic or intermittent conditions. These studies are usually in rodents and last for 2 years.
Special toxicity studies	These are studies appropriate for specific formulations, route of administration, or conducted in particular animal models relevant to a human condition, disease or age. They include immunotoxicity studies.
Reproductive toxicity studies	For drugs to be used in women of childbearing potential. They include fertility and general reproductive performance, teratology and perinatal/postnatal development.
Genotoxicity studies	Mutagenicity studies. Are used to assess the likelihood of the drug causing genomic damage that could induce cancer development.
Toxicokinetic studies	Used to describe the systemic exposure achieved in animals and its relationship to the drug concentration, dose and time course of the toxic effect. The purpose is to contribute in the assessment of the relevance of these findings to clinical safety, and support the choice of species and dose regimen in other nonclinical studies as well as the design of subsequent nonclinical studies.

Table 1. Basic Nonclinical Toxicology Studies.

approval period. These are not mandates that determine the specific structure or design of clinical trials. Although these phases are in general to be conducted sequentially, frequently they overlap. Clinical development programs commonly proceed in the following stages:

5.1 Phase I
This is the phase where initial introduction of an investigational product to humans. The drug is administered cautiously to a few patients or normal human volunteers (usually less than 80), to gain an understanding of the pharmacology, and basic safety of the drug, including tolerability, activity, pharmacodynamics, pharmacokinetics, mechanism of action in humans and optimal route of administration. Drug metabolism, structure-activity relationships and studies in which investigational drugs are used as research tools to explore biological phenomena or disease processes are also included in this phase. The first evidence of the drug efficacy in humans may also be observed in these phase. Subjects are monitored very closely. The studies in Phase II are designed based on the results obtained during this phase.

5.2 Phase II
In this phase a small group of patients are tested, usually 100 to 200, who suffer from the condition the drug is intended to treat or diagnose. The studies include well controlled, closely monitored trials. The investigational product is administered with the objective of increasing the understanding of the safety profile and the initial observations on the efficacy of the drug in the proposed disease. In this phase the aim is to establish a foundation for the phase III trials. The information gathered includes dose, dose regimen and fine tuning of the target population.

5.3 Phase III
The drug in this phase is used in much larger groups of patients, several hundred or thousands, who suffer from the condition that the compound is supposed to treat. This phase includes controlled and uncontrolled studies. The idea is to gather additional safety and efficacy information to determine the benefit-risk ratio of the drug. In this phase the trials follow more rigorous standards since they will serve as the primary basis for the approval of the drug to be marketed.

5.4 Phase IV
In addition to these 3 phases, regulatory authorities may require additional studies after approval to clarify some finding observed during the development program or to produce additional safety data, or treat special populations (e.g. the elderly, patients with renal function impairment, children, etc.). In a general sense the clinical development process continues long after the drug has been approved for marketing. Collection and evaluation of adverse experiences and other information collected while the drug is in the market provides the sponsor and regulatory authorities of a continuous flow of data that allows ongoing review and reassessment of safety and efficacy of the drug. The concept of risk minimization action plans have been introduced recently. Risk minimization action plans are strategic plans to minimize a drug's known risks and for the regulatory agency to monitor the sponsor's implementation of the plan. These postmarketing commitments range from comprehensive literature reviews to large controlled trials. These post marketing

studies are usually called post approval trials or phase IV trials. Phase IV trials can be undertaken at the request of the regulatory agency as part of a postapproval commitment, as a specific regulatory agency requirement, or at a company's own decision to learn more about their product.

6. Quality systems in clinical research[12]

Many aspects of Good Manufacturing Practice (GMP) apply to the drug development process. Quality is a measure of the ability of a product, process, or service to satisfy stated or implied needs. A high quality product is one that meets these needs. For human experimentation, quality may apply to data generation and management, or, the processes involved in the implementation of the trials. Quality systems for human experimentation are the formalized practices, e.g. monitoring programs, auditing programs, complaint handling systems, etc., for periodically reviewing the adequacy of the activities and practices during human experimentation, and for revising such activities and practices so that data and process quality are maintained. For human experimentation GCPs are the basis for implementation of quality systems through quality management. This is done through the coordination of activities by the sponsors of the experiments, the clinical investigators and their staff, the institutional review boards and independent ethics committees, and by regulators to direct and control the operations with respect to quality. Quality management consists of three components: quality control, quality assurance, and quality improvement.

In the case of human experimentation, **Quality Control** is the steps taken during the implementation of the clinical trial to ensure the quality of the data generated and the processes involved. These include investigator supervision, sponsor monitoring, and any review by the regulatory authorities, to ensure that the trial meets the protocol and procedural requirements and is reproducible. **Quality Assurance** is the systematic process to determine whether the quality control system is working and effective. In clinical trials this is usually done by the sponsor through independent auditing of quality control activities, and also by the regulatory authorities through inspection of the quality systems and activities.

With the knowledge obtained from the quality assurance, audits and activities changes are made to the systems and activities with the purpose of increasing the ability to fulfill the quality requirements for the moment and in the future. This process can be called **Quality Improvement**.

Another activity central to maintaining and improving quality in clinical trial is the process of monitoring. Monitoring is a quality control activity conducted by the sponsor or a representative of the sponsor. The purpose is to ensure that the research data are accurate, complete, and verifiable from source documents. GCP guidelines (ICH E6)[9] defines monitoring as "the act of overseeing the progress of a clinical trial, and of ensuring that it is conducted, recorded, and reported in accordance with the protocol, standard operating procedures, good clinical practices, and the applicable regulatory requirements." Monitors usually compare the data in the case report forms designed for the study and the source documents, i.e., with the medical chart of the patient, physician notes, laboratory results, etc. Monitors also make sure that the activities related to protecting the rights and welfare of the study subjects were carried out appropriately. On the other hand, auditing is an independent quality assurance activity used by the sponsor to evaluate the effectiveness of the monitoring program. Auditing procedures are similar to the monitoring activities. The difference is that monitoring occurs only during the execution of the clinical study, auditing occurs at any time during or after the clinical study is completed. In addition to quality audits there are inspections conducted by the regulatory authority(ies). An inspection is the act of conducting an official review of documents, facilities, records, and any other resources the authorities deem related to the clinical study. The inspection may be at the clinical trial site, at the sponsor's facilities, and/or at the Contract Research Organization (CRO) facilities, or at any other establishment the authorities deem appropriate. CROs are organizations which are normally contracted by sponsors to monitor their clinical studies. CROs may also conduct a complete development program for a sponsor on occasions, or deliver part of the activities related to the development of the investigational product. The purpose of monitoring is to determine if the research was conducted in compliance with national and local laws and regulations for the conduct of research and the protection of human subjects.

All parties involved in human experimentation (sponsors, clinical investigators, Institutional Review Boards/Independent Ethics Committees, and regulatory authorities) need to adopt and implement quality systems for the processes and activities they are responsible for.

This includes clinical research facilities. Clinical research should include **Quality Systems** to measure the quality of clinical research through the use of standard operating procedures (SOPs), study protocol compliance, internal monitoring and the sponsor's monitoring activities. This is accomplished through training of the personnel involved in clinical trial activities, internal and external audits, and accountability of the personnel.

A Typical quality system would include production and process control, equipment and facilities control; records, documents and change controls; material controls, design controls and corrective and preventive action. (Figure 1). This system can easily be adapted for the development of medical devices.

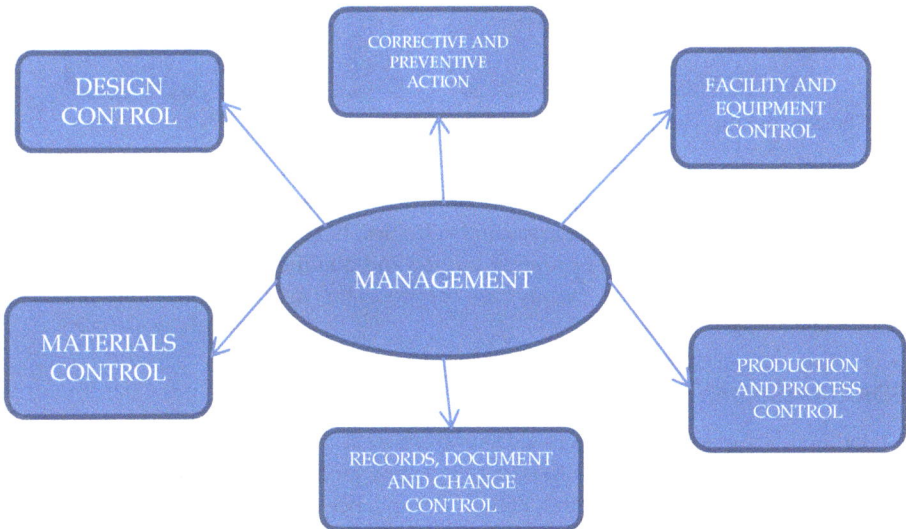

Fig. 1. Typical quality system. This system can easily be adapted to a medical device development facility.

A **Quality System** for an investigational clinical center may be also adapted from this diagram to include the following areas under the control of the clinical investigator (Figure 2):
- Facility and Equipment Evaluation and Documentation
- Source Documentation Generation, Integrity and Retention
- Consent Process and Documentation
- Safety Management and Reporting Processes and Documentation
- Investigational Product Accountability and Integrity and Documentation
- Site Staff Qualifications, Training, and Documentation
- Corrective and Preventive Action Development and Implementation Facility

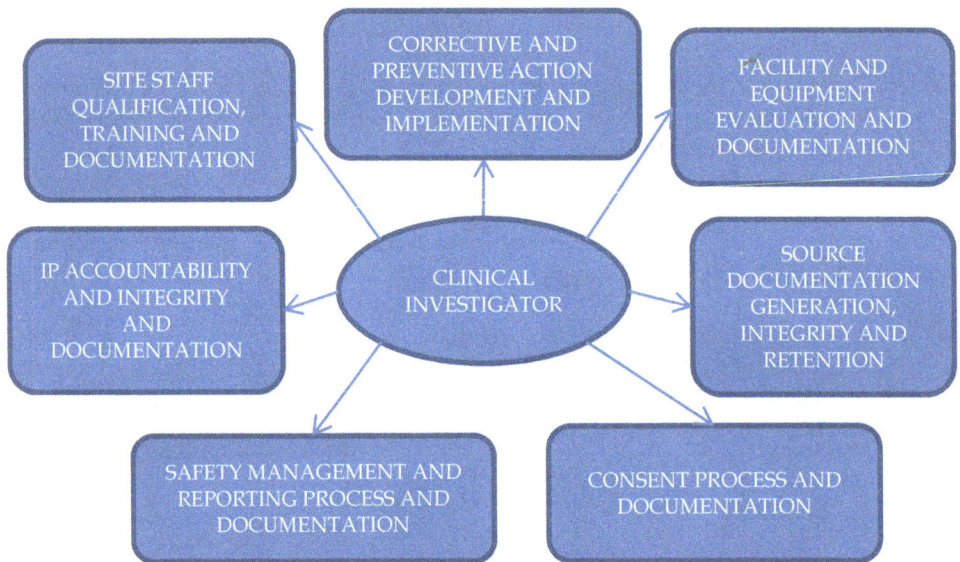

Fig. 2. The organization of a clinical investigational site.

These represent the activities required in a well run clinical investigational site. The investigator is responsible for all activities. The site should have guidelines and/or standard operating procedures for each these areas and activities. In addition, the investigator should have sufficient personnel who are properly trained and qualified to conduct these activities. It is also important that the facility are appropriate in size and configuration to accommodate all these areas.

7. FDA Bioresearch monitoring [12]

The Food and Drug Administration's (FDA) bioresearch monitoring program (BIMO) was established in 1977 with input from the drug, biologics, medical device, veterinary medicine, and food areas. Chapter 48 of the FDA's Compliance Program Guidance Manual is dedicated to Bioresearch Monitoring and delineates the inspection and reporting procedures for studies under FDA jurisdiction. The stated objectives of the bioresearch monitoring program are: to protect the rights, safety, and welfare of subjects involved in FDA-regulated clinical trials; verify the accuracy and reliability of clinical trial data submitted to FDA in support of research or marketing applications; and assess compliance with FDA's regulations governing the conduct of clinical trials. The purpose of the program is to provide instructions for FDA's field personnel for conducting such inspections.

BIMO developed compliance programs to provide uniform guidance and specific instructions for inspections of clinical investigators, sponsors and monitors, in-vivo bioequivalence facilities, Institutional Review Boards, and nonclinical laboratories involved in the testing of investigational products. The purpose of these programs is to adapt a Quality System framework for the oversight and management of clinical studies.

The most useful elements of a quality system that applies to clinical studies are: corrective and preventive action (CAPA) and management controls. CAPA procedures can be adapted to ensure effective and efficient clinical study management.

The application of CAPA to clinical research activities involve:

- Identification of non-conformances, e.g. protocol deviations, errors of omission or transcription.
- Investigation of the cause of the problem identified
- Identification of the actions needed to correct and prevent recurrence of the problem
- Verification that the corrective action is effective
- Making sure that the information is appropriately disseminated
- Submission of the information for management review on problems identified and actions taken
- Documentation of the process

Management controls involve the appointment of a management representative responsible for the research, in this case the investigator or sub Investigator, and to conduct management reviews.

Figure 3 shows the relationship between CAPA, management reviews and audits, external (sponsor monitoring, third party or FDA) and internal through monitoring internal activities.

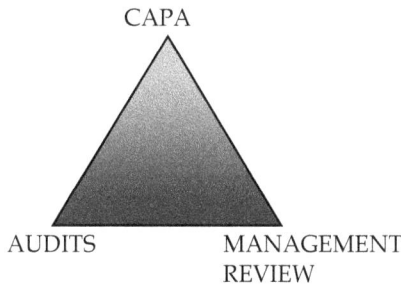

Fig. 3 Relationship between CAPA, Management Reviews and Audits.

Although FDA inspections are focused on clinical investigators, they are of great importance to sponsors. The inspections are designed to determine how well sponsors performed their responsibilities for the conduct of the study; should the inspections uncover serious problems it may result in rejection of the data essential for drug approval. As a result the sponsor may face inspections and compliance actions if it is found to have worked with noncompliant investigators and did not take corrective action.

8. Good clinical practice (E6)[9]

Good Clinical Practices (GCP) is not a set of instructions on how to develop a product or how to design human experiments. GCP is a series of general principles that must be observed during the conduct of human experimentation. This GCP guideline provides a unified standard for designing, conducting, recording, and reporting clinical trials that involve human subjects. Compliance with this guideline provides public assurance that the rights, well-being, and confidentiality of the trial subjects are protected and that the results of the study are credible. GCP are part of the quality systems to cover testing of medicinal products and devices, and conducting clinical studies in human beings. Their objective is to provide a unified standard for the European Union, Japan, and the United States, with consideration to existing GCPs of Australia, Canada, the Nordic countries as well as the World Health Organization, and to facilitate the mutual acceptance of clinical data by the regulatory authorities. It includes also the minimum information that should be included in the information to the investigator, which are the documents considered essential, their purpose, and how to file them. Many countries around the world have adopted these guidelines as their own. The ICH guideline on GCP (E6) outlines the 13 principles of good clinical practices. These principles are in line with the Nüremberg Code, the Declaration of Helsinki and the Belmont Report. These guidelines should be adopted by IRBs/IECs, sponsors, and clinical investigators as well as regulatory authorities who oversee or conduct clinical trials.

8.1 Principles of GCP

1. Clinical studies should be conducted according to ethical principles
2. Foreseeable risks and inconveniences should be weighed against anticipated benefit to the subjects
3. The rights, safety and well-being of the trial subjects should be the most important consideration and prevail over scientific or societal interests.
4. Available preclinical and clinical information on a product should be adequate to support the proposed trial
5. A clinical trial should be scientifically sound and described in a clear detailed protocol.
6. A clinical trial should be conducted in compliance with a protocol previously approved by an IRB/IEC.
7. Medical care given to, and decisions made on behalf of, trial subjects should be always the responsibility of a qualified physician or qualified dentist.
8. Each individual involved in conducting the trial should be qualified by education, training, and experience to perform his/her respective tasks.
9. Freely given informed consent should be obtained from every study subject prior to clinical trial participation.
10. All clinical trial information should be recorded, handled, and stored in a way that allows accurate reporting, interpretation, and verification.
11. The confidentiality of the records should be protected
12. The investigational products should be manufactured, handled, and stored in accordance with applicable good manufacturing practices (GMP), and used in accordance with the approved protocol.
13. Systems with procedures that assure the quality of every aspect of the trial should be implemented.

The GCP outline the duties of the IRBs/IECs, sponsors, and the clinical investigators.

8.2 Independent Ethics Committees (IECs) and Institutional Review Boards (IRBs)

Their main responsibility is to safeguard the rights, safety, and wellbeing of all trial subjects, with special attention to the inclusion of vulnerable subjects to the trial. The IRB/IEC is required to have standard operating procedures and maintain written records of their meeting and decisions. The composition and authority under which the IRB was established should be documented in writing. All meetings, notification to members, and schedules should be disseminated in writing. In summary all information and documentation of activities should be documented and transparent.

The IRB/IEC should consider the qualifications of the investigator, ensure that all subjects have freely provided their informed consent to be included in the study, ensure that payments to the subject for participation in the trial are not coercive or exercise undue influence, and continuously review the progress of the experimentation at intervals appropriate to the degree of risk to human subjects, but at least once per year. The IRB can request additional information if in the judgment of the IRB members the additional information would add meaningfully to the protection of rights, safety and/or well-being of the trial subjects. The IRB should always determine that a protocol or the information provided adequately addresses relevant ethical concerns and meets applicable regulatory requirements.

The IRB is usually composed of at least 5 members, at least one member whose primary area of interest is nonscientific, and at least one member who is independent of the trial site. The investigator may provide information on any aspect of the trial but should not participate in deliberations or in the vote, or opinion of the IRB.

8.3 The informed consent process

For the implementation of the informed consent the investigator should comply with all regulatory requirements and adhere to GCP and the ethical principles originating in the Declaration of Helsinki. The subject should be thoroughly informed of the experiment to be conducted, the risks and the potential benefits. Ample time should be given to the subject to make his/her decision to participate in the study. It should be very clear what the experimentation is all about. No coercion should be applied on the potential study subject.

The informed consent document should include explanations of the following:

- The trial involves research, and some parts of it are experimental.
- The purpose of the trial and the voluntary nature of the subject's participation.
- The treatment and probability for random assignment to treatment. That is in a double blinded study neither the patient nor the investigator may know which treatment is being administered.
- The trial procedures. Including potential for risky procedures and their potential consequences.
- The subject's responsibility to follow the indications from the investigator.
- Reasonable foreseeable risks or inconveniences to the subject, and embryo, fetus or nursing infant, if applicable.
- Reasonable expected benefits.
- Alternative forms of treatment for the condition under investigation and their potential risks and benefits.
- Compensation for trial related injury.
- Payment to the subject, if any.

- Expenses for the subject, if any.
- That the subject's original medical records may be accessed by regulatory authorities, IRB/IEC, the monitor and auditors for verification of the information.
- That confidentiality will be maintained and not be made public.
- All new information related to the trial that becomes available that may be relevant to the subject's willingness to continue to participate in the trial will be forwarded to the subject.
- Who to contact with questions or in the event of a trial related injury
- Foreseeable circumstances or reasons under wich the subject's participation in the trial may be terminated.
- The duration of the trial and approximate number of subjects involved.

8.4 The investigator

The investigators supervise the study staff to ensure they follow established procedures for the conduct of the study. They should be qualified by training, education and experience to conduct clinical trials. The investigators should be thoroughly familiar with GCP, the product under investigation and the study protocol. Investigators are responsible for all medical decisions. In their role they obtain approval to conduct the study from the IRB/IEC; ensure that informed consent is obtained freely and without coercion before the study starts; establish and maintain the subjects' case histories; transcribe the subjects' medical data from the medical files to a case report form for the sponsor; ensure the accuracy, completeness, legibility, and timeliness of the data reported; promptly report all adverse events and other problems; document and explain any deviations from the study protocol; be responsible for the accountability and proper storage as well as the use according to the protocol of the investigational product; and provide all required reports at the end of the study to the sponsor.

Investigators should be in contact with the IRB/IEC and the sponsor frequently. Communications involve,

- Before initiating the trial, obtain a written and dated approval/favorable opinion from the IRB/IEC to start the study
- Provide the IRB/IEC with a copy of the information on the product under investigation (the Investigator's Brochure, IB) and any amendments to the IB during the study.
- Report promptly any serious adverse event or laboratory abnormality immediately to the sponsor, the regulatory authorities and the IRB/IEC, and follow up with a detailed written report with any additional information requested.
- For patients who die during the study the investigator should supply the sponsors, regulatory authorities and the IRB/IEC with all pertinent information on the event.

Upon completion of the trial the investigator should inform the sponsor, the IRB/IEC and the regulatory authorities with a summary of the trial outcome, and any other report required by applicable regulation.

8.5 The sponsor

Sponsors are responsible implementing and maintaining quality assurance and quality control systems with written SOPs to ensure that the trials are conducted, and data are generated, documented and reported in compliance with the protocol, GCP, and applicable regulatory requirements. Sponsors are also responsible for securing agreement from all involved parties to ensure direct access to clinical trial related sites, source documents, and reports for the purpose of monitoring and auditing by the sponsor, CRO and regulatory authorities. Agreement with the investigators or any other party involved in the study should be in writing.

Quality control should be applied to each stage of data handling to ensure that all data are reliable and have been properly processed, for securing the services of monitors to ensure compliance of clinical investigators and verify that the study is carried out according to the approved study protocol. Sponsors also audit the monitor's performance, other quality control activities and systems to ensure performance. The monitors hired by the sponsor to review the records at the clinical centers, and report their finding to the sponsor in written reports of all visits and trial related communications.

Sponsors may transfer in writing any or all their obligations to a contract research organization (CRO), but the ultimate responsibility for the quality and integrity of the data always resides with the sponsor. CROs have the same obligations as the sponsor.

The sponsor is responsible for the medical expertise. Qualified medical personnel should be readily available to advise on trial related matters. An external consultant may be appointed for this function.

Sponsors are responsible for the trial design, trial management, trial data handling, and retention of documents for the specified period required by law and regulations. They are also responsible for the selection of qualified investigators and to apply with the regulatory authorities to conduct the trial.

Finally, the sponsor is responsible to provide insurance or indemnification to the investigator against claims arising from the trials, except for claims arising from malpractice and/or negligence.

8.6 Regulators

The regulators may inspect all parties who conduct or oversee clinical research and verify the information submitted to the regulatory authorities. Regulatory agencies inspect specifically clinical investigators, pharmaceutical companies, device companies, CROs, IRBs/IECs, as well as nonclinical laboratories, to ensure the accuracy and validity of the data generated, and to ensure that the rights and welfare of the research subjects are protected. The regulatory inspectors evaluate how well sponsors, monitors, clinical and non clinical investigators, CROs, and IRBs/IECs comply with the regulations. They may require certain conditions for a study to proceed. They develop policies and procedures for reviewing product applications and for the conduct of GCP inspections as exemplified by the FDA's BIMO compliance programs.

9. Conclusion

Over the last century the scientific community has developed a better understanding of how to protect and respect the rights, safety and wellbeing of research subjects. For centuries the Hippocratic Oath was the only ethical guidance for physicians and scientists on how to treat subjects, and specifically research subjects. The development of Good Clinical Practice was the result of various incidents that resulted in the Nuremberg Code, the Declaration of Geneva, the Declaration of Helsinki and the Belmont Report. ICH is an attempt to harmonize GCP in the most advanced democracies. Today, many regulatory agencies around the globe use these principles to regulate human experimentation in their countries.

The responsibility of GCP is shared by all parties involved in human experimentation, investigators, sponsors, ethics commitees, regulatory authorities, and research subjects. To guarantee the quality and accuracy of the data generated during human experimentation Quality Systems have been developed and are applied around the world.

10. References

[1] Markel, H. (2004) Perspective. Becoming a Physician. "I Swear by Apollo"- On Taking the Hippocratic Oath. N Engl J Med 2004; 350:2026-2029

[2] Shuster, E. (1997) Fifty Years Later: The Significance of the Nuremberg Code. N   Engl   J Med 1997; 337: 1436-1440, Nov 13, 1997

[3] Jones, D.A. (2006) The Hippocratic Oath II. The Declaration of Geneva and other modern adaptation of the classical doctor's oath. Catholic Medical Quarterly, February 2006

[4] WMA Declaration of Helsinki – Ethical Principles for Medical Research Involving Human Subjects. October 2008 available from
http://www.wma.net/en/30publications/10policies/b3/index.html

[5] The Belmont Report. (1979) Ethical Principles and Guidelines for the Protection of Human Subjects of Research. Department of Health, Education, and Welfare available in http://ohrs.od.nih.gov/guidelines/belmont.html

[6] U.S. Food and Drug Administration available in http://www.FDA.gov

[7] European Medicines Agency available in http://www.EMA.europa.eu

[8] Japanese Regulatory Agency: Pharmaceutical and Medical Devices Agency available in http://www.pmda.go.jp/english/index.html

[9] International Conference on Harmonization and Technical Requirements for the Registration of Pharmaceuticals for Human Use (ICH) available in http://www. ICH.org/home.html

[10] Good Laboratory Practice (GLP), 21CFR58 available in http:// www.fda.gov

[11] Phases of Drug Development, 21CFR312.21 available in http:// www.fda.gov

[12] Bioresearch Monitoring in Compliance Program Guidance Manual, Chapter 48 available in: http://www.fda.gov

2

Quality and Quality Indicators in Peritoneal Dialysis

Javier Arrieta
Hospital Universitario de Basurto – Bilbao
Spain

1. Introduction

Physicians have historically shared an intuitive concept of Quality, concerning the care we provide to our patients. Our academic education and practice have been focused on Quality as a technical concept, assessable only by technicians and with no strong correlation with outcomes. The concept of Medicine as an Art is related to the values of vocation, dedication and good practice, recognizing that results can after all be negative

In the XXI Century, we all now accept the scientific nature of Medicine and, therefore, its dependence on the objective assessment of outcomes. In contrast, the patient's perception of Quality strongly depends on the culture and the environment. The current definition of disease given by the World Health Organization (WHO) focuses on self-perceived health and wellbeing. In this context, quality-based medicine should also be oriented towards the health and welfare as perceived by the patient.

Quality is one of the strategic elements on which the transformation and improvement of modern health systems is based. The effort made in recent years towards quality assurance in this field –including in the particular case of nephrology-, entails recognition of the need for objective and standardized measurement tools for health activities: Quality is not just good intentions.

2. Definitions of quality

There are many definitions of Quality, which in itself suggests that none of them are comprehensive. Definitions focused on Quality in Health, mainly date from the 1980's, when Palmer, Donabedian (Donabedian, 1980), the American Medical Association and many other authors tried to develop an adequate definition. As early as 1990, the Institute of Medicine adapted the definition given by the ISO (International Organization for Standardization), which does not specifically refer to Health: "Quality is the degree to which the characteristics of a product or service meet the objectives for which it was created", defining Quality in Medicine as "the degree to which health services for individuals and populations increase the likelihood of desired health outcomes and are consistent with current professional knowledge"(Lohr, 1990). This is the current philosophy of Medical Quality, which assesses the results, both objective and perceived by the patient (Committee on Quality of Health Care in America, 2001), assuming a degree of uncertainty with respect to processes and the final outcomes.

Quality in patient care depends on a large number of factors, but doctors tend to consider only a few, such as efficacy and effectiveness, more recently including accessibility, efficiency, privacy, and safety, among others as respect for the environment. Some factors are of great interest to Society as a whole - like those just listed - while others may be interesting primarily for patients, such as timeliness, convenience, patient participation, etc. The restrictive view of Quality used by doctors explains the differences we find between the technical quality and the quality perceived by the patients (JCAHO: Agenda for change, 1989).

2.1 Quality models

Although all models of quality are based on common ideas, such as reducing the variability in medical practice through standardization -using standards and indicators-, two types of quality models can be distinguished with respect to their underlying purpose. On the one hand, some models pursue standardization. This involves assessment by a qualified and independent entity that will accredit or certify us for providing high quality medical care. On the other hand, there are models that aim for continuous quality improvement based on self-monitoring. These produce continuous feedback that should help eliminate errors and lead to improvements in outcomes.

These two types of models are by no means exclusive. In Europe, public hospitals commonly use the European Foundation for Quality Management (EFQM) model, which not health specific:

(http://www.efqm.org/en/Home/aboutEFQM/tabid/108/Default.aspx); while, providers in the United Stated and private centers in Europe have chosen accreditations based on the standards of the Joint Commission on Accreditation of Healthcare Organizations (JCAHO, 1989; JCAHO update, 1990; JCAHO, 2011).   Meanwhile, in Latin America the EFQM model is adopted more widely through the Latin American Foundation for Quality Management (FUNDIBEQ; www.fundibeq.org).

The Joint Commission uses a wide range of indicators and standards from the National Institute of Standards and Technology (NIST) -355 in the international version, of which 171 are mandatory for accreditation-, divided into medical and organizational indicators. They can be accessed from www.jointcommission.org or www.quality.nist/gov/.

The EFQM model allows centers to choose their indicators and standards -as long are they are logical and supported by scientific evidence- and pays greater attention to the evolution of the indicator towards "Excellence", than to the achievement of a standard at a given time. In other words, centers are not valued for their good work, but for their year-on-year improvement.

Another important difference from an operational point of view is that the Joint Commission certifies Centers -although it may also test Units- and assesses both clinical and other organizational, structural, plant safety and accessibility indicators. On the other hand, the European model can readily be applied to processes. For example, it is possible to apply the EFQM to a chronic hemodialysis process, peritoneal dialysis, nephrology hospitalization ward or kidney transplantation unit. Therefore, we can first apply it to one of the processes in our Service or Hospital, and within a few years extend it to other processes. It should be noted that processes are only one part of the EFQM (Figure 1), a useful aspect of the model is that we can start by applying it into individual processes, based on the priorities of clients and employees. Later on, the analysis can evolve to address common issues for the Hospital, such as leadership and strategy.

Finally, the European model can theoretically be implemented with no additional economic cost (the only requirement being training in use of the model), while those who pursue international accreditation need to pay the evaluators. There is a cost calculator on the Joint Commission's website, and it should be noted that the center's accreditations have to be renewed every three years.

In this context, we note that dialysis is a high cost therapy that can rarely be paid for by the patient. Funders have the authority -and obligation- to monitor the quality of the Healthcare for which they pay. Therefore, they increasingly demand the accreditation of Dialysis Units. Evaluators are usually independent from the payer, and they act as intermediaries between the payer and the health provider. Nevertheless, even in the accreditation systems, evaluation is considered as an element to guide these units in making improvements.

In this chapter, we will consider a quality system focused on a continuous improvement (rather than quality accreditation) that every dialysis unit could adopt if so desired.

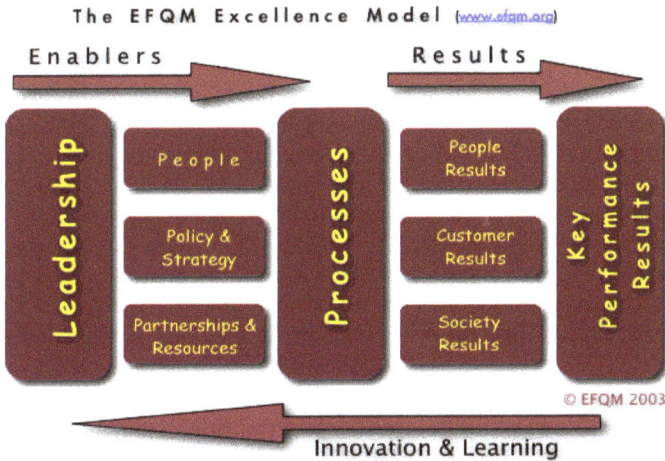

Fig. 1. European Foundation for Quality Management. Model of Excellence.

2.1.1 Quality systems in dialysis

Initially, quality systems have been used in acute care processes (mainly surgical), as well as general services such as laboratories, radiology units, etc. Quality indicators in these cases are derived from different patients who undergo a procedure at different points in time. However in dialysis, patients continue treatment over periods of months and years, and this implies several conceptual changes. It is clear that dialysis is not a curative procedure, but rather a life support technique. Its purpose is then to prolong life and improve its quality. Accordingly, indicators that seek to measure the quality of a certain dialysis therapy should be correlated with those two endpoints: survival and quality of life.

2.1.2 Quality systems in peritoneal dialysis

Quality systems in hemodialysis have been implemented for two decades, fundamentally, due to accreditation requirements. However, peritoneal dialysis (PD) is performed at the patient's home under clinical guidance depending on the general hospital, itself already under global assessment and accreditations. That dependence explains why quality systems in PD have not been prioritized. The EFQM model can be applied to isolated processes, so it can be used in Peritoneal Dialysis Units.

The full EFQM model (Total Quality Management) includes the assessment of multiple criteria, grouped into facilitators (5 types) and results (4 types).

In this chapter, we will only describe the PD process (as a part of the Dialysis Process) and the most appropriate indicators and standards for its evaluation..

3. The peritoneal dialysis process

The process includes information concerning the alternative techniques of dialysis offered to patients from which they can choose, and withdrawal from the PD program due to death, transplantation, changing to hemodialysis or recovery of renal function. As hemodialysis and PD have a similar start and end, and the same therapeutic purpose, we have grouped them under a single process of chronic dialysis, with its two main variants (Figure 2). Logically, the dialysis process is part of a series of support processes including those of the laboratory, pharmacy, maintenance, etc. The description of each activity in the process (Table 1) should not be exhaustive but rather refer to specific protocols that need to be written, accessible to all staff and regularly updated. However, it is important that there is a designated person in charge of each activity in the process and a record of the activity that could be consulted if necessary (Lopez-Revuelta et al., 2002; Arenas, 2006).

The process of peritoneal dialysis is a part of a more complex dialysis process that includes all the renal replacement therapies (Figure 2). Patients' opinions and medical contraindications determine the decision between the three main alternatives for dialysis, whether as a definitive therapy or as life support waiting for kidney transplantation. In this chapter we consider only the indicators of quality for the home peritoneal dialysis option.

3.1 Characteristics of clinical quality indicators

A clinical indicator is a quantitative measure that can help us monitor and evaluate quality in care activities and support services. It is not a direct marker of quality, but rather can serve to alert to areas which require specific action within a healthcare organization

Indicators express information as numbers of events or ratios. In the latter case, the denominator is the number of patients among whom the event could potentially occur. Although the event selected is undesirable, in general it should occur commonly enough to be used as an index. There is, however, a special kind of indicator that cannot be expressed as an index or a ratio: the Sentinel Event Indicator, that measures events which are undesirable, preventable, rare and have serious outcomes. When detected, such indicators warrant a thorough investigation and urgent action (even if there is just one case).

Indicators can measure either processes or results. The best process indicators are those closely linked to patient outcomes, and for which there is scientific evidence that indicates that the care provided will lead to a specific result. In the event that the result of a process cannot be measured, or there is an excessive delay for corrective action, process indicators are the only type that can be used.

Further, indicators can measure desirable or undesirable results. In the former case, the objective is that the vast majority of patients meet the criteria; while in the latter case, the aim is just the opposite. Ideally, the quality systems used by a given Unit should select only those indicators that represent desirable objectives, in order to avoid confusion. For instance,

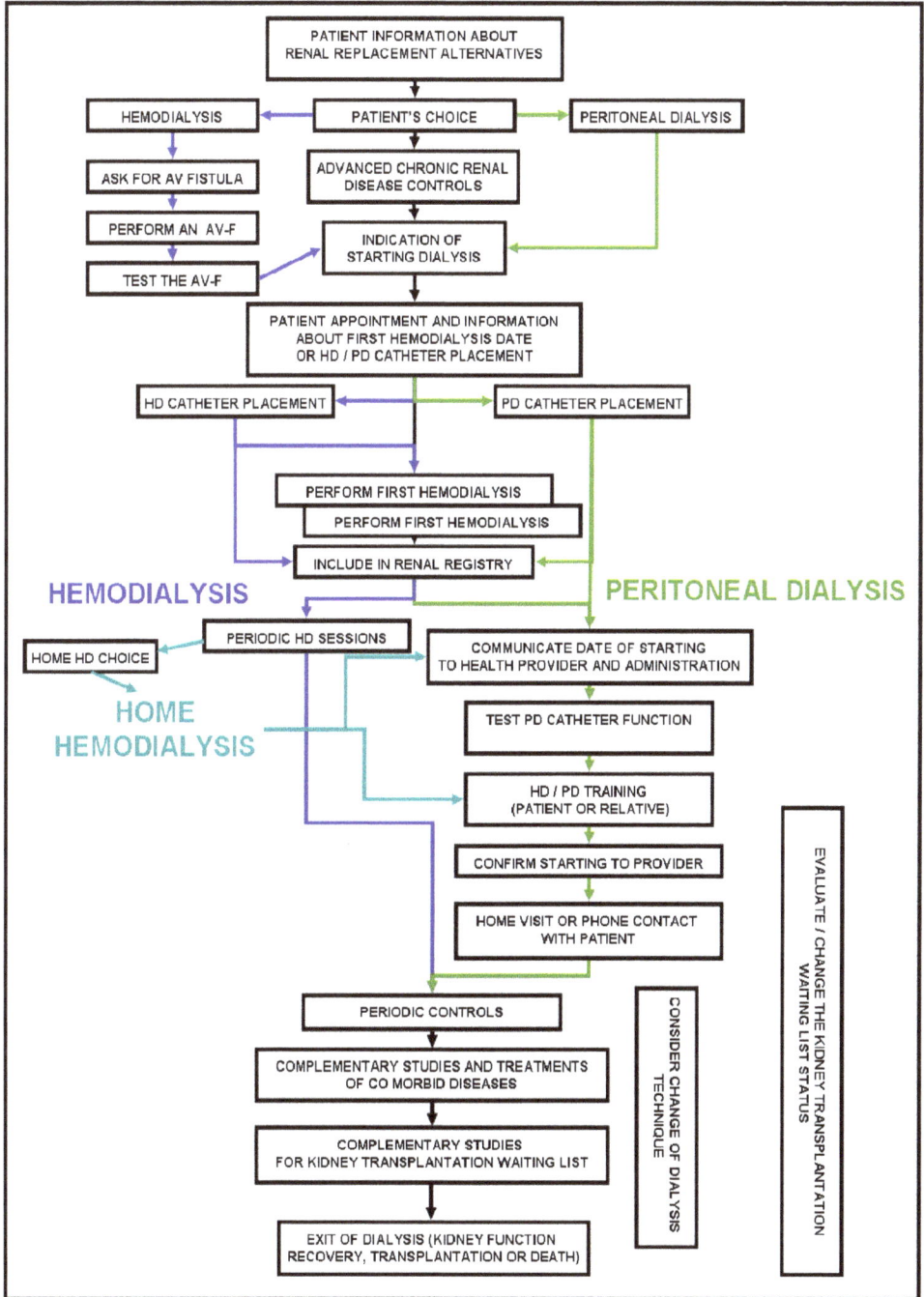

Fig. 2. Dialysis Process.

ACTIVITY	DESCRIPTION	RESPONSIBLE	REGISTRY
Patient information about RRT techniques	DIALYSIS GENERAL PROTOCOL	Nephrologist PD nurse	Clinical Record Nursing record Consent Form
Indication for starting PD	DIALYSIS GENERAL PROTOCOL	Nephrologist PD physician	Clinical Record
Appointment for PD catheter insertion	Written appointment, with date, time and premedication	PD nurse	Nursing record
Deliver to patient Information about appointment for catheter insertion	PERITONEAL DIALYSIS PROTOCOL	PD nurse	Nursing record
Catheter insertion	PERITONEAL DIALYSIS PROTOCOL	PD nephrologist or surgeon	Lab Reports Clinical Record
Incorporate in the RRT Registry and waiting list for renal transplant	On line data communication to RRT Registry Database, including Identification, Clinical and Serological Data, and required data for Kidney Waiting List	Nephrologist	RRT Registry. Waiting List for kidney transplant
Convey to Administration and PD material provider about the patient	To be done by email about patient information and chosen PD technique (CAPD or APD)	Nephrologist	E-mail and letter of approval
Check PD catheter permeability	PERITONEAL DIALYSIS PROTOCOL	PD nurse	Nursing record
Patient PD training	PERITONEAL DIALYSIS PROTOCOL	PD nurse	Nursing record
Call the PD provider	To finalize the supply of PD equipment at patient's home from specific date	PD nurse	Nursing record
Home visit or phone call to patient's home	After starting PD at home, some contacts asking for possible problems or doubts.	PD nurse	Nursing record
Regular controls at hospital, or by phone, mail, web-cam, etc.	PERITONEAL DIALYSIS PROTOCOL	PD nephrologist PD nurse	PD Graphics Nursing record Clinical Record
To consider change in dialysis technique	In case of patient's decision or unsolvable problems. PD PROTOCOL	PD nephrologist	PD Graphics Nursing record Clinical Record
Regular controls about studies and treatments of associated illnesses	PERITONEAL DIALYSIS PROTOCOL	PD nephrologist	PD Graphics Clinical Record
Regular control's studies about Waiting List for kidney transplant	PRE-TRASPLANT STUDIES PROTOCOL	Nephrologist	PD Graphics Clinical Record
Reconsider situation in Waiting List for kidney transplant	PRE-TRASPLANT STUDIES PROTOCOL	Nephrologist	PD Graphics Clinical Record RRT Registry

ACTIVITY	DESCRIPTION	RESPONSIBLE	REGISTRY
Discharge from PD due to partial improvement of renal function, change to hemodialysis, transplant or patient's death.	PERITONEAL DIALYSIS PROTOCOL	Nephrologist	Lab Reports Clinical Record

Table 1. Peritoneal Dialysis: Activities in the Process.

"peritonitis rate with a negative culture" is an indicator of low quality, a high rate suggesting poor quality in sample collection, transport or laboratory processing of peritoneal effluent. As this may not be intuitive, it is preferable to use the "peritonitis rate with a positive culture", aiming for this indicator to exceed 80% of cases.

In addition, indicators must be valid. This means they should identify situations in which quality in the healthcare provided can be improved (reflected in final outcomes). Validity is often only apparent and the indicator has to be "validated" afterwards. Lastly, indicators must be sensitive, able to identify real problems with care, avoiding "false positives", and they must also be specific, so that they detect only these real problems.

The selection of a set of indicators is a complicated task. It is preferable to select only a few, avoiding an increase in workload related to maintaining the database that would have no direct translation into improvements. From the selected recommendations, the quality indicators were drawn up according to a format which includes: their definition, criterion, equation, units, frequency of the assessment, standard, bibliographical references and comments. The methodologies proposed by the Joint Commission and the Standing Committee of the Hospitals of the European Union were followed by systems for monitoring healthcare processes. These have been complemented by the specific HD methodology that is followed by the American ESRD Special Project and implemented by the Centres for Medicare and Medicaid Services (CMS) -such as in the ESRD Clinical Performance Measures (CPM) Project-. Initially, quality criteria were selected from each recommendation for measurement of performance. The indicator is a quantitative measurement to evaluate a criterion. A "standard" was set for each indicator (namely, the required degree of performance to ensure an acceptable level of quality) based on scientific evidence or, in its absence, by consensus. On many occasions, sufficient scientific evidence was not available, but experience derived from the follow-up of indicators will help us adjust and redefine them in the future. Those interested in understanding the subject more deeply, should consult the 1989 and 1990 JCAHO references.

3.2 Quality indicators in peritoneal dialysis

Traditionally (Donabedian 1980; JCAHO, 1989), we distinguish between structure, process and outcome indicators. Variations in the quality of the structure or the process tend to influence the outcomes. Structure Indicators are highly valued for accreditation, as adverse results caused by structural defects imply a greater responsibility if patients file lawsuits. However, we assume process indicators are a more accurate reflection of quality than those directly related to outcomes, as they detect systematic errors and their correction more commonly produces improved outcomes (Williams et al, 2006; Ballard, 2003).

Indicators must monitor quality. Therefore, they should be correlated to survival or quality of life of the patients, and be based on scientific evidence. In our case, we based them on the Clinical Practice Guidelines in PD, recently published by the Spanish Society of Nephrology (Arrieta, 2006). Following the publication of these guidelines, a panel of peritoneal dialysis, with the support of the Quality Management in Nephrology Group (a working group of Spanish Society of Nephrology), designed a definition for quality indicators and standards that can be used by all the nephrology community -especially those dedicated to PD-. The new definitions would also serve as a framework or terms of reference for future areas of improvement, filling the gap between the development of guidelines and subsequent monitoring.

Often, we found that there was not sufficient scientific evidence to define a standard. In these cases, we proposed a provisional framework that should be evaluated later. Earlier in this chapter, we have explained that continuous improvement objectives should be set by each unit, based on local outcomes.

Whatever the result of applying an indicator in a given unit, what is important is that they guide improvement activities, and there will then be ongoing monitoring of whether such measures are effective. In fact, indicators are basically an internal tool that permits comparisons with our own previous performance and helps us assess our own improvement. In the future, the pooling of results from different institutions would determine the appropriate quality standards in peritoneal dialysis for the Spanish population.

Having similar quality criteria in all centers is a medium-term objective, as we are all interested in comparing our results and assessing whether variations in clinical practice lead to different final outcomes (Jha et al., 2005).

On the other hand, it has been shown that regular measurement of quality indicators -and the fact of having set up targets and standards- encourages monitoring and improves the outcomes of the process (Williams, 2005; Fink, 2002).

The initial list of indicators, standards and objectives selected includes a large number of indicators that have been already established for hemodialysis. As the most prevalent renal replacement technique, many Quality Systems have already been developed in that field (Lopez-Revuelta et al., 2007). Nevertheless, we should always consider those indicators or standards that have not been specifically validated for PD patients as provisional, and focus on the survival and quality of life outcomes instead.

There are usually too many indicators. Each unit should select those that seem most relevant to its daily routine. In addition, data management technologies become a priority. A wide range of computer software (Renalsoft®, Nefrolink®, Nefrosoft®, Versia® etc.) is used in peritoneal dialysis and hemodialysis units in Spain. In some cases, more advanced programs are being developed and adopted than enable quality indicators to be estimated automatically and rapidly.

In the following sections, we will describe the initial selection of Quality Indicators used by the Spanish Society of Nephrology (currently, at the evaluation stage). They are Clinical Indicators, so they have to be supplemented with Structure Indicators, Satisfaction Surveys and Quality of Life Questionnaires for patients. From a business point of view, and in order to obtain Accreditations of our units, it is also a good idea to carry out Satisfaction Surveys of our staff and suppliers.

3.3 Classification of peritoneal dialysis indicators

We use Global Indicators and Comorbidity Indicators to describe patients (Table 2). Most of these are not quality indicators but Registry data, local practice frameworks or terms of reference which enable us to identify certain patient and PD unit characteristics that may influence outcomes and modify other indicators. It is interesting to see how their evolution pans out over time. In some cases, they do indicate aspects of the quality of medical attention before starting PD, but our intention is to use them to adjust the evaluation of Outcome Indicators. The modified Charlson Index (Bedhu et al., 2000) extends the item "Documented History of Myocardial Infarction" to include another one namely "Ischemic Heart Disease (CHD)", which includes all forms of coronary heart disease (angina, myocardial infarction, angiographic evidence of coronary artery disease history of angioplasty or bypass surgery). For this reason, we consider it more appropriate for the usual profile of PD patients. Global and Comorbidity Indicators are collected annually, as they are not used to make improvements. The Charlson Index is measured at the start of PD and, as it can only increase, it is reassessed once or twice a year.

Outcome Indicators (Table 3) (Arrieta et al., 2009; Bajo et al., 2010) include more up-to-date data, such as the rate of infections associated with the technique, the adequacy of the dialysis dose, test results and medications taken. These Indicators can alert us to deficiencies in the initial stages of treatment, and early correction can rapidly improve outcomes. Usually, they are compiled twice a year, but with a good IT system they can be calculated and consulted as often as is agreed to be appropriate in each unit, though clearly this involves additional work.

Other indicators such as rates of hospitalization or withdrawals from DP should be explored more carefully, as they are influenced by local characteristics, the socio-cultural context and the availability of replacement therapy.

GLOBAL INDICATORS
PD Incidence
Period Prevalence (prevalents at begin of period + incidents)
Mean age of incidents
Mean age of prevalents
Sex rate of incidents y prevalents
Mean time in PD treatment of prevalents
Percent of diabetics among incidents
" of incidents not dialyzed before
" " coming from HD
" " coming from TX
" of incidents with a signed Informed Consent about all RRT techniques
" of prevalents on CAPD (vs total in PD techniques)
COMORBIDITY INDICATORS
Median of Modified Charlson Index in incidents
Median of Modified Charlson Index in prevalents

Table 2. Quality Indicators at starting PD.

OUTCOME INDICATORS (1) (ANNUAL INDICATORS)	
Hospitalization	
admissions	
average days by admission	
Exits from PD	
totals	
change to HD	
deaths	
transplants	
Transplants	
percent of patients in Kidney Tx Waiting List (WL) (among prevalents in PD)	
mean time in PD before inclusion in WL	
annual rate of transplants in PD patients (among patients in WL)	
mean time in PD before kidney Tx	
mean time between Tx and PD catheter extraction ¿< 1-3 months?	
OUTCOME INDICATORS (2) (SEMESTER INDICATORS)	
Infections (limited to PD technique)	
rate of peritonitis ¿< 0.5 / pte / yr?	
partial rates in APD and CAPD	
percent of peritonitis with a positive culture (identified germ)	
" of peritonitis by Gram +	
by Gram –	
by fungus	
" of peritonitis "catheter dependent"	
rate of infections of catheter exit site	
rate of patients with nasal cultures (positive or not)	>80%
Adequacy and membrane function	
percent of patients with an urea KT/V measured in the semester	>80%
" of patients with urea KT/V > 1.7	>80%
" of patients not anurics with Renal Residual Function measured	>80%
" of patients with a daily UF rate > 1000ml/ day	>80%
" of patients using daily one or more hypertonic bags (3.86 / 4.25%)	<20%
" of patients with a PET performed in the 3 months alter starting PD	>80%
" of patients with a PET performed annually	>80%
" of patients resulting High Absorbers in PET. (D/P Cr 4h > 0.81)	
Analysis and medication	
percent of patients within Hb objective (11 to 13)	>80%
" of patients with serum ferritin > 100	>80%
" of patients with Index of Resistance to EPO < 9	>80%
" of patients with I.R. to darboepoetin < 0.045	>80%
" of patients with serum cholesterol LDL < 100	>80%
" of patients with serum albumin > 3.5	>80%
" of patients with serum phosphate < 5.5	>80%
" of patients with serum corrected calcium > 8.4 and < 9.5	>80%
" of patients with Ca x P < 55 (in mg/dL)	>80%
" of patients with PTH < 300	>80%

Table 3. Quality Indicators of Outcomes.

Calculation of the rate of occurrence of a certain outcome may present problems in units with few patients. We recommend estimating the prevalence of "at-risk" patients per month, to determine the "real" total number of patients to be used in the denominator of the ratios (Jager et al., 2007).

3.4 Standards and objectives of quality indicators

Every indicator should have a clear definition, a target or objective (threshold or range), and a standard for assessing compliance. We have defined objectives when there is a reasonable amount of scientific evidence to support them. However, such evidence is often not sufficiently tested in PD (though it may have been tested in HD patients or in the general population, as is the case of LDL cholesterol levels). The original standard is commonly set at the percentage of patients who meet the target. For clarity, we prefer to express the degree of compliance than the rate of "non-compliance".

It is important that targets are always to be established based on scientific evidence. For instance, the hemoglobin target is set at 11 mg/dL or above because the Guidelines for Good Clinical Practice (based on hemodialysis) agree on this level; nevertheless; PD patients may have Hb higher than 13 in the absence of EPO. Accordingly, we will not set a maximum target as we do in HD. The standard is a given rate of compliance with objectives -usually 80% to 85%-, and is later adapted to the real results obtained and the real possibilities of achieving the Standard in our healthcare context.

When we initially apply an indicator in our units, we may find that our compliance rates are very low. This could mean that the target was too high, the indicator was not appropriate or, even, that the sample of patients on which the assessment if based are really ill. The objective must be based on high-grade evidence. If it is well established, we must strive to achieve it over time and accept a low compliance rate, re-evaluating the rate once or twice a year.

I insist that a good progress is more important than a good result. Evidence is often drawn from clinical trials involving highly selected patients, with a high rates of adherence to prescribed medication   (which is often free during the trial) and under close medical supervision. These results would be very difficult to achieve in routine practice. In any case, it is absolutely not permissible for the threshold for compliance with an objective to be lowered as a means of achieving a better rate of compliance, unless on reconsideration it is judged that the target is not supported by current evidence, or that the effort required to achieve the target is not justified by real improvements in the final outcome measures (namely, survival and quality of life).

Finally, we must remember that just measuring outcomes tends to produce an improvement in clinical practice (Williams et al., 2005; Fink et al., 2002). It has also been proven that, in hemodialysis, the level of compliance with quality standards is directly related to mortality and morbidity, although most of the standards applied have not yet been validated (Rocco et al., 2006; Plantiga et al., 2007). From a theoretical standpoint, this introduces a bias towards the validation of an Indicator or a Standard, but it should also encourage doctors to use the quality control systems as tools for continuous improvement of our daily practice, rather than consider them as management tools with little relevance to medical practice.

4. Conclusion

It has already been demonstrated that the regular measurement of quality indicators –as well as having standards and establishing objectives-, helps to improve the monitoring and results of the dialysis process, and contributes to improving outcomes in terms of patient morbidity and mortality. Access to management software becomes a priority. A Quality System should be focused on achieving Continuous Improvement of Quality expressed in terms of Survival and Quality of Life. Patients' opinion about self-perceived health and wellbeing and about quality of health care must be considered. Accreditation of the Unit should not be a final objective.

5. Acknowledgment

Groups of Quality in Hemodialysis and Peritoneal Dialysis of Spanish Society of Nephrology have played an essential role in the process of selecting indicators and testing the suitability of proposed standards of Quality in PD.

6. References

Arenas, MD.; Lorenzo, S.; Alvarez-Ude, F.; Angoso, M.; López- Revuelta, K. &Aranaz, J. (2006). Quality control systems implementation in the Spanish Dialysis Units. *Nefrología,* Vol 26, No.2, pp. 234-245. Online ISSN: 2013-2514. Print ISSN: 0211-6995.

Arrieta, J.; Bajo, MA.; Caravaca, F.; Coronel, F.; García-Perez, H.; Gonzalez-Parra, E.; et al. (2006). Guidelines of the Spanish Society of Nephrology. Clinical practice guidelines for peritoneal dialysis. *Nefrología.* Vol 26. Suppl 4, pp 1-184. Online ISSN: 2013-2514. Print ISSN: 0211-6995.

Arrieta, J. (2009). Calidad en Diálisis Peritoneal. In: *Tratado de Diálisis Peritoneal.* (Chapter 31). Montenegro, J.; Correa-Rotter, R & Riella, MC., pp 573-582. Elsevier. ISBN: 978-84-8086-394-0, Madrid.

Bajo, MA.; Selgas, R.; Remón, C.; Arrieta, J.; Alvarez-Ude, F.; Arenas, MD.; Borrás, M.; Coronel, F.; García-Ramón, R.; Minguela, I.; Pérez-Bañasco, V.; Pérez-Contreras, J.; Fontán, MP.; Teixidó, J.; Tornero, F. & Vega N. (2010). Scientific-technical quality and ongoing quality improvement plan in peritoneal dialysis. *Nefrologia.* Vol 30, No. 1, pp. 28-45. : 2013-2514. Print ISSN: 0211-6995.

Ballard, DJ. (2003). Indicators to improve clinical quality across an integrated health care system. International Journal of Quality in Health Care. Vol. 15, Suppl 1, pp i13-i23, Online ISSN 1464-3677 - Print ISSN 1353-4505.

Bedhu, S.; Bruns, FJ.; Saul, M.; Seddon, P. & Zeidel, ML. (2000). A simple comorbidity scale predicts clinical outcomes and costs in dialysis patients. *American Journal of Medicine,* Vol 108, pp 609-613, ISSN 0002-9343.

Committee on Quality of Health Care in America. (2001). Crossing the quality chasm: a new health system for the 21st Century. Washington, DC: National Academy Press. ISBN: 0-309-07280-8.

Donabedian, A. (1980). Explorations in quality assessment and monitoring. Ann Arbor, MI: Health Administration Press. ISBN: 0914904477, 0914904485

Fink, JC.; Zhan, M.; Blahut, SA.; Soucie, M. & McClellan, WM. (2002). Measuring the Efficacy of a Quality Improvement Program in Dialysis Adequacy with Changes in Center Effects. Journal of American Society of Nephrology, Vol 13, pp. 2338-2344. Online ISSN: 1533-3450, Print ISSN: 1046-6673.

Jager, KJ.; Zoccali, C.; Kramar, R. & Dekker, FW. (2007). ABC of Epidemiology (1): Measuring disease occurrence. Kidney International, Vol 72, No 4, pp. 412-415. Online ISSN: 1523-1755, Print ISSN: 0085-2538

Jha, AK.; Li, Z.; Orav, ÈJ. & Epstein, AM. (2005). Care in U.S. hospitals--the Hospital Quality Alliance program. New England Journal of Medicine, Vol 353, pp. 265-274. On line ISSN 1533-4406 Print ISSN 0028-4793.

JCAHO, 1989: Joint Commission on Accreditation of Healthcare Organizations: Agenda for change. (1989). Characteristics of clinical indicators. Quality Review Bulletin. Vol 15, No 11, pp 330-339. ISSN: 0097-5990.

JCAHO, 1990: Update: clinical indicators. (1990). Hospital Food & Nutrition Focus. Vol 6, No 11, pp. 6-7. ISSN: 0747-7376.

JCAHO, 2011: Joint Commission on Accreditation of Healthcare Organizations: Agenda for change (2011). Advanced Chronic Kidney Disease Certification. http://www.jointcommission.org/certification/chronic_kidney_disease.aspx

Lohr, KN. (editor). Committee to Design a Strategy for Quality Review and Assurance in Medicare. Institute of Medicine. (1990). A strategy for quality assurance. Washington, DC: National Academy Press. ISBN: 0-309-04230-5

López-Revuelta, K.; Lorenzo, S.; Gruss, E.; Garrido, MV. & Moreno-Barbas, JA. (2002). Application of process management in nephrology. Hemodialysis process management. Nefrología, Vol 22, No 4, pp. 329-339. Online ISSN: 2013-2514. Print ISSN: 0211-6995.

López-Revuelta, K.; Barril, G.; Caramelo, C.; Delgado, R.; García-López, F.; García-Valdecasas, J. et al. (2007). Developing a Clinical Performance Measures System for hemodialysis. Quality Group, Spanish Society of Nephrology. Nefrología, Vol 27, No 5, pp. 542-559. Online ISSN: 2013-2514. Print ISSN: 0211-6995.

Platinga, LC.; Fink, NE.; Jaar, BG.; Sadler, JH.; Levin, NW.; Coresh, JK.; et al. (2007). Attainment of clinical performance targets and improvement in clinical outcomes and resource use in hemodialysis care: a prospective cohort study. BMC Health Services Research, Vol 7, pp. 5-18. On line ISSN: 1472-6963

Rocco, MV.; Frankenfield, MV.; Hopson, SD. & McClellan, VM. (2006). Relationship between Clinical Performance Measures and Outcomes among Patients Receiving long-term Hemodialysis. Annals of Internal Medicine, Vol 145, pp. 512-519. On line ISSN: 1539-3704. Print ISSN: 0003-4819.

Williams, SC.; Schmaltz, SP.; Morton, DJ.; Koss, RG. & Loeb, JM. (2005). Quality of care in U.S. hospitals as reflected by standardized measures, 2002-2004. New England Journal of Medicine, Vol 353, pp. 255-264. On line ISSN 1533-4406 Print ISSN 0028-4793.

Williams, SC.; Watt, A.; Schmaltz, SP.; Koss, RG. & Loeb, JM. (2006). Assessing the reliability of standardized performance indicators. *International Journal of Quality in Health Care,* Vol 18, pp. 246-255.   Online ISSN 1464-3677 - Print ISSN 1353-4505.

3

Procedures for Validation of Diagnostic Methods in Clinical Laboratory Accredited by ISO 15189

Silvia Izquierdo Álvarez and Francisco A. Bernabeu Andreu
Servicio de Bioquímica Clínica, Hospital Universitario Miguel Servet, Zaragoza
Hospital Universitario Príncipe de Asturias, Alcalá de Henares, Madrid,
Spain

1. Introduction

Actually, each clinical and/or biochemical laboratory has responsibility for demonstrating its competence and therefore must obtain results of good quality. Medical laboratories provide vital medical services to different clients: clinicians requesting a test, patients from whom the sample was collected, public health and medical-legal instances, referral laboratories and authoritative bodies. All expect results that are accurate and obtained in an effective manner, within a suitable time frame and at acceptable cost. There are different ways of achieving the end results, but compliance with International Organization for Standardization (ISO) 15189, the international standard for the accreditation of medical laboratories, is becoming progressively accepted as the optimal approach to assuring quality in medical testing. As result, the accreditation of clinical laboratories is shifting from being a "recommendation" to becoming a "requirement" in many countries throughout Europe and in the other countries around the world (Berwouts, 2010). *Accreditation* is defined by ISO as the "Procedure by which an authoritative body gives formal recognition that a body or person is competent to carry out specific tasks". Although accreditation also considers the quality management system (QMS), it has additional formal requirements of technical competence, including initial and continuous training of personnel, *validation of methods and instruments*, and internal and external quality control.

A good QMS in the laboratory has a lot of advantages such as increased transparency, traceability, uniformity, work satisfaction and better focus on critical points. On the contrary, it will require extra time on aspects such as document control and there is a danger of losing critical attitude and curbing innovation and changes. Therefore, a formal accreditation and the linked periodical audits are stimulant for keeping the quality system (QS) alive. Without accreditation, there is a danger of giving less attention to quality improvement. In addition, accreditation is a good way to demonstrate and attest competence and a worldwide tool to recognize laboratories. Finally, all parties (patients, families, the laboratory and clinicians) are benefited through better processes and quality of results (Berwouts, 2010).

All essential elements of QS are covered by the ISO 15189 accreditation standard in two distinct chapters: management requirements and technical requirements. Technical elements enclose personnel and training, accommodation, equipment, *validation* and assuring quality of examination procedures by internal quality control (IQC), external quality control (EQA), maintenance and calibration.

ISO 15189 standard emphasizes so in the quality of contributions to patient care as in laboratory and management procedures and specifies the quality management system requirements, in particular to medical laboratories and stages:

<<The laboratory shall use only validated procedures for confirming that the examination procedures are suitable for intended use>>, <<The validation shall be as extensive as are necessary to meet the needs in the given application or field of application>>, and <<Procedures need to be periodically revalidated in view if changing conditions and technical advances>>.

IQC is an internal verification that the test yields consistent results day after day; in the other words, the identification measure of *precision*, but not necessarily of *accuracy*. ISO 15189 requires that "the laboratory shall design IQC systems that verify the attainment of the intended quality of results". On the hand, the laboratory should avoid mistakes (ISO 15189, 5.6.1.) in the process of handling samples, requests, examinations, reports and so on; on the other, the laboratory should determine *uncertainty* (ISO 15189, 5.6.2) where relevant and possible. For each test, the laboratory should identify and define potential errors, risks and challenges (typically, during the validation phase); subsequently, specific IQC should be defined to assure each risk and potential problem.

EQA is an important complement to IQC in which a large number of laboratories are provided with the same material and required to return results to a coordinating centre. The results are compared to determine the *accuracy* of the individual laboratory. In addition, EQA provides continuous education and training for laboratories as well. Accredited laboratories are required to "participate in interlaboratory comparisons such as those organized by EQA schemes" (ISO 15189, 5.6.4). EQA should, as far as possible, cover the entire range of tests, and the entire examination process, from sample reception, preparation and analysis to interpretation and reporting (ISO 15189 5.6.5). For some specific tests, no EQA scheme exists. ISO 15189 (5.6.5) states "whenever a formal laboratory comparison programme is not available, the laboratory shall develop a mechanism for determining the acceptability of procedures nor otherwise evaluated"; examples include reference materials or interlaboratory exchange. Interlaboratory comparisons should cover the scope of services offered and there should be a formal mechanism of review and comparison of results.

Used together, IQC and EQA provide a method of ensuring accuracy and consistency of results and are vital tools in the laboratory. The relation between precision and accuracy may be illustrated by the familiar example of shooting arrows at a target (Berwouts, 2010; Burnett, 2006) (figure 1).

The results provided by the clinical/medical laboratory must be accurate to allow a correct clinical interpretation and to be comparable with earlier or later and between laboratories. So the purpose of this chapter is to establish a set of guidelines and recommendations to help personnel carry out their work in clinical/medical laboratories that are accredited or under accreditation by ISO 15189. It is necessary to establish and define the different procedures validation, the fundamental guidelines for the proper design of the validation, the recommendations to validate an established method in the laboratory, and the different parameters to be assessed.

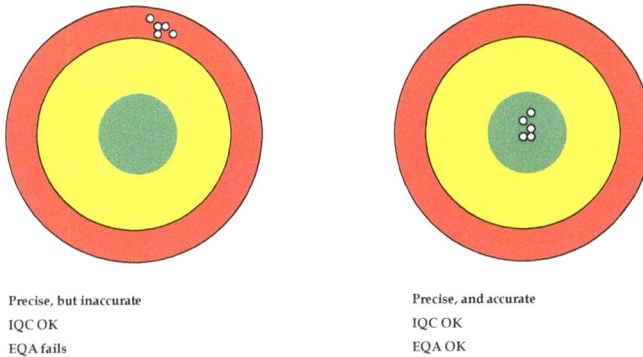

Precise, but inaccurate
IQC OK
EQA fails

Precise, and accurate
IQC OK
EQA OK

IQC: Internal Quality Control
EQA: External Quality assessment

Fig. 1. Accuracy and precision.

2. Validation design of a method

Diagnostic validation is a formal requirement of accreditation standards, including ISO 17025 and ISO 15189, those tests/methods and instruments must be validated before diagnostic use to ensure reliable results for patients, clinicians or referring laboratories and their quality must be maintained throughout use. In other words, the laboratory must demonstrate that their tests/methods are fit for the intended use before application to patient samples. Figure 2 shows a summary of what ISO 15189 states with regard to validation (Berwouts, 2010; Burnett & C. Blair, 2001, Burnett et al., 2002; Burnett, 2006).

Although the concept of validation makes explicit reference to the purely analytical aspects may also include preanalytical and sampling procedures, handling and transport. At a minimum, the techniques used to determine the performance of a method should be one or more of the following:

- Calibration using reference standards or reference materials or traceable to these
- Comparison of results obtained with other methods
- Interlaboratory comparisons
- Systematic evaluation of the factors that influence outcomes
- Estimate of the uncertainty of the results based on scientific knowledge of theoretical principles of the method and practical experience.

In any case, analytical methods must be those that meet customer requirements, that is, those that provide clinically useful information. Thus, an analytical method for determination of aluminium in serum based on the complexation of this element with 8-hydroxyquinoline and quantification by fluorimetry can have high reliability but its detection limit is at least an order of magnitude above the upper limit of the reference element, which makes this method, does not meet customer requirements.

There are publications that provide general methods of evaluation of analytical methods including the following:

1. Definition of the evaluation protocol that registers the results of the measurements.
2. Determining the range of application and dilution mechanisms, if any.
3. Identify the components of precision in the day, every day.
4. To determine the accuracy through recovery studies before a definitive method or reference, if any.
5. To determine the sensitivity.
6. Estimate the limit of detection and quantification (and others if applicable).
7. To study the specificity of the method, checking for interference.
8. Establishing the reference range.
9. Document the validity of the method.

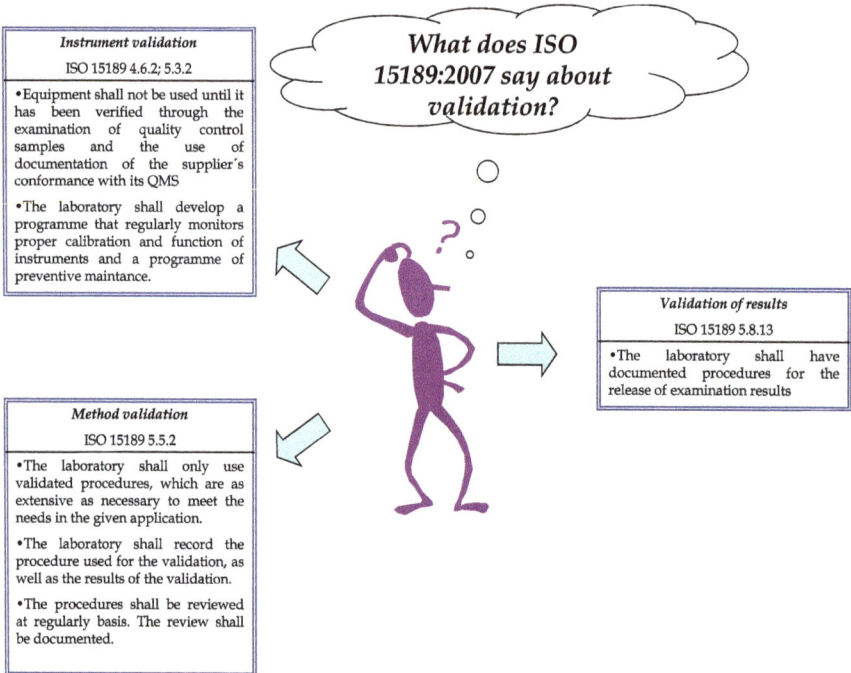

Fig. 2. Validation requirements according ISO 15189:2007.

Regarding the validation and control of analytical procedures, paragraph 5.5 of ISO 15189: 2007 specifies to be used those procedures "that have been published by experts or international guidelines, national or regional". Own procedures "must be properly validated for its intended use and fully documented".

So, ISO 15189: 2007 says "as appropriate, the documentation should include the following: technical specifications (e.g., linearity, precision, accuracy, expressed as a measurement uncertainty limit of detection, measurement, sensitivity and specificity, and interference)" (ISO 15189, 5.5.3).

When a unit or section of a clinical/medical laboratory chooses to engage in the accreditation ISO 15189, must be aware that although the analytical methods which usually works have been validated in its implementation must be validated in time.

A validation process like any other requires a series of planning, execution and control to ensure that the results come to fruition.

a.   Planning:
   • Definition responsible for performing the validation process.
   • Definition of objectives and internal requirements applied the method to validate (purpose, parameters to measure in the matrix or matrices to be determined).
   • Definition and documentation of the method (procedure for validation).

b.   Implementation: implementation of activities, results obtained and recorded (date, operators and results).

c.   Control:
   • Verification of compliance with targets.
   • Final Declaration of the appropriateness of the procedure defined.

a. *Planning* consists of the following phases:

1) Assign responsibility. In this phase it will be defined a person responsible for carrying out the validation process and deciding the outcome. This person can count on help from others, but he is responsible for making decisions so, he must have a proper qualification; 2) Definition of the characteristics and requirements applied to the method: The definition of requirements has to do with the intended use of the method (i.e. as property or analyte, the matrix or matrices in which they will determine the use that will make the test results and legal requirements or economic policy to be applied to test results), from the specified requirements and based on a literature search using other standards, etc. There is a design and optimization phase of the procedure that is performed by laboratory. This is the stage where, for example for an instrumental method, you must establish a priori the linearity of the method, the working range, the limit of detection and quantification is desired, the accuracy and precision fit. In short what features the laboratory can apply the method to the intended use; 3) Description documented procedure: It should be sufficiently detailed to ensure its proper performance and repeatability. This ensures that all laboratory personnel that are qualified can do just as the method with comparable results.

To accomplish this phase can be helpful in establishing a suitable index of the case as the reference standard.

b. *Implementation*: Outcome is based on the realization of a series of tests and experiments that occur as a result values for the parameters defined in the requirements. These parameters can be variable depending on the type of method applied and the requirements and can include accuracy, precision, limit of detection, limit of quantization, selectivity, etc.

c. *Control*: The control is the verification of compliance and the final declaration. 1) Verification of compliance: As a result of the implementation of activities will be decided whether the values meet the specified requirements, in which case proceed to establish which checks should be made to the method as regular monitoring to confirm that remain requirements requested at the time of validation, e.g. using a control pattern periodically check the parameters of the regression line, etc., proceeding to their inclusion in the proceedings and preparing a final edition of the same. Otherwise you may be assessed if an amendment to the previously established requirements. 2) Final Statement: All the validation process should conclude with a formal statement of the adequacy of the procedure defined as stated is suitable for their intended use, according to specified requirements (Burnett & C. Blair, 2001; Burnett et al., 2001; Burnett, 2006).

2.1 Types and methods of validation

The laboratory shall validate examination procedures from non standard methods, laboratory designed or development methods, standard methods used outside their intended scope and modified validated methods.

When examination procedures have been validated by the method developer (i.e., the manufacturer or author of a published procedure), the laboratory shall obtain information from the method developer to confirm that the performance characteristics of the method are appropriate for its intended use. When changes are made to a validated examination procedure, the influence of such changes shall be documented and, if appropriate, a new validation shall be carried out.

Examination procedures from method developers that used without modification shall be subject to verification before being introduced into routine use. The verification shall confirm, through provision of objective evidence (performance characteristics), that the performance claims for the examination method have been met. The performance claims for the examination method confirmed during the verification process shall be those relevant to the intended use of the examination results.

Verification and validation are two slightly different procedures (figure 3). By default, all new laboratory procedures must be validated before application to clinical testing. In addition, a validation is necessary when major technical modifications to existing methods are carried out or when the performance of existing methods has been shown to be unsatisfactory (Berwouts, 2010; Hauck et al., 2008).

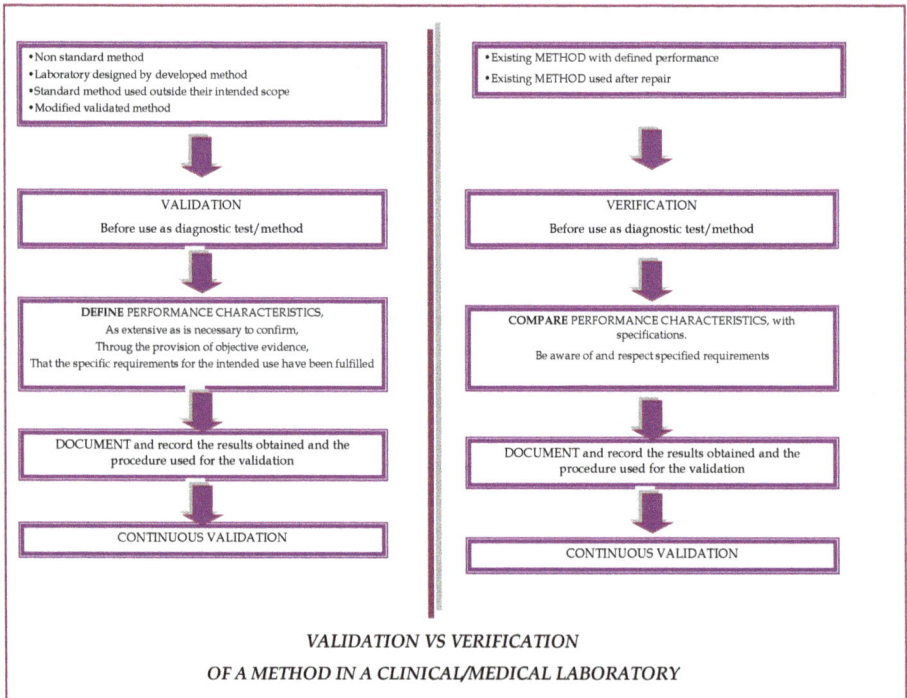

Fig. 3. Validation vs verification in diagnostic methods.

The validation or verification of methods, as defined in figure 4, is a normal requirement for the accreditation of laboratories according to the two major international standards applicable to clinical/medical laboratories, ISO 15189 and ISO 17025. Although the general requirements are clearly stablished (figure 4), the standards provide very little guidance about the detailed requirements or procedures.

Before a test/method can be validated, it is necessary to establish (a) that the particular measurements are diagnostically useful and (b) that the correct analyte(s), and only the correct analyte(s), are measured.

Full validation is required when is no suitable performance specification available, for example, with novel tests/methods or technologies. This process involves assessing the performance of the test/method in comparison with a "gold standard" or reference test/method that is capable of assigning the sample status without error. In simple terms, validation can be seen as a process to determine whether the laboratory is "performing the correct test/method". Validation data can be used to assess the accuracy of either the technology or the specific test/method. Generally speaking, the generic validation of a novel technology should be performed on a larger scale, ideally in multiple laboratories (interlaboratory validation), and should include a much more comprehensive investigation of the critical parameters relevant to the specific technology to provide the highest chance of detecting sources of variation and interference (Berwouts, 2010; Burnett, 2006).

Principle requirements of ISO 157025:2005	Principle requirements od ISO 15189:2007
5.4.2 "Laboratory-developed methods or methods adopted by the laboratory may also be used if they are appropiate for the intended use and if they are validated". 5.4.5.2 "The laboratory shall validate non-standard methods, laboratory-designed/developed methods, standard methods used outside their intended scope, and amplifications and modifications of standard methods to confirm that methods are fit for the intended use. The validation shall be as extensive as is necessary to meet the needs of the given application or field or application.The laboratory shall record the results obtained, the procedure used for validation, and a statement as to wether the method is fit for the intended use". 5.4.5.3 "NOTE 1 validation includes specification of the requirements, determination of the characteristics of the methods, a check that the requirements can be fulfilled by using the method, and a statement on the validity. NOTE 3 Validation is always a balance between costs, risks and technical possibilities, There are many cases in which the range and uncertainty of the values (eg accuracy, detection limit, selectivity, linearity, repeatability, robutness and cross-sensivity) can only be given in a simplified way due to lack of information".	5.5.1 "[....]If in-house procedures are used, they shall be appropriately validated for their intended use and fully documented". 5.5.2 "The methods and procedures selected for use shall be evaluated and found to give satisfactory results before used for medical examinations. A review of procedures by the laboratory director or designated person shall be understaken initially and at defined intervals". 5.6.2 "The laboratory shall determine the uncertainty of results, where relevant and possible".

Fig. 4. Principle requirements of ISO 15189: 2007 and ISO 17025:2005 about validation and verification.

2.2 Recommendations to validate a method developed in the laboratory
2.2.1 Quantitative methods

Two components of analytical accuracy are required to characterize a quantitative method: trueness and precision. Trueness expresses how close the methods result is to the reference value. Typically, multiple measurements are made for each point and the rest method result is taken to be the mean of the replicate results (excluding outliers if necessary). As quantitative assays measure a continuous variable, mean results are often represented by a regression of data (a regression line is a linear average). Any deviation of this regression from the reference indicates a systematic error, which expressed as a bias (i.e., a number indicating the size and direction of the deviation from the true result). There are two general forms of bias. With constant bias, method results deviate from the reference value by the same amount, regardless of that value. With proportional bias, the deviation is proportional to the reference value. Both forms of bias can exist simultaneously. Although measurement of bias is useful, it is only one component of the measurement uncertainty and gives no indication of how dispersed the replicate results are. This dispersal is called precision and can be measured by imprecision, that provides an indication of how well a single method results is representative of a number of replicates or repetitions. Imprecision is commonly expressed as the standard deviation of the replicate results, but is often more informative to describe a confidence interval (CI) around the mean result. Precision is subdivided according to how replicate analyses are handled an evaluated.

Repeatability refers to the closeness of agreement between results of test performed on the same method items, by the same analyst, on the same instrument, under the same conditions in the same location and repeated over a short period of time. Repeatability represents "within-run precision".

Intermediate precision refers to closeness of agreement between results of methods performed on the same method items in a single laboratory but over an extended period of time, taking account of normal variation in laboratory conditions such as different operators, different equipment and different days. Intermediate precision therefore represents "within-laboratory, between-run precision" and is therefore a useful measure for inclusion in ongoing validation.

Reproducibility refers to closeness of agreement between results of methods carried out on the same method items, taking into account the broadest range of variables encountered in real laboratory conditions, including different laboratories. Reproducibility therefore represents "inter-laboratory precision".

In practical terms, internal laboratory validation will only be concerned with repeatability and intermediate precision and in many cases both can be investigated in a single series of well-designed experiments. Reduced precision indicates the presence of random error. The relationship between the components of analytical accuracy, types of error and the metrics used to describe them is illustrated in figure 5.

Any validation should also consider robustness, which, in the context of a quantitative method, could be considered as a measure of precision. However, robustness expresses how well a method maintains precision when faced by a specific designed "challenge", in the form of precision does not represent random error. Typical variables in the laboratory include sample type, sample handling, sample quality, instrument make and model, reagent lots and environmental conditions (e.g., humidity, temperature). Appropriate variables should be considered and tested for each specific method. The principle of purposefully challenging methods is also applicable to both categorical and qualitative methods and should be considered in the validation as well. Robustness can be considered as a useful prediction of expected intermediate precision (Berwouts, 2010).

As trueness and precision represent two different forms of error, they need to be treated in different ways. In practice, systematic error or bias can often be resolved by using a correction factor; constant bias requires an additive correction factor, whereas proportional bias requires a multiplicative correction factor.

For quantitative methods, particularly those requiring absolute quantification, it is most effective to estimate analytical accuracy on an ongoing basis by running a set of calibration standards (standard curve) with each batch or run. In this case, it is important that linearity be evaluated and that the lower and upper standards are respectively below and above the expected range of the results as precision cannot be assessed on extrapolated results. Where possible, calibration standards should be traceable to absolute numbers or to recognized international units.

Other factors that may need to be evaluated include the limit of detection defined as the lowest quantity of analyte that can be reliably detected above background noise levels and the limits of quantification that define the extremities at which the measurement response to changes in the analyte remains linear (Berwouts, 2010).

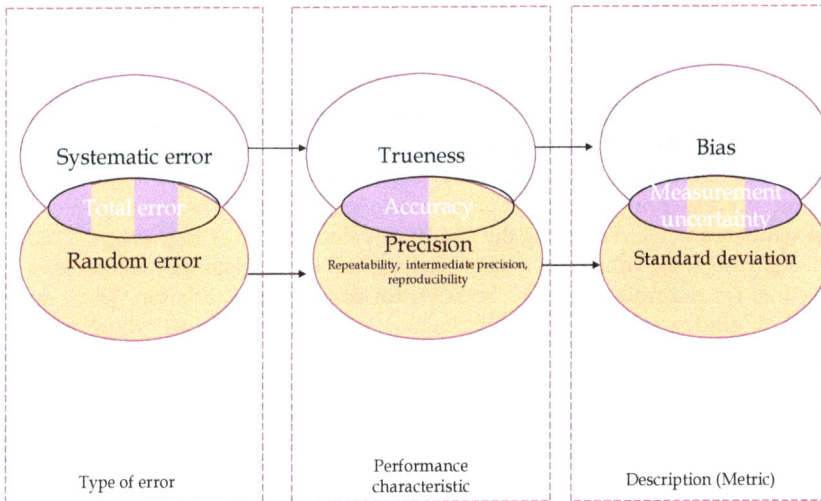

Fig. 5. Performance characteristics, error types and measurement metrics uses for qualitative methods (adapted from Menditto et al., 2007).

2.2.2 Qualitative methods

This is an extreme form of a categorical test/method, in which there are only two result categories, positive and negative. This binary categorization can be based either on a cut-off to a quantitative result. The diagnostic accuracy of a qualitative method can be characterized by two components, both of which can be calculated: sensitivity (the proportion of positive results correctly identified by the method), and specificity (the proportion of negative results correctly identified by the method).

3. Parameters required for considering in the validation or revalidation of a method

There are several measurable parameters that should be taken into account during validation or verification. The estimation of *accuracy* is a key parameter. Accuracy consists of both precision and trueness for quantitative and semiquantitative test/method. *Precision* or "closeness of agreement between results of replicate measurements" includes the following:
Repeatability: within-run variation (same sample, same conditions).
Intermediate precision: between-run variation within a single laboratory (different samples, operators, equipments).
Reproducibility: between-run variation in different laboratories (different samples, operator).
Robustness: variation when confronted with relevant challenges (e.g., sample type, environmental conditions and so on).
Trueness is the "closeness of agreement with a reference value". Appropriate reference materials are, therefore, essential and could include positive and negative/normal controls, certified reference materials, EQA materials, synthetic samples or material characterized by another technique.
The components of accuracy for quality tests/methods are *sensitivity* and *specificity*. *Sensitivity* is a measure of how well the test/method detects positive results, whereas *specificity* describes how well negatives are detected.
Thorough documentation during a validation process is essential, especially in the context of accreditation process pragmatic approaches, reconciling the formal requirements of accreditation standards while respecting the aim that "validation must be practical", such as the design of IQC based on validation results, making full use of data that laboratories are already collecting, for example from IQC or EQA, for continuous validation. There are no detailed practical guides for validation of diagnostic methods in medical laboratories; moreover, the accreditation standards have no specific details about how to fulfil their requirements. The laboratory must decide on this, on the basis of their experience and performance requirements; it is duty of the laboratory to provide evidence that tests results provided are reliable, and that the performance claims are correct (Burnett, 2006; Hauck et al., 2008).
In the end, validation is never finished. The implementation of quality indicators for systematically monitoring and evaluating the laboratory's contribution to patient care is good way to continuously validate diagnostic tests/methods, apart from IQC, EQA and other data (Eurachem, 2010; Hauck et al., 2008).
But when the method has already implemented several years and with performance and optimal quality specifications, we recommend periodic revalidation through the information provided by the treatment program reports or international interlaboratory comparison or testing fitness, programs or external quality control, in which the laboratory has been involved for years. The advantage of this validation methodology described above, is to estimate the imprecision, bias and uncertainty, without much effort and without having to make a lot of trials and experimental trials that would mean stopping the routine work in

the laboratory. You can validate the micro-range or specific concentration range, applicable as diagnostic daily reality for clinical laboratories.

3.1 Estimation of the accuracy using reference materials

ISO 15189 accreditation for clinical laboratories require a verification of the accuracy of the measurement procedures. The study of the accuracy, by estimating the systematic error, should be in the validation of the measurement. To study the accuracy of a measurement procedure is necessary to compare average values obtained with a conventional true value. In the clinical laboratory can be used as true values considered, mainly 3 types: value assigned to a reference material, the consensus value obtained in a program of external quality assessment, the value obtained with a reference measurement procedure.

The accuracy is expressed numerically by systematic error, which is the difference between the average measurement results obtained and a conventional true value. The values assigned to some reference materials may be considered conventionally true values.

The reference materials used in the study must have a value assigned to the magnitude that is measured and the corresponding value of uncertainty. You should also know the traceability of the assigned value. It is preferable that the material has a matrix similar to human samples. The main types of materials: certified reference materials, prepared by metrology institutes or other organizations related to metrology and reference materials business (controls to the truth).

The reference material manufacturer must provide the traceability and uncertainty of assigned values; the latter expressed as standard uncertainty or expanded uncertainty. Along with the uncertainty value must also specify the coverage factor used.

The results of accuracy studies should be used to validate the accuracy of the measurement procedures, ensures the absence of relevant and introduce systematic errors in the calculation of uncertainty of measurement uncertainty components associated with any correction factors.

3.2 Estimation of the accuracy from participation in external quality assessment programs

The external quality control includes different activities aimed at assessing the accuracy of the results through the intervention of an organization outside the laboratory. The most common form of external quality control is comparisons between laboratories or programs of external quality assessment.

These programs are organized by professional associations, government agencies or manufacturers of control materials that have a similar function. Participating laboratories measured once a magnitude of a control material of unknown value. Organization of the program collects the results of laboratory and a study of the data then forwarded to each participating laboratory, informing about the error of its outcome. The duration of the program, the number of measurements that are performed and the number of different materials are used, depending on the different programs. To study the accuracy is recommended that the program in which you participate fulfill the following conditions: high number of participants, the laboratory has a minimum of 12 results for participation and that you know the standard deviation characterizes the dispersion of results among participating laboratories.

3.3 Estimation of measurement uncertainty

The results provided by the clinical laboratory must be accurate (true and precise) to allow a correct clinical interpretation and to be comparable with earlier or later and between laboratories.

The error of measurement of clinical laboratory results is almost always unknown. Instead, it is possible to ascribe a measurement uncertainty and metrological traceability of each result. The uncertainty is a numerical expression of the degree of doubt of the result. Traceability relates the result with reference values established allowing reproducibility over time and between laboratories (Eurachem, 2010).

In the estimation of measurement uncertainty is assumed that any systematic error is eliminated, corrected or ignored, random effects are assessed on the outcome of an action and establishing a range within which lies the true value of the measured magnitude a certain level of confidence. The standard for laboratory accreditation ISO 15189 requires an estimate of the uncertainty of the results. The appropriate methodology for estimating the uncertainty described in the Guide to the Expression of Uncertainty in Measurement (GUM). The GUM was developed jointly by several international organizations for standardization and metrology for use in calibration and testing laboratories and measures applied to physical or chemical analysis. Currently, the GUM is difficult to apply to measures that are performed in clinical laboratories, although they maintained their principles. Moreover, the complexity and cost of obtaining an estimate of the uncertainty of measurement must be commensurate with the quality requirements applicable to the clinical use of the results.

Sources are contributing to the uncertainty of a result as follows: sample collection, sample preparation, calibrators or reference materials, input quantities (e.g., absorbance), computer equipment used, environmental conditions, sample stability and changes in workers.

The uncertainty associated with the collection and sample preparation is difficult to estimate and should be reduced through rigorous standardization of procedures. In this paper only consider the sources of uncertainty in the analytical phase, which begins when the sample interacts with the first technical step of the measurement (for example, placing the sample into an analyzer) and ends with obtaining a value numerical measurement result.

The main components of the uncertainty of the analytical phase correspond to the uncertainty of the measured, the stability of the sample in the measurement system calibration, the volume dispensed, the batch of reagents, instrumentation equipment, operators and environmental conditions. In the following paragraphs, are discussed in more detail the main components.

Measurement uncertainty is a parameter that is specifically associated with each outcome. In clinical laboratories, it is impossible to estimate particular measurement uncertainty for each measurand of each sample, so it makes a rough estimate of the uncertainty of measurement for a measurand defined and values of the same close to decision clinic. Measurement uncertainty does not apply to qualitative tests, in which the result is a numeric value.

3.3.1 Definition of measurand

The measurand is defined by the following parameters:

a.   Analyte to be measured. For example, protein, sodium ion, cholesterol, ASO, hemoglobin, white blood cell counts, etc.

b.   System. For example, serum, urine, venous blood, pleural fluid, etc.

c.   Type size and unity. For example, substance concentration (mmol/L), mass concentration (g/L), catalyst concentration (nkat/L), etc.
d.   Measurement procedure.

The existence of different molecular forms of the analyte can introduce uncertainty in the results. This source of uncertainty can be reduced or eliminated by careful definition of the measurement, so they may react differently to some or other molecular forms.

Another source of uncertainty regarding the definition of the measurand are possible cross-reactions and interference that can occur with some samples and must be identified and documented to prevent, where possible, their influence.

In short, uncertainty caused by the uncertainty of the measurand can not be quantified, but may be reduced or eliminated by detailed specification of the measurand.

3.3.2 Imprecision

Most of the components of measurement uncertainty of the analytical phase are contained in the estimation of imprecision (CV_{id}). It is usually obtained using control materials.

This assessment should be a sufficient number of data to collect the different sources of uncertainty apply, i.e., a minimum of six months of data and new estimation every year. In the period of data collection should include several calibrations to collect the uncertainty generated by the calibration process. Moreover it is necessary to use different batches of calibrator if you have the uncertainty of the assigned value.

The estimate of CV_{id} is made for a measurement value close to the values of clinical decision.

3.3.3 Value assigned to the calibrator

The clinical laboratory must know the uncertainty and metrological traceability of values assigned to calibration materials used. As usual it is commercial material the manufacturer must provide such data (Directive 98/79/EC). Along with the uncertainty value must also specify the coverage factor used. Typically, uncertainty is expressed as expanded uncertainty (U) for a confidence level of 95% (coverage factor = 2).

The standard uncertainty (u) is calculated by dividing U by the coverage factor. U on (%) of the value assigned to the gauge should not vary excessively batch to batch and should generally be lower than CV_{id}.

3.3.4 Systematic error (bias)

The estimation of measurement uncertainty is assumed that any significant systematic error of the measurement procedure has been deleted, corrected or ignored. The identification of a possible systematic error should be done during the validation of the measurement procedure.

When systematic error is corrected by a factor, the correction has an associated uncertainty (u_{cf}) that should be considered in calculating the combined measurement uncertainty.

Systematic errors caused in the routine use of the measurement by the inevitable differences between different calibrations behave randomly in the long term, so this component of uncertainty is reflected in CV_{id}.

3.3.5 Uncertainty calculation

The uncertainty is calculated by combining various sources. For this reason, clinical laboratories should identify each measurand, specifying the measurement procedure, and calculate for each of them calculate the combined uncertainty from the data of internal quality control and other data, using the following equation:

$$u_c = \sqrt{CV_{id}^2 + u_{cal}^2 + u_{cf}^2}$$

Where:

u_c: relative combined standard uncertainty (%); CV_{id}: imprecision (coefficient of variation) interday; U_{cal}: relative standard uncertainty (%) of the value assigned to the calibrator; u_{cf}: relative standard uncertainty (%) of the factor used to correct a systematic error. It is recommended to express the combined uncertainty for a confidence level of 95% (expanded uncertainty, U_c). To do this, multiply the value of u_c for k = 2.

$$U_c = 2 \times \sqrt{CV_{id}^2 + u_{cal}^2 + u_{fc}^2}$$

The relative expanded uncertainty should be expressed to two significant figures, for example: 4.2%, 16%.

3.3.6 Interpretation

The estimation of measurement uncertainty provides a quantitative indication of the level of doubt that the laboratory has in each result and is therefore a key element in the system of analytical quality in clinical laboratories.

The relative expanded uncertainty of a measurand should be less than one third of the Maximum Permitted Error (MPE). If it was superior, should be studied in greater detail the different sources of uncertainty, identify the most significant and perform the appropriate actions to reduce them.

3.3.7 Applications

The uncertainty of measurement should be used primarily for:

- Selection of measurement procedures that fulfill the specifications of accuracy.
- Strict interpretation of the significance of a change between two consecutive values of magnitude biochemistry.
- Strict interpretation of the significance of a result compared with a value of clinical decision.

3.3.8 Limitations

The value of the measurement uncertainty varies with the concentration of the measurand and may be substantially different for very low or very high analyte. For this reason it is recommended that the estimate for a concentration closes to clinical decision values.

3.4 Estimation of precision

Precision is one of the most important metrological characteristics to be considered for selection and implementation of a measurement procedure in the clinical laboratory. In addition, the quantitative understanding of this feature is essential for establishing tolerance intervals of internal control materials for the objective interpretation of the significance of a change between two consecutive values of magnitude biochemistry, and the calculation of uncertainty.

The accuracy can be studied under conditions of repeatability, reproducibility and intermediate. The study conditions that are more interested in the clinical laboratory are the kind of repeatability and intermediate terms, which vary from day to day.

Before starting the study a period of familiarization with the measurement procedure and the operator's experience are recommended.

Also for this study is recommended that samples for this study were commercial materials or control samples. We recommend using at least two samples with different concentrations of the magnitude under study, with a value within the physiological range or close to a discriminate value, and one with a pathological value. When it sees fit, they can not be tested with values close to the limits of the measuring range of the procedure. Samples should be stable during the duration of the study. If using commercial control materials, whenever possible, they should be interchangeable with samples of human origin.

The run imprecision were obtained under the same conditions of repeatability, ie the same samples, the same operator, the same components of the measurement system for a short time and without calibrations between measurements. A minimum of 30 measurements are required for fulfilling statistical criteria.

If the results of the run imprecision are not consistent with previous results, either supplied by the manufacturer or obtained in the literature, the study should be stopped to find and correct the cause of the discrepancy.

Imprecision is obtained under certain conditions. Each laboratory should perform the estimation with standard calibration frequency (daily weekly, etc), and changes of operator, calibrator lot, reagent lot, etc, which are common in everyday work. Following the statistical criteria are recommended to estimate a minimum of 30 days.

Before calculating the mean, standard deviation and coefficient of variation of the results must be detected the presence of possible outliers. An abnormal result will be removed provided that it is related to a documented error or has demonstrated statistically that is an outlier. After the removal of outliers, if any, imprecision is calculated by the coefficient of variation.

4. Documentation: procedures and instructions needed for the validation of a method in the clinical laboratory

International Standard ISO 15189:2007 clearly identifies the documentation requirements necessary to determine compliance with the requirements referred to for quality and competence of clinical laboratories.

Standard clinical laboratory means (paragraph 3.8) that "laboratory devoted to biological, microbiological, immunological, chemical, immunohematological, biophysical, cytological, pathological or other material derived from the human body in order to provide diagnostic information, prevention and treatment of diseases or the assessment of human health and can provide a consultant advisory service covering all aspects of laboratory analysis, including interpretation of the findings and recommendations on any proper analysis additional".

The implications of documentary that suggests the validation of a method, it follows that it must develop a set of documents or records.

Registration means that documentary evidence of a fact that has occurred and is understood by documentary evidence to document that describes how the activities should be conducted.

According to these definitions it is able to state that the laboratory should have a defined overall validation procedure (document) that describes: What activities will be performed; responsibility to perform; records to retain; how to be performed.

To confirm the verification of compliance, apply the method to real matrices, records to keep are: 1. Requirements applied to the method (Must be defined prior to conducting the tests, indicating preserved based on what have been defined); 2.Records of previous tests. (Straight calibration standards used, results obtained from different computers, etc.); 3. Written procedure (approved by qualified personnel); 4. Results of tests for checking compliance with requirements (The laboratory must clearly indicate the results of the parameters and the comparison with the specified requirements); 5. Statement by the head of the validation of the procedure is suitable for their intended use based on the evidence (All these records should include dates, personnel and equipment used in ways that can be reconstructed).

It must have an overall validation procedure describing the activities undertaken; those responsible for conducting, records to keep (the method established requirements, records of tests: calibration lines, patterns, etc.).

Lab logo	Registration	"Name of the Laboratory"	
	Design/planning of the method validation	Date:	
		CODE:	
		Review:	
		Page X of Y	
Method:			
Responsible for validation:			
Used:			
References:			
Procedure/codification:			
Objectives:			
Parameter	Value	Observations	

Fig. 6. Template record: Design / planning of the method validation.

It must be developed and used a generic template so that it is not necessary to have to develop a validation process for each method, but simply change the data in the template. Thus, for any method in the laboratory which will continue to want to validate one of the two ways described: the classical or from the results of inter-comparison programs. It is used for the models listed in the Annexes to this case: report validation, design / planning of the method validation, quality control plan method. They detail steps for each procedure to be followed in the validation.

Here are a few examples of formats and / or templates of records that are considered necessary for the validation of a method:

- Template record: Design / planning of the method validation (figure 6).
- Template record: Report of validation (figure 7).
- Registration: Plan quality control method (figure 8).
- Title page of a validation procedure in clinical laboratory methods (figure 9).
- Technical Registration. Spreadsheets (Excel) (figure 10).

LAB LOGO	VALIDATION REPORT		"Name of the laboratory"		
			Date:		
			Code:		
			Review		
			Page X of Y		
Date of procedure of validation	Register	Observations	Personnel		
Additional data					
Parameter	Result / value	Observations			
Declaration: ❑Validated method ❑No validated method ❑Validated method with restrictions					
Validation criteria:					
Date of validation of the method:		Signature: (Responsible of validation)			

Fig. 7. Template record: Report of validation.

Lab logo	Registration	"Name of the Laboratory"	
	Quality control plan method	Date:	
		CODE:	
		Review:	
		Page X of Y	
Method for validation (Description)	**Quality Control**	*Periodicity*	*Acceptance criteria*
	Accuracy		
	Trueness		
	Precision		
	Repeateability		
	Reproducibility		
	Uncertainty		
	Robutness		
	Specificity		
	LOQ		

Fig. 8. Registration: Plan quality control method.

LAB LOGO	MANUAL OF PROCEDURES	"Name of the laboratory"
		Date:
	PROCEDURE FOR VALIDATION METHODS	Code:
		Review
		Page X of Y

PROCEDURE FOR VALIDATION METHODS

Developed	*Revised*	*Approved*
Date		
Technical Manager	*Quality Manager*	*Lab Director*

Fig. 9. Title page of a validation procedure in clinical laboratory methods

LAB LOGO	MANUAL OF PROCEDURES	"Name of the laboratory"
		Date:
	PROCEDURE FOR VALIDATION METHODS	Code:
		Review
		Page X of Y

CONTENTS

1. OBJECTIVE AND SCOPE

2. DEFINITIONS

3. REFERENCE DOCUMENTS

4. PROCEDURE

      4.1. Accuracy

      4.2.Precision

      4.3.Trueness

      4.4. Repeateability

      4.5.Reproducibility

      4.6. Uncertainty

      4.7. Another validation parameters

5. ANNEXES

Fig. 10. Contents in a validation procedure in clinical laboratory.

5. Conclusion

It is important to have documented procedures for the validation of different diagnostic methods available within a clinical laboratory. There is a need to develop practical guidelines for method validation procedures in clinical laboratories through the various tools available to the laboratory. There is no single way to validate a diagnostic and clinical laboratory validate and verify the validation of their methods over time to meet the requirements of the existing accreditation standards and to demonstrate the laboratory's technical competence to offer quality results.

6. References

Berwouts, S.; Morris, M.A. & Dequeker, E. (2010).Approaches to quality management and accreditation in a genetic testing laboratory. *European Journal of Human Genetics.* No.18, pp. 1-19.

Burnett, D. (2006). ISO 15189:2003-quality management, evaluation and continual improvement. *Clin Chem Lab Med.* No. 44, pp. 133-739.

Burnett, D. & Blair, C. (2001). Standards for the medical laboratory-harmonization and subsidiarity. *Clin Chem Acta.* No. 309, pp. 137-145.

Burnett, D.; Blair C.; Haeney, M.R.; Jeffcoate, S.L.; Scott, K.W. & Williams, D.L. (2002). Clinical pathology accreditation: standards for the medical laboratory. *J Clin Pathol.* No. 55, pp. 729-733.

Eurachem. (May 2010). The firness for purpose of analytical methods a laboratory guide to
    method validation and related topics, Avaliable from:
    http://www.eurachem.org/guides/valid.pdf
Hauck, W.W.; Kock, W.; Abernethy, D. &Williams, R.L. (2008). Making sense of trueness,
    precision, accuracy and uncertainty. *Pharmacopeial Forum.* No.34, pp.838-842.
International   Organization   for   Standardization.   Medical   laboratories-Particular
    requirements for quality and competence. ISO 15189:2007.
International Organization for Standardization: General requirements for the competence of
    testing and calibration laboratories.ISO/IEC 17025:2005.
Menditto, A.; Patriarca, M. & Magnusson, B. (2007). Understanding the meaning of
    accuracy, trueness and precision. *Accred Qual Assur.* No. 12, pp.45-47.

4

From a Quality Assurance and Control System for Medical Processes, Through Epidemiological Trends of Medical Conditions, to a Nationwide Health Project

Yossy Machluf[1,2], Amir Navon[1], Avi Yona[1], Avinoam Pirogovsky[1,3], Elio Palma[1,4], Orna Tal[5], Nachman Ash[6], Avi Cohen[1] and Yoram Chaiter[1]

[1]*Quality Assurance and Control Committee, Medical Corps, IDF,*
[2]*Weizmann Institute of Science, Rehovot,*
[3]*Head of Standards and Regulation Department in the Division of Community Medicine, Ministry of Health, Tel Aviv,*
[4]*Head of Department of Occupational Medicine, Clalit Health Services, Afula,*
[5]*Israeli Center for Technology Assessment in Health Care; The Gertner Institute for Epidemiology and Health Policy Research, Head of Emerging Technologies Unit, Tel Aviv,*
[6]*Chief Medical Officer, Medical Corps, IDF,*
Israel

1. Introduction

A health policy is an integral part of the general welfare policy in every state. According to Dye (1987), interior public policy constitutes the same action as a government that chooses to "do or not do" (Dye, 1987), and thus it is assumed that any governmental action is derived from its will to preserve and provide quality services to the public. Quality of patient health care is defined by two principal dimensions: access and effectiveness (Campbell *et al.*, 2000). Over the last few decades, quality control has been gaining a central place in public organizations in order to improve the quality of services and treatment (Blumenthal, 1996; Landon, *et al.*, 2003; Mandel, *et al.*, 2003; 2004). Quality assurance is a key component in the processes aimed at improving the quality of service and medical care.

1.1 Quality control and the health system

The quality index in the health system is a criterion that shows measurable values in morbidity and service levels (Campbell *et al.*, 2000). Since the 1960s, the quality indices in health systems were divided into three main levels: infrastructure and structure, process, and outcome. The last two are usually included as preferred measures (Donabedian, 2005). Infrastructure and structure indices are related to organizational issues of the health services, the attending person's nature, and the procedures and medical policy that are being implemented by the organization on both the private and public levels. The assumption is that any health organization should be capable of providing quality health services according to its resources, which are made up of human, economical, and infrastructure components. Process indices are aimed at examining the extent of the medical actions that are taken to achieve the desired target, assuming that well performed operations increase the chance of accomplishing the desired effects. By contrast, outcome indices focus on the individual level rather than on the organization level. Outcome measurements provide an indirect measure of the overall quality assessment and may provide a benchmark for tracking progress. In general, the more the first two indices (the infrastructure and structure and the process indices) are involved in measuring quality, the greater the reliability of the outcome measures (Donabedian, 2005).

Many approaches were developed to assess the quality of health care and to improve both medical and patient processes, such as the Six Sigma, which utilizes the DMAIC model (Define, Measure, Analyze, Improve, and Control); ISO (International Organization for Standardization); BOS (Business Operating System); CI (Continuous Improvement); TQM (Total Quality Management); etc. (Donabedian, 2003; Munro, 2009; Ovretveit, 1992; Ransom *et al.*, 2008). Many of these approaches are derived and adapted from quality assurance systems in industry, where processes are straightforward and the implementation of such methods is easier. Yet, the medical process represents an intricately interwoven and dynamic process (Donabedian, 2003, Ovretveit, 1992) where many variables are interconnected. For instance, in medical committees, managing authorities, health care medical professionals, technical and administrative personnel, and patients, as well as medical policy, regulations, and goals are all part of medical processes. Therefore, difficulties may arise while applying these methods to medical procedures.

Although there are many models of health care quality control and assurance, most focus on specific issues and are a tool for managerial decisions and not for the day-to-day surveillance of the processes on a clinical level. They may address a specific issue at a local medical facility and may try to improve the specific circumstances, such as the high rates of certain infections at a specific ward or at a specific medical center. Most of the quality assurance processes measure outcomes rather than the completeness and intactness of the continuous process and use a set of tracers or specific indicators, such as Hemoglobin A1C levels in diabetics, cholesterol levels, blood pressure, etc. There are also many economical issues addressed by such quality assurance systems and only rarely do they achieve an optimal resolution of the processes that they aim to control. Many decisions are made in an effort to save expenses and by that to achieve control, while the clinical issues of the processes are not addressed in detail. Most quality systems deal with a limited aspect of the health care process, rather than dealing with the whole process and its various components, including both personnel (policy makers, managers, medical professionals, technical staff, administrative shell, patients) and non-personnel (policy, regulations, infrastructure, economy). In this chapter, the design, application, and outcomes of a unique quality control and assurance program within the framework of medical committees will be described.

1.2 Overview of medical processes at recruitment centers

In Israel, adolescents aged 16-19 are obligated by law to enlist for military service and are examined by medical committees at conscription centers in order to determine their medical status. The medical process at the recruitment centers necessitates the coordinated action of medical professionals, technical and administrative personnel, and managing authorities, on both the local and national levels. The medical process is based mainly on a medical interview and an examination by a medical committee and is supplemented and supported by information from the family physician, medical consultants, and experts (Fig. 1).

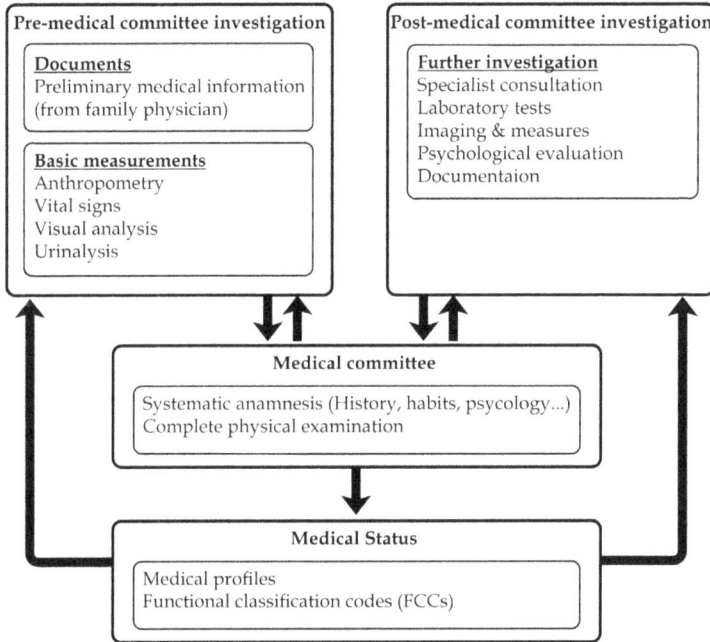

Fig. 1. Schematic representation of the medical process at the recruitment centers.

The administrative medical department employees at the recruitment centers constitute the administrative shell of the medical process, are thought of as medical facility workers, and serve as an extension of the medical process. They are exposed to medical information and follow strict laws of keeping medical confidentiality. They are in charge of the appointments of the recruits for various medical procedures at the recruitment centers and at medical facilities outside the recruitment centers; queue regulations; exchanges of letters and documents, and submitting requests from clinics, hospitals, and government facilities.

The technical medical assistants are part of the medical committee. Prior to medical committee examinations, they measure vital signs and anthropometric values (such as height, weight, blood pressure, and pulse) and check visual acuity and color vision according to Ishihara color tables. Urinalysis testing is done by a laboratory technician. They also collect all previously sent documents, add all required forms, and provide all of this accumulated medical documentation and forms to the medical committee.

The physician's work on the medical committee consists of a thorough medical anamnesis, including family history, habits, and a psychological evaluation. A systematic and comprehensive physical examination is performed. According to the findings, the committee chairman decides whether additional tests, such as a specialist consultation, laboratory tests, imaging, or other measures, are required or if the information is sufficient to apply functional classification codes (FCCs) and complete a recruit's medical profile. The medical profile reflects the recruit's current health status and is used by the Personnel Directorate to assign a military position that is consistent with that health status. Similar to coding systems like the International Classification of Diseases (ICD), Medical FCCs describe disorders, their severity, and determine the profile (Chaiter *et al.*, 2010).

For a complete and successful medical process, a close collaboration between the medical, administrative and technical personnel and harmony in their working relationship are needed.

1.3 Complex and dynamic medical processes at recruitment centers

Based on our annual analysis of the last decade (Conscription Administration Data, 2001-9), each year about 100,000 new recruits start the medical processes, and more than 400,000 medical encounters are performed at the recruitment centers. About 50% and approximately 13% of the recruits have ≥ 2 and ≥ 4 encounters, respectively, mostly with specialists or in hospitals. Most recruits ($\geq 75\%$) have at least one FCC, while many (~20%) have 3 or more co-existent medical conditions, in spite of their age. Furthermore, it is not just the prevalence of medical conditions but also their severity and morbidity. For instance, ~8% of recruits suffer from chronic asthma, while ~1.5% and ~10% suffer from various cardiac anomalies and mental disorders, respectively (Machluf *et al.*, 2011). In about 10% of the recruits, the discovery of new conditions (following accidents, operations, or a change in the severity of the medical conditions from the first check-up) leads to the modification, addition, or cancellation of FCCs. This may result in the determination of a new medical profile. The thorough examination at a recruitment center might reveal a new, previously unknown, medical problem that warrants further follow-up or treatments at a primary clinic by the primary care physician. Among these conditions, we have revealed even cases of severe disorders, such cardiac anomalies, nephropathies, cancer, etc. In these cases, the diagnostic-treatment loops need to be closed.

1.4 Databases

The information acquired during the medical process in stored, organized, and archived in a database. Computer-based tools allow analysis and visualization of data. With regard to the administrative aspects of the medical process, the computerized system consists of three main components (Machluf *et al.*, 2011):

A status system: a specific status code is assigned to each file/recruit using code numbers reflecting the specific specialist, test, or documentation needed. A status code and its beginning and end dates allow the administrative medical department to actively and dynamically follow up and manage the medical and administrative processes on both the individual and collective levels.

An appointment system: this system is used to assign a specific date for an appointment to a certain specialist or medical procedure, to generate invitations, and to document the appointment outcome. Its design principles allow better control and management of human and medical resources according to the capacities and limitations of the medical system.

A directing, monitoring, and controlling system: a local smart card-based system for real-time follow-up and regulation of waiting lists of patients at each station. It automatically directs recruits either to an available station or according to priorities pre-set by the medical personnel that the specific recruit is required to pass. It provides the medical administrative personnel and the medical committee members with relevant information on both the individual (recruit) and collective (queues) levels.

Each component system is directed toward answering a particular need and, although each is independent, they are all compatible with each other and provide the user with a comprehensive view of the medical process and information. The combined computerized system improves the control and management of the medical processes and informatics from the point of view of both the patients and the system operators. Implementation and automation of medical regulations and procedures within the computerized system also make this system play a key role and serve as a control tool during the decision-making process (Machluf *et al.*, 2011).

2. Quality assurance and control system

During 1997, a quality assurance and control system was set up (Chaiter *et al.*, 2008; 2010; 2011; The State Comptroller and Ombudsman Office, 2002). This unique system, unlike most policies or systems for quality assurance of medical care, originated from within the medical profession, rather than from industry or academia, and its regulation and modes of action reflect the real daily health care activity in medical departments.

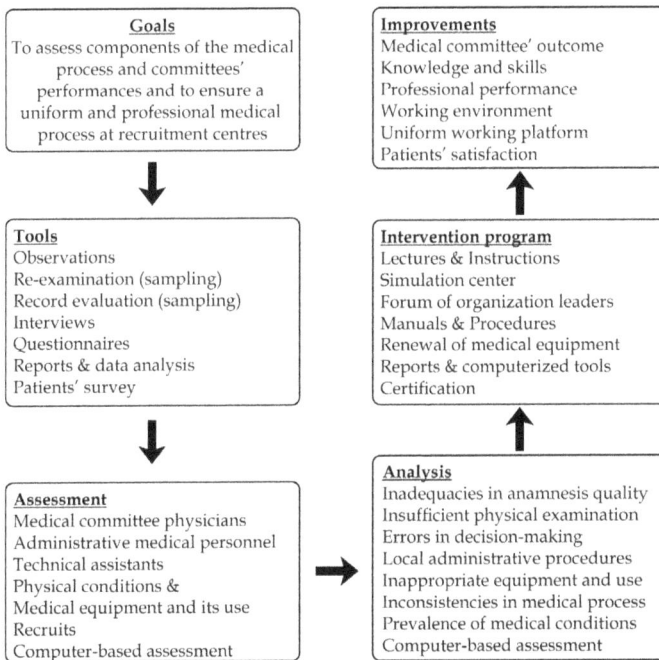

Fig. 2. Scheme of the quality assurance and control project of medical processes at the recruitment centers.

The goals of the quality assurance and control committee are not solely to assess and improve the medical committee outcomes. They are also to identify the limitations and needs of whole medical process regarding the medical, technical, administrative, procedural, and physical aspects and to evaluate knowledge, skills, judgement and working patterns, which count for quality care (Holmboe *et al.*, 2008), among all medical personnel; to develop and implement intervention programs to answer the needs of all of the parties participating in the medical process; and to achieve a higher quality of patient care and patient satisfaction (Fig. 2). Accordingly, the quality assurance and control system evaluates and analyzes in detail the various facets of the activities of the medical policy makers, managers, administrative staff, medical professionals (physicians, experts, consultants), and technical assistants, utilizing different complementary methods (Fig. 2). Experts from all these fields (and others), with their high skills and experience, are incorporated into the quality assurance and control system. Of note, performance measures are evidence-based and valid, feasible to collect, applicable to a large enough population of patients, attributed to the performance of individuals, adjustable to the patient level, and representative of the activities of a specialty (Landon *et al.*, 2003).

2.1 Approach

Two main approaches are deployed by the quality assurance and control committee: (i) physical visits to recruitment centers, during which the procedures, work, decision-making processes, and outcomes are directly assessed, and (ii) data mining and processing from the computerized databases. The different means used in each approach to assess the work and results of the different medical process-related staff are summarized in Table 1. Incorporation of such complementary assessment methodologies provides both quantitative and qualitative analyses of daily activities and practices in the medical departments.

Methodology		Components of the medical process			
Approach	Tools	Medical physicians	Technical assistants	Administrative staff	Recruits
Direct assessment	Observation	✓	✓	✓	-
	Re-examination	✓*	-	-	-
	Record evaluation	✓	✓	✓	-
	Interviews	-	✓	✓	✓
	Questionnaires	✓*	✓#	✓#	✓
Data mining	Reports-QC	✓	✓	✓	-
	Epidemiology-like	✓	✓	-	✓

*-distributed to recruits; #-distributed to soldiers

Table 1. The various methodologies used to assess components of the medical process.

2.2 Direct assessment

Direct assessment during the physical visits to the recruitment centers provides an opportunity to evaluate the work of the medical process as a whole. It also enables direct interaction and brainstorming with local medical department members, from administrative staff though technicians, physicians and experts to managers.

2.2.1 Direct assessment of medical committee performance

The medical committee is central to the medical process. Direct assessment of the medical committee's performance consists of direct observation and clinical evaluation of the physicians' work, random samplings, re-examination of recruits that were examined by the medical committee, distribution of questionnaires to recruits following their examination by the medical committee, and analysis of random samples of completed medical files.

Observation and clinical assessment (clinical audit): Upon receiving a recruit's consent, a physician from the quality and assurance system joins the medical committee as an observer and evaluates the completeness and adequacy of the anamnesis, physical examination, and decision-making process (using a pre-designed form), as well as the quality of the communication and service to the recruits. For each of these components, quantitative (a numerical scale following specific criteria) and qualitative (description and comments) assessments are used. For example, for each issue in the anamnesis or physical examination, codes are used to determine whether and how the issue was addressed (according to regulations and clinical merits as determined in the medical literature). The decision-making process is also assessed for the quality of referrals to further investigation and the quality of decision making according to clinical criteria and regulations.

For each medical committee, 5-10 cases are observed. Findings are shared and discussed with the observed physicians, and proper instructions and guidance are provided, if necessary.

Sampling for recruits' re-examination and record evaluation: With a recruits' consent, physicians from the quality and assurance system re-examine a random sampling (8-10 on average) of recruits immediately following their examination by the medical committee. The findings of the complete re-examination, including the "assigning" of a medical profile and FCCs, are compared to those of the local medical committee, and discrepancies are recorded and discussed for each case. In addition, a random sampling of files (30-50 files from each recruitment center), where the profiling process was completed, are re-checked by a physician from the quality and assurance system to assess the anamnesis, the medical findings and documents, the decision-making process, and the assignment of profiles and FCCs.

Questionnaire (patients' survey): In each recruitment center, questionnaires are randomly distributed to recruits (15-20 on average) following their examination by the medical committee to gain more insight regarding the medical processes (basic measurements, anamnesis, and physical examination), preserving medical confidentiality and right of privacy, as well as the recruits' rates of satisfaction with the service, during the medical process. Recruits are also asked to express their general impression of the medical process.

2.2.2 Direct assessment of technical and administrative medical staff

There are many similarities that are shared by the work of administrative staff and the technical assistants and also by their quality control and assessment. Direct assessment of each person's performance consists of direct observation and evaluation of their work, distribution of questionnaires to professional personnel, interviews, and analysis of random samples of completed medical files.

Observation (audit): The work of all of the technical medical staff is assessed prior to, during, and after the medical committee examinations. First, medical equipment (scales, altimeters, chart tables for the visual acuity examination, sphyngomanometers, and Ishihara color tables) and its use are examined during the process of taking basic measurements. Also, documentation handling, data recording, and proper directing to further processes are checked. When required, correct instructions are provided to prevent future mistakes. Similarly, observation of the administrative staff involves the same parameters, except for an evaluation of basic measurement techniques.

Questionnaire: A detailed questionnaire is used in order to evaluate the skills of all technical assistants in measurement techniques (weight, height, etc.), knowledge of the normal range and abnormal findings regarding different measurements and the corresponding FCCs (blood pressure, pulse, urinalysis, etc.), and administrative issues (recording of measurements and medical history, signatures, etc.). By contrast, questionnaires are given to the administrative staff to evaluate their knowledge of regulations (status, appointments, and the smart card-based system) and administrative issues (special populations, reports, etc.). Correct instructions are given and even demonstrations are done to promote an increase in knowledge and skills.

Interviews: The technical assistants and administrative personnel are interviewed about their work, the findings of the control system are discussed individually, and ideas of how to improve the infrastructure, process, and procedures are shared.

Sampling for records evaluation: a random sampling of files (10 on average of each kind, at each recruitment center), where the profiling process was completed, are re-checked by members of the quality and assurance system to assess the documentation and administrative processes.

2.3 Data mining and processing

The demographic information and medical-administrative data are stored, organized, and archived in a database. This data can be visualized and retrieved. Computerized tools, such as reports and regulation-based automated procedures, allow in-depth data analysis. Computerized databases and tools, when integrated into the medical process, allow efficient follow-up and management of medical processes and informatics (Machluf *et al.*, 2011). These databases and reports can also serve as quality control and assessment means (Chaiter *et al.*, 2008; 2010; Machluf *et al.*, 2011; Navon *et al.*, 2011). Using the reports, one can assess the work of medical professionals (physicians and experts, for example), technical assistants, and administrative staff members. For example, reports are aimed at identifying discrepancies between the medical information (such as anthropometric and basic measurement data) and FCCs or medical profiles, inconsistencies in the medical-administrative information between the status and appointment systems, inadequate or incomplete medical processes (deviations from defined regulations), etc. (Chaiter *et al.*, 2008; Machluf *et al.*, 2011; Navon *et al.*, 2011). Such populations, at the individual level, are monitored, and reports are distributed monthly to the relevant personnel at each recruitment center and to the managing authorities.

Reports also support the design, planning, monitoring, and use of human and medical resources and are a component of the decision making that is made by the medical-administrative managers. For instance, various aspects of the availability and need for medical services, such as specialists, consultants, medical procedures, waiting queues, and various causes of congestion that need to be taken care of, are all assessed regularly by reports. The findings are distributed to the relevant personnel at the recruitment centers and to the managing authorities.

Reports also allow a comparison between the performance of the medical personnel within and between the recruitment centers and across a longitudinal time axis. Furthermore, since medical profiles and FCCs are indicators of medical conditions among recruits, an epidemiological investigation of the profile distribution and prevalence of FCCs can be performed, providing inter- and intra-recruitment center analysis (Chaiter *et al.*, 2010). This valuable information can be related to or crossed with gender, geographical area, country of origin, ethnicity, socio-economic background, education, and morbidity trends in the general population in Israel and other countries.

3. Findings

Using these tools, we analyzed key parameters related to the performance, integrity, and completeness of the medical processes and procedures. Furthermore, we were able to identify difficulties concerning physical conditions, administrative or medical procedural deficiencies, and insufficient knowledge or skills of medical/technical/administrative personnel within the medical departments.

3.1 Medical committee performance

In general, during the years 1997-2010, more than sixty visits to the different recruitment centers were carried out. Regarding the performance of the medical committees, the work of more than 110 chairmen (of medical committees) was assessed by different means. Six hundred and fifty-five recruits were re-examined, 452 filled out questionnaires, and approximately 1700 records (medical files) were examined and evaluated, providing a 3-pronged approach to observational findings. For the sake of clarity, we will present the findings from the first years and from the period after the intervention programs, focusing on recruitment centers with bigger populations.

During the first years of the program, we found that at most of the recruitment centers there was an inadequate amount or a lack of medical equipment, and often it was found that the equipment was used inappropriately (Chaiter *et al.*, 2008). The following were among the more common medical equipment-related faults that were found: broken altimeters, inadequately balanced scales, unsuitable Ishihara books for color vision examination, incomplete regulation manuals, and inadequate equipment for visual acuity examinations. Following our visits, a major effort was made to improve the physical conditions and equipment at the medical committee examinations. Examination rooms were redesigned according to the needs of the physician and the patients. Inadequate and old medical equipment were replaced by new machines, complete manuals and regulations were provided, and attention was paid to the physical surroundings and environment. After a gradual improvement in this area throughout the recruitment centers, such faults, even minor ones, are now only rarely detected.

Among the physicians, we observed inadequacies in anamnesis quality, insufficient physical examinations, and errors in decision making (Chaiter *et al.*, 2008). Furthermore, significant differences were observed between recruitment centers.

Anamnesis includes both specific tracer questions and general systematic queries. The tracer questions are all obligatory and refer to night vision, refractive eye surgery, dyslexia, drug use, gynaecological issues among female recruits, prolonged staying abroad at specific areas that are endemic to HIV, and psychological/psychiatric treatments. The main findings show a lack of anamnesis about most of these tracers, except the last two. Yet, differences were observed between recruitment centers. For instance (Chaiter *et al.*, 2008), in 2002, complete anamnesis for refractive surgery was found to range between 12.5% of recruits at center 2 to 90% of recruits at center 5 ($p<0.001$). Anamnesis of drug use was found to be only 25% of the cases at center 2 as compared to 100% of the recruits at center 5 ($p=0.04$). Similarly, at center 3, 10% of recruits there were asked about dyslexia, while at center 2 the rate was 100% ($p<0.001$). During and especially following the intervention program, a clear improvement in the anamnesis process was evident by the completion of anamneses and specific tracer questions at all of the recruitment centers. For example (Chaiter *et al.*, 2008), at center 1, the rates of recruits who were asked about the night vision trace and about prolonged stays abroad at specific areas that are endemic to HIV significantly increased (from 4.35% to 50% ($p=0.003$) and from 8.7% to 50% ($p=0.01$) during the years 2006 and 2007).

Noted findings concerning physical examinations included a lack of examination of the lateral motion range of the spine and an incomplete examination of heart auscultation (at one point only - left sternal border), lung auscultation (one or two points on each side), the abdomen (partial palpation of one or two quadrants), the lower extremities (especially the feet), and male genitalia (lack of examination of the inguinal canal for possible inguinal hernia or incomplete examination of testicles) (Chaiter *et al.*, 2008). Moreover, similar to anamnesis, clear differences were observed between recruitment centers. In 2002, the rate of complete examination of the abdomen ranged from 25% (center 2) to 90% (center 5) of recruits (p=0.048), while complete foot examinations were carried out on 38% (center 4) to 70% (center 5) of recruits. During the years following the intervention program, noticeable progress was observed in the physical examinations. For example, at center 1, a significant improvement was noted from 2006 to 2007 in abdominal examinations (from 8.7% to 75% (p<0.001)), foot examinations (from 47.8% to 100% (p=0.01)), and in male genitalia examinations (from 4.35% to 75% (p<0.01)).

Decision-making processes are integral and central to the medical process. Over the years, no clear trends were observed in the rates of correct decision-making procedures (Chaiter *et al.*, 2008). This may be attributed to the turnover of human resources (chairmen, physicians), and the persistence of local medical procedures at recruitment centers which are not in line with the general guidance. Nevertheless, an improvement in the decision-making processes at specific time points, or over several years, at specific centers was associated with cooperation with the quality control and assurance system.

Together, better medical history recording, physical examinations, and decision-making processes by the physicians of the medical committees were noted.

Data mining and processing by computerized reports revealed a major decrease in the rate of discrepancies between the medical information (such as anthropometric and basic measurements data) and FCCs or medical profiles (Chaiter *et al.*, 2008).

Analysis of medical profile distribution and the prevalence of FCCs, the indicators of medical conditions among recruits, uncovered significant differences between recruitment centers (Chaiter *et al.*, 2010). Analyzing all of the FCCs revealed the 26 most common FCCs, which comprised approximately 90% of all assigned FCCs. Almost 90% of these common FCCs (23 out of 26) were found to vary significantly between different recruitment centers. Data stratification according to ethnic origin did not affect the results (Conscription Administration Data, 2001-9). These 26 FCCs include overweight, underweight, anemia, asthma, cardiac anomalies (either valvular or non-valvular), hypertension, varicocele/hydrocele, epilepsy, mental illness conditions (personality disorders, neurosis, psychosis, depression, mental retardation, and autism), hernia, visual acuity, allergic rhinitis/sinusitis, and flat feet. Some of these FCCs (such as bee sting allergy; anemia; valvular and non-valvular cardiac anomalies, including mitral valve prolapse; hypertension; hydrocele/varicocele; flat feet; hernia; hearing loss; visual acuity problems; and color blindness) were found not to significantly affect the final profile outcome. Yet, a set of only 8 FCCs (those indicating recruits who were underweight or suffered from asthma, chronic headache/migraine, mental illness, scoliosis/kyphosis, chronic back pain, knee joint disorders, or rhinitis/sinusitis) accounted for 90% of the medical profiling differences between recruitment centers (Conscription Administration Data, 2001-9). Of these key profile-affecting FCCs, the prevalence of all of them except scoliosis/kyphosis and mental conditions was found to be 1.5 to 2.5 times higher at recruitment center 1 as compared to the other centers (Chaiter *et al.*, 2010).

Over the years, significant trends were observed: the prevalence of chronic headaches (increased), symptomatic scoliosis/kyphosis (increased), and active asthma (decreased) in recruitment center 1 (Table 2). Interestingly, a trend toward a significant increase in prevalence until the year 2005 and then a significant decrease in prevalence until the year 2009 was common to all other FCCs, except for that of underweight (Table 2).

Medical condition	Year									Correlation
	2001	2002	2003	2004	2005	2006	2007	2008	2009	p value
Chronic headache	4.31	4.82	4.88	5.77	6.66	6.21	6.06	6.24	6.17	0.820, <0.01
Symptomatic scoliosis/kyphosis	1.90	3.41	3.93	5.35	5.52	4.88	4.91	5.16	4.72	0.709, <0.05
Active asthma	11.29	9.29	8.66	9.62	9.94	9.54	8.52	8.94	8.12	0.684, <0.05
Under-weight	4.44	4.56	4.91	5.49	4.79	3.57	3.44	3.85	4.14	0.558, n.s.
Chronic back pain	8.32	7.99	8.27	8.52	9.03	8.50	7.28	7.63	7.50	0.515, n.s.
Allergic rhinitis/ sinusitis	11.89	11.75	11.19	12.39	12.86	12.31	11.18	11.19	10.30	0.471, n.s.
Knee joint disorders	3.62	3.19	3.79	4.37	4.94	4.16	4.24	4.02	3.09	0.082, n.s.
Mental conditions	9.83	8.17	9.14	8.94	11.13	10.20	10.03	9.75	8.10	0.040, n.s.

n.s. – not significant

Table 2. Percentages of main medical conditions contributing to a difference in profiling at recruitment center 1 from 2001 to 2009 (Modified with permission from Chaiter et al., 2010).

Some of these 26 FCCs were found to vary significantly between sub-districts in all of the recruitment centers (data not shown and Chaiter et al., 2010). This may be attributed to both demographic-environmental parameters and professional-human causes. Across recruitment centers, and over the years, clear and significant disparities were found in the prevalence of how a majority of these FCCs were assigned by different medical committee chairmen in the year 2006 (data not shown and Chaiter et al., 2010). This further strengthens the supposition that there are differences among recruitment centers and among the chairmen's reporting of medical histories, performing physical examinations, and interpreting various medical conditions, all of which lead to differential assignments of FCCs. Differences in FCC assignment by different chairmen at the same recruitment center indicate decision-making disparities or lack of experience and expertise in specific fields. The intervention program led to an improvement in certain aspects of the chairmen's knowledge and skills and, as a result, to an improvement in the performance of the committees; while variations between chairmen and between recruitment centers still exist, it is to a slightly lesser extent. The impact of professional development and understanding of the whole medical process (see below), as well as the establishment of a uniform working platform, resulted in decreased variability of medical conditions in the various conscription centers and among physicians. However, it could have been greater if the turnover of professional medical human resources would have been lower.

It is noteworthy that a subset of the 26 most commonly assigned FCCs were found to vary significantly among all three stratification criteria (recruitment centers, sub-districts, and chairmen assigning the profiles) (data not shown and Chaiter *et al.*, 2010). The common characteristic of these FCCs (such as underweight, asthma, chronic headache, symptomatic scoliosis/hypnosis, chronic back pain, knee joint disorders, and allergic rhinitis/sinusitis) is that their assignment procedure is prone to a relatively high degree of variation in anamnesis, examination, chairman discretion, and interpretation.

3.2 Technical assistant staff performance

The work of more than 110 technician assistants and laboratory staff members was assessed by questionnaires, interviews, and direct observation of their performance before, during, and after the medical committee examinations (Conscription Administration Data, 2006-9; Chaiter *et al.*, 2010; 2011). Prior to the intervention program, insufficient knowledge was revealed mainly with regard to the normal range of values for blood pressure and pulse, urinalysis, visual acuity, and color vision and to the interpretation of abnormal values of these measurements. Furthermore, some technician soldiers suffered from inadequate execution of the techniques, such as incorrect weight and height measurements. Inadequate knowledge regarding the relationship of all of the above mentioned measurements to specific medical FCCs was found. Some of these findings (insufficient knowledge, technical skill, and their relationships and meanings) were common to the technical staff at all of the recruitment centers, while other aspects were evident at specific recruitment centers.

After the intervention program, a higher level of expertise, increased skills in measuring medical parameters, and a more accurate interpretation of these values were observed among the technical staff. Improvements were found in the measurement techniques of weight, height, color vision tests, and determining of visual acuity, as well as an increased understanding of the normal parameters of these and other measurements, such as pulse and urinalysis, and their interpretation and relationship to medical FCCs (Chaiter *et al.*, 2011). For example, in recruitment center 2, a significant and sustained improvement was observed in the interpretation of low systolic or diastolic blood pressure and in the determination of color blindness and other issues, and in recruitment center 1 there was a higher rate of correct measurements for weight, height, and visual acuity. The number of medical inconsistencies was progressively and dramatically reduced.

3.3 Administrative staff

During the years 2007 to 2009, 23 visits were carried out to assess the work of the administrative medical departments at the recruitment centers. During these visits, the work of almost 200 medical administrative personnel and managers was analyzed. The main findings (Navon *et al.*, 2011) include incomplete knowledge of medical-administrative processes (such as appointments and statuses), a lack of professional collaboration between medical departments at different recruitment centers that was inevitably caused by differences in working patterns and operational procedures at all of the centers, a partial management of diaries with abnormal laboratory results, and local procedures which deviated and were not in line with regulations and instructions (such as those related to recruit identification or the management of medical questionnaires that were received from primary care physicians). Furthermore, in some cases, the managers of the medical administrative departments were only partially or inappropriately trained for their duty, and therefore their performance was far from optimal during the first period with regard to the professional-administrative-medical aspects and the management of human resources.

As a result of the intervention program (Navon *et al.*, 2011), improvements were found in all of these aspects of knowledge, skills, working procedures, professional collaboration, and management. This led to a considerable decrease in the rates of errors in records, such as inconsistencies in medical-administrative information between the status and appointment systems, and a significant increase in the rate of proper and complete administrative medical processes and profiling processes. Nevertheless, a significant increase was observed in the rate of recruits with medical profile that did not take into account the information written in the medical questionnaires that were filled out by their primary care physicians.

3.4 Computer-based tools – not just for quality control and assurance
The quality control and assurance system and the local medical departments at the recruitment centers continuously analyze data from the computerized system (mainly by means of reports) in order to assess and evaluate the performance of the medical and administrative processes, as well as to identify errors and discrepancies in individual medical files. In light of the findings, the medical committees and departments take action to correct mistakes, follow up specific populations, etc. Moreover, reports can uncover difficulties and vulnerabilities in global processes, which in turn lead to further improvements or modifications of either the medical procedure or the computerized medical database and system.

The combined computerized system (including the appointment system, the status system, and the directing, monitoring, and controlling system) improves the control and management of the medical processes and informatics from the point of view of both the patients and the system operators (Machluf *et al.*, 2011). Different parameters of quality control regarding the medical and administrative processes are assessed (such as efficiency), and solutions are sought. Computerized system-based design and re-allocation of human and medical resources are implemented according to the capacities and limitations of the medical system. For instance, at recruitment center 1, reports during late '90s revealed a significant number of recruits waiting for specific medical processes (ophthalmologist, cardiologist, pulmonologist, orthopede, neurologist, etc.), and the intervention led to a reduction of at least 50% in the number of recruits holding a specific status over a period of 16 months; this reduction was achieved for most (12) of these statuses (15) (Machluf *et al.*, 2011). In addition, analysis revealed the specialists in each recruitment center for which recruits wait long periods of time for successful completion of the process. In all of the recruitment centers, a higher number of recruits who were waiting for more than 3 months was found with recruits who were in the process of medical documentation (centers 1 and 5) or who were waiting to see a neurologist (centers 2-4). Also in this case, computer-aided planning and re-allocation of human and medical resources played a key role in the intervention and in the solutions found for the specific specialists and recruits at each center (Machluf *et al.*, 2011). Another parameter regarding both quality control of the medical process and quality service to the recruits is the number of attendances at a recruitment center for each recruit until the final profile is assigned. Among the recruits who received their medical profile during 2010, approximately two-thirds of the recruits were required to report to the medical departments up to two times (Fig. 3). This rate was better in recruitment centers 2, 4, and 5 (>70% of recruits). On the other hand, a considerable fraction of recruits were obligated to attend six times or more, especially in recruitment centers 1 and 5 (2.5% and 2.3% of recruits, respectively) (Fig. 3). In some cases, the profile was assigned but no arrival to a recruitment center was recorded. These findings, and others, are the basis to analyze the medical and administrative aspects of the medical process, such as coordinated invitations, unnecessary or insufficient investigations, and the proper recording of arrivals.

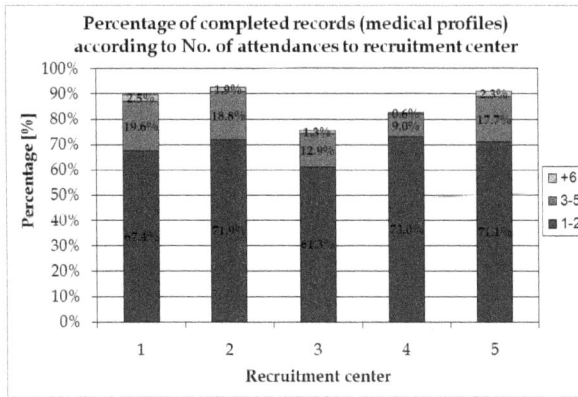

Fig. 3. Analysis of the number of attendances to a recruitment center in completed records during 2010.

The combined computerized system together with reports and automated tools allows the management of populations with special needs, such as mentally retarded and cerebral palsy patients. According to the regulations, these special populations are exempt from reporting to the recruitment centers, and their files are managed with maximum discretion and sensitivity to the individual, thus respecting them and their families' wishes.

Based on data mining findings, modifications were also introduced in working procedures. A reduction in the daily number of invited recruits improved the quality of the medical encounters. Specific combined status codes were introduced for the efficient planning of the medical encounters. Implementation and automation of medical regulations and procedures within the computerized system cause this system to play a key role and serve as a control tool during the decision-making process.

4. Intervention

The quality control and assurance system operates via an analysis→ design→ implementation→ evaluation→ modification loop.

First, all medical processes were mapped and analyzed, and particular infrastructure, medical, technical, management, and administrative needs were characterized by experts. The design principles of the intervention program with regard to the components, mode of action, time lines, etc. were determined to address those needs, in line with the general medical processes and goals. Then, a small-scale pilot program was launched in one recruitment center and was systematically analyzed. Throughout these steps, the quality control and assurance system personnel collaborated with the experts and experienced staff members from the medical departments and with policy makers. This collaboration allowed for better intervention planning, an efficient feedback process, contributed to the participants' sense of ownership and commitment to the process, increased their confidence, and made them receptive to the intervention program. After the required modifications

From a Quality Assurance and Control System for Medical Processes, Through Epidemiological Trends
of Medical Conditions, to a Nationwide Health Project

67

were made, the same procedure was carried out as a large-scale intervention program that was gradually implemented at other recruitment centers, until full implementation of the system was achieved throughout the whole organization. The gradual implementation process allowed for both the full support of each recruitment center and an assimilation of adjustments to specific local needs.

It is noteworthy that acceptance of and cooperation with the quality control and assurance system were not trivial. During the first years of the program, physicians and policy makers, and to a lesser extent technical assistants and administrative staff members, were reluctant to comply with the quality control and assurance activities, during all of the stages and particularly during the intervention. The main reasons for that were professional autonomy and time/procedural requirements. This phenomenon and its causes were found also in France (Giraud-Roufast & Chabot, 2008). To build trust and cooperation, except for the fact that it was an obligatory process, few key actions were taken. First, the main goals of performance assessment were to improve professional competence, rather than to take sanctions. Medical personnel were considered partners, taking part in activities and brainstorming (mainly during the design and implementation stages). During intensive educational activity, the benefits (professional development, money, time, resources, quality care, etc.) to the individual service-providing staff (physicians, managers, etc.) and to the patients (quality of service and care, safety, etc.), to the medical departments, and to the organization were highlighted. Gradually, quality control and assessment became an integral component of the medical department's work, both routinely by the medical personnel and occasionally by the quality control and assessment system.

Except for the physical condition issues, three main needs were acknowledged in every aspect of the medical processes, in the medical department: uniform medical and administrative processes at all recruitment centers; proper and comprehensive acquisition of medical, technical, and administrative knowledge, expertise, and skills; and sharing between all conscription centers. In light of these needs, an intervention program was designed. It utilizes various means:

Forums of organization leaders: A forum was established, which includes leading physicians, medical administrative personnel, and the managers of the medical departments. Meetings are held on a regular basis in order to update all involved personnel on the novelties introduced into the medical system, to exchange ideas, and to formulate recommendations (and thereafter their implementation) for ongoing quality assessment and improved working procedures. This led to the development and production of manuals for administrative and technician soldiers, the printing of a catalogue of medical equipment, and the initiation of theoretical and practical training programs.

Training and simulation center for physicians: A training program was implemented for physicians, which includes lectures, clinical training at a simulation center, and continuous observation and feedback on their work. Each chairman physician is invited once a year to participate in a workshop that provides simulated scenarios of patient-physician encounters. Each scenario is played by specially trained actors performing the role of the recruits. A detailed anamnesis, a recording of the findings, and further investigation is performed by the physician. Each encounter is recorded on video and is then discussed in detail with the special training team to point out mistakes and inconsistencies in the process. Each physician receives a personal feedback summary and accreditation is given for participation in the course.

Instruction of physicians: After being observed, the physicians are instructed and trained in all issues assessed by the quality control and assurance system. Proper physical examination, partial or complete, is demonstrated upon request or, if necessary, is based on the findings.

Manual for technical medical staff performance: A comprehensive manual was written describing measurement techniques that are carried out by the technical assistants (and laboratory staff) at the medical committee examinations. It was distributed to relevant personnel and is used on a daily basis. The manual also contains information about the normal range of systolic and diastolic blood pressure and pulse measurements and correct interpretation of visual acuity and color vision tests (Ishihara and D15), as well as about the interpretation of abnormal values with instructions of how to act if an abnormal value is encountered during measurement or is written in the medical committee protocol. The manual contains information that involves the technician soldiers in the coordination and quality assurance processes of the medical committee examinations. For instance, the technician soldiers are instructed to return a file to the physician if they find an abnormal blood pressure value recorded in the protocol of the medical committee without any instructions from the examining physician on how to proceed or if the physician determines the profile of the recruit without assigning a FCC that indicates hypertension.

Frontal lectures: All technical medical personnel and administrative staff at all of the recruitment centers were given lectures at each recruitment center and in special meetings arranged at the medical assessment branch of the IDF Medical Corps. The issues discussed in the lectures further stressed what was described in the written manuals and also emphasized cases of risk management in order to strengthen the notion about the importance of the technical medical staff's work and the administrative processes as part of the quality assurance of the medical committee examinations and as assistants of the physicians in the process of medical profiling. During some of the lectures at the recruitment centers, training in measurement techniques, such as blood pressure measurements, was performed. In addition, the relationships and links between the medical, technical, and administrative processes were highlighted.

Instruction of technical medical staff: After the observations, questionnaires, and interviews were conducted, the technician soldiers were given detailed feedback and were instructed and trained in all of the issues that were assessed by the quality control and assurance system. In addition, all problematic areas that were identified by the system were discussed with the medical department managers.

Instruction of administrative staff: After the observations, questionnaires, and interviews were completed, the administrative staff were given detailed feedback and were instructed and trained in all of the issues that were assessed by the quality control and assurance system. In addition, all problematic areas that were identified by the system were discussed with the administrative staff's managers.

Written reports: Written reports summarized the findings with an emphasis on the recommendations required to make improvements and to correct the mistakes found in each recruitment center. The reports were distributed to the relevant medical and management authorities at the local recruitment centers and at headquarters. These reports also allowed a comparative overview between medical departments at different recruitment centers and between different time periods or specific assessments.

Computerized tools: The quality control and assurance system plays a key role in the design and planning of computerized systems or in their modifications so that the medical department's needs, mainly with regard to procedures and regulations, will be answered.

The quality control and assurance system also gives complementary support to the medical department with implementation, adjustments, and instruction in the proper use of the computerized tools in the medical processes for the purpose of higher quality of control, management, and service. The computerized system, through its implementation and automation of medical regulations and procedures, plays a central role and serves as a control tool during the decision-making process and as a way to prevent inconsistencies in the medical information. In addition, electronic medical files were incorporated into the computerized system.

Certification: Certification is virtually awarded for the completion of requisite training/instruction by accredited authorities, including the quality control and assurance system. It is important to note that, unlike most licensures and board certifications (Landon *et al.*, 2003), the qualification process and examinations are periodic and specific to particular medical fields and ensure professional competency according to pre-determined standards.

5. Effects of intervention

During the years of intervention, benefits were observed in the quality of all of the areas that were examined. As these effects were described before (see the section entitled Findings) or earlier (Chaiter *et al.*, 2008; 2010; 2011; Machluf *et al.*, 2011; Navon *et al.*, 2011), we can summarize them in a few main categories: 1) infrastructure and physical conditions; 2) consistent improvement in the knowledge, skills, and performance of the physicians (anamnesis, examination, decision-making processes, etc.), technicians (measurements, recording, interpretation, etc.), and administrative staff (procedures, regulations, etc.); 3) uniform working platforms and procedures in terms of the medical and administrative processes at all of the recruitment centers; and 4) a launching of the framework and forums for sharing knowledge and skills between all of the conscription centers.

One important contribution of the quality control and assurance system is in terms of the diagnosis, design, and implementation of the intervention program and its analysis. The impact on the medical department's performance is clear and evident. Furthermore, after the intervention program's implementation, the satisfaction rate, sense of belonging, and responsibility were higher among all of the medical department personnel. It led to the increased perception of the administrative and technical medical staff to feel that their work was part of the medical profiling process, acting as case managers and part of quality assurance aimed at providing the best medical service for recruits. Therefore, it is only natural that this would consequently lead to a consistent, major increase in the satisfaction level reported by recruits.

As mentioned above, not all of the goals were achieved fully, and in some respects there was a disparity between what was achieved and the desired objectives. Clearly, some changes are time-dependent, and the benefits of current efforts will be evident in the near future. Other challenges may require different solutions. We are always looking for new and better tools and modes of action to achieve these goals.

6. Epidemiological aspects

The medical processes for the adolescent population of Israel present a unique opportunity to assess the health status of the young Israeli population on a nationwide level and to identify risk factors that can affect present and future morbidity.

A pilot study, including over 105,000 adolescents, has been carried out using data from one of the recruitment centers to assess trends in weight, height, and other medical parameters (e.g., blood pressure) and conditions (e.g., congenital heart disease) among 16-19 year olds in Israel (born in 1971-1992). Our preliminary analysis suggests that variability between recruitment centers may affect the consistency and reliability of nationwide analysis, as opposed to information from a single recruitment center. Furthermore, in one selected center, the differences between chairmen in assigning profiles and FCCs were much smaller than in other centers. Clear trends were demonstrated, and their association with demographic variables was examined. The findings are out of the scope of this manuscript, yet we wish to illustrate two possible insights drawn from such analysis.

6.1 Anthropometric values

Among the young (16-19 year olds) Israeli population born between 1971 and 1992, a significant increase in average body weight was demonstrated in females and mainly in males (Fig. 4, upper panel). In both genders, the increase in average weight is more dramatic in teenagers who were born during 1982-1992. On the other hand, no dramatic change was observed in average height in females or in males (Fig. 4, lower panel). This might be an indication of an increase in the body-mass index (BMI).

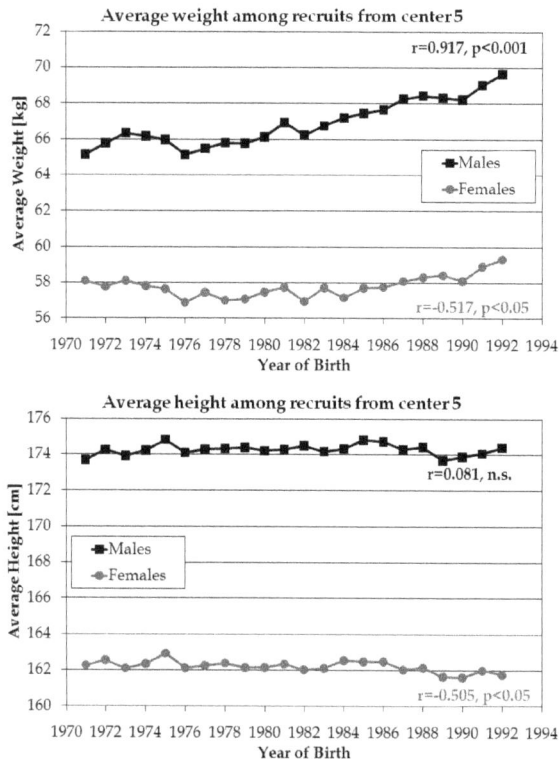

Fig. 4. Trends of average weight and height among the 16-19 year old Israeli population.

General population data and military records suggest that US heights essentially remained stable after World War II (mainly in birth cohorts of 1955-74), which is concurrent with continual rapid increases in height in Western and Northern Europe (Komlos & Lauderdale, 2007). A historical retrospect of German military male recruits found an increase in both the average body weight and height, where the changes in body weight were greater than those of body height (Jaeger *et al.*, 2001).

The implications of these findings with regard to different demographic parameters, socioeconomic status, anthropometric indices, medical conditions, and other risk factors are now under investigation.

6.2 Medical conditions

Analysis of the prevalence of specific FCCs provided the opportunity to gain knowledge about the trends of different medical conditions. Such an analysis in recruits from center 5 uncovered an increase in the prevalence of valvular and non-valvular congenital heart anomalies among male recruits (Fig. 5. upper panel). Referring to cardiac diseases, the prevalence of congenital valvular heart disorders was higher than the prevalence of non-valvular heart disorders. Furthermore, an increase in the prevalence of solid tumors in both males and females was also demonstrated (Fig. 5. lower panel).

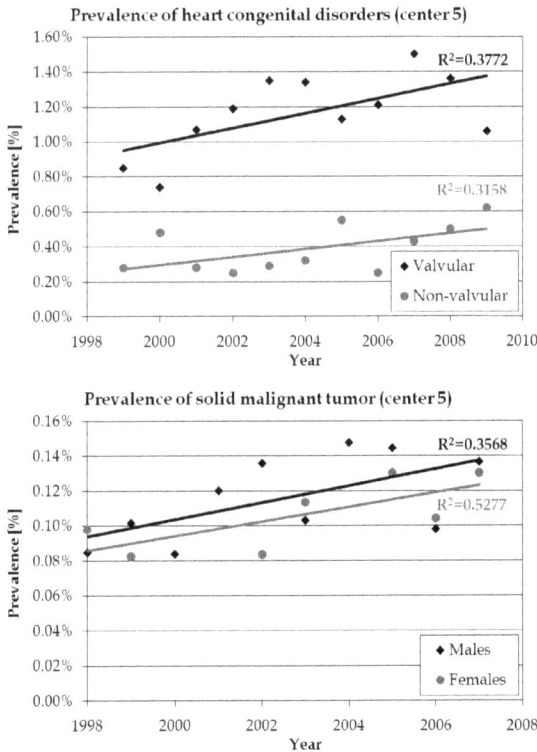

Fig. 5. Prevalence of solid tumors and congenital heart anomalies among the 16-19 year old Israeli population.

Previous analyses of specific morbidity prevalence in Israeli conscripts from all recruitment centers also revealed an increase in the prevalence of heart defects among male recruits (Farfel *et al.*, 2007), as well as a higher prevalence of congenital valvular heart disease compared to non-valvular heart disease (Bar-Dayan *et al.*, 2005).

7. A nationwide program

These findings demonstrate the need for a nationwide intervention program to reduce morbidity, future illness, and even mortality. Furthermore, a project of information sharing and cooperation was established with family physicians at primary clinics on a nationwide basis, the Ministry of Health, the National Insurance Institute, the Israeli National Cancer Registry, and the Ministry of Social Affairs and Social Services. This national project (Fig. 6) is aimed at education, prevention, and early intervention in target populations.

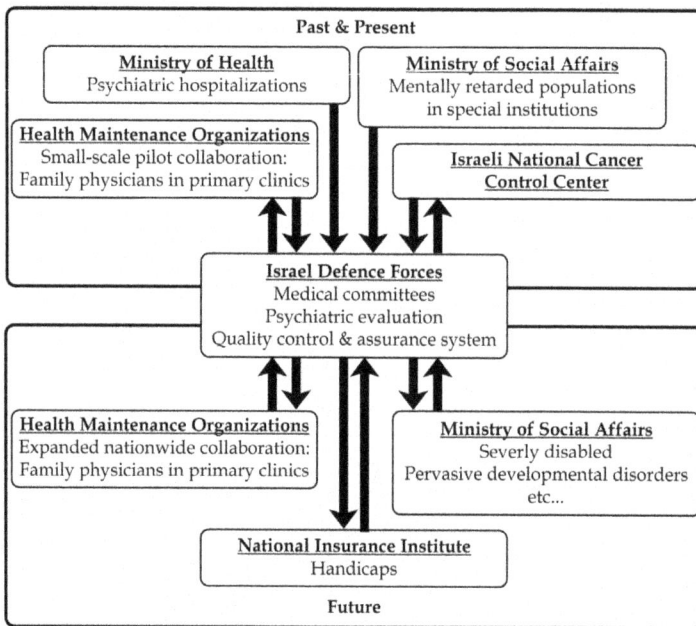

Fig. 6. Scheme of national project for information sharing and collaboration.

So far, although this information-sharing project was only initiated on a small-scale pilot format, covering only a minor fraction of the relevant population, bi-directional benefits are evident. For example, medical investigations in recruitment centers uncover medical conditions that were unknown before to the civilian authorities and vice-versa. These medical conditions include vision and hearing problems, essential hypertension, asthma, cardiac anomalies, tumors, urological conditions (hernia, varicocele, hydrocele, etc.), nephrological disorders (nephropathies, microscopic hematoria, severe proteinuria, etc.), orthopaedic problems, neurological problems, and mental disorders. Such findings are then reported to the primary care physician for further investigation with a referral to specialists and/or for treatment. The medical departments at the recruitment centers are then informed about further findings and about any change in the medical condition and status of the recruit by the primary physicians. Beyond the issue of medical informatics and process, recruits are provided with better medical service, which might contribute to early diagnosis, quick and appropriate treatment, and better and faster recovery. In the future, if this information sharing is combined with an epidemiological study, we believe it may also contribute to the prevention of health conditions, either via the medical education of teenagers or by active medical intervention.

8. Concluding remarks

Quality assessment, control, and improvement in a health system should include the infrastructure and structure, process, and outcome levels. The quality control and assurance system for medical committees at recruitment centers operates via an analysis→ design→ implementation→ evaluation→ modification loop. It first relies upon the identification of the limitations and needs of the whole medical process and of each department with regard to the medical, technical, administrative, procedural, and physical aspects and with regard to official policy, including the systematic evaluation of the knowledge, skills, judgement, and working patterns of all of the medical personnel. To this end, complementary methodologies are utilized to provide both quantitative and qualitative analyses of daily activities and practices. Among the main tools utilized are observations and assessment, a sampling of recruits for re-examination and of records for evaluation, questionnaires, interviews, data mining and analysis by reports, and patients' surveys. Then, intervention programs are designed and implemented to answer the needs of all of the parties participating in the medical process in order to improve their quality and performance; to increase the quality of patient care; and to achieve a higher patient satisfaction rate. Intervention programs include the establishment of a training and simulation center, lectures and instruction to all of the medical department's personnel, a forum of organization leaders, production of manuals, certification, written reports, and the design of computerized tools. These intervention programs and their impact need to be continuously evaluated and modified according to the specific needs and effects in particular recruitment centers. Significant improvements have been observed in various key parameters, such as the knowledge, skills, and judgement of the personnel and their professional performance, the conditions of their working environment, uniform working platforms, and the patients' satisfaction rate. Incessant monitoring and intervention are important to maintain quality in a medical organization.

Quality improvement at the individual (physicians, assistant technicians, administrative staff, managers, etc.) and global levels (performance, outcomes, physical conditions, procedures and regulations, etc.) is a goal in and of itself but is also a means of improving patient care and safety. Furthermore, the collaboration of all the participants in the process, on all levels – medical professionals (physicians, experts, etc.), technical assistants, administrative staff, and the active support and involvement of the managing authorities and policy makers – is a critical determinant for a successful outcome.

The successful application of a quality control and assurance project can lay the foundation for a population-based investigation, namely an assessment of the health status of the young Israeli population, by measuring anthropometric values and the prevalence of various medical conditions among recruits on a nationwide level. They may result in the identification of risk factors that can affect present and future morbidity. A national project of information sharing and cooperation was set up, aimed at screening, education, prevention, and early intervention in target populations.

There are multi-directional effects and relationships between the quality assurance and control process and its implications on improving the quality of health care and nationwide projects of preventative medicine through collaboration and information sharing (Fig. 7).

Recently, Zalmanovitch and Vashdi (2010) proposed an inherent trilemma, a trade-off between three desirable objectives (Iverson & Wren, 1998), in any debate on health care policy. In this context, the critical broad objectives are quality, funding, and coverage. In this context, quality refers to the efficiency and effectiveness of the health care services provided; funding refers to the public expenditures for health care that are incurred by taxpayers; and coverage refers to the percentage of a country's population eligible for state health care services and the comprehensiveness of these services. A trade-off means that, at most, only two of the three objectives can be satisfied simultaneously, and satisfying any two will always come at the expense of the third (Zalmanovitch & Vashdi, 2010). The cumulative experience of the quality assurance and control system and its effects suggest that it is central to the successful balancing between the three objectives, at least in the context of the medical departments at the recruitment centers. Considering the common characteristics of medical processes at the recruitment centers and at primary medical facilities, implementation of multi-armed quality assurance and control systems at clinics and hospitals holds great promise in finding the best solution to the trilemma, beyond the direct and clear impact on the medical staff's performance, medical process outcomes, service quality, and patient safety. Together with an epidemiologic investigation and preventive action, this system may contribute to the identification of risk factors and a reduction in future morbidity.

Fig. 7. Establishment of a quality control and assurance system for medical processes in recruitment centers led to an epidemiological study and a nationwide health project.

It remains to be seen in future studies whether the system can efficiently address these issues.

9. References

Blumenthal, D. (1996). Quality of health care part 1: Quality of health care - what is it?. *The New England Journal of Medicine*, Vol.335, No.12, pp. 891-4.

Bar-Dayan, Y.; Elishkevits, K.; Goldstein, L.; Goldberg, A.; Ohana, N.; Onn, E.; Levi, Y. &
     Bar-Dayan, Y. (2005). The prevalence of common cardiovascular diseases among
     17-year-old Israeli conscripts. *Cardiology*, Vol.104, No.1, pp. 6-9.

Campbell, S.M.; Roland, M.O. & Buetow, S.A. (2000). Defining quality of care. *Social Science
     and Medicine*, Vol.51, No.11, pp. 1611-25.

Chaiter, Y.; Machluf, Y.; Pirogovsky. A.; Palma, E.; Yona, A.; Shohat, T.; Yitzak, A.; Tal, O. &
     Ash, N. (2010). Quality control and quality assurance of medical committee
     performance in the Israel Defense Forces. *International Journal of Health Care Quality
     Assurance*, Vol.23 No.5, pp. 507-15.

Chaiter, Y., Palma, E., Machluf, Y., Yona, A.; Cohen, A.; Pirogovsky, A.; Shohat, T.; Ytzhak,
     A. & Ash, N. (2011). Quality assuring intervention for technical medical staff at
     medical committees. *International Journal of Health Care Quality Assurance*, Vol.24,
     No.1, pp. 19-30.

Chaiter, Y.; Pirogovsky, A.; Palma, E.; Yona, A.; Machluf, Y.; Shohat, T.; Farraj, N.; Tal, O.,
     Campino-Abbebe, G. & Levy, Y. (2008). Medical quality control in conscription
     centers- ten years of activity. *Journal of Israeli Military Medicine*, Vol.5 No.2, pp. 75-9.

Conscription Administration Data. (2001-9), *Biannual control on medical profiles in recruitment
     centers*, Personnel Directorate, IDF, Israel (unpublished data analysis).

Donabedian, A. (2003). *An introduction to quality assurance in health care*, Oxford University
     Press, ISBN 0195158091, New York, USA.

Donabedian, A. (2005). Evaluating the quality of medical care. *Milbank Quarterly*, Vol.83,
     No.4, pp. 691–729.

Dye, T.R. (1987). *Understanding Public Policy* (6th edition), Prentice-Hall, ISBN 0139369732,
     Englewood Cliffs, New Jersey, USA.

Farfel, A.; Green, M.S.; Shochat, T.; Noyman.; Levy, Y. & Afek, A. (2007). Trends in specific
     morbidity prevalence in male adolescents in Israel over a 50 year period and the
     impact of recent immigration. *The Israel Medical Association Journal*, Vol.9, No.3, pp.
     149-52.

Giraud-Roufast, A. & Chabot, J.M. (2008). Medical acceptance of quality assurance in health
     care, *The Journal of the American Medical Association*, Vol.300, No.22, pp. 2663-5.

Holmboe, E.S.; Lipner, R. & Greiner, A. (2008). Assessing quality of care: knowledge
     matters, *The Journal of the American Medical Association*, Vol.299, NO.3, pp. 338-40.

Iverson, T. & Wren, A. (1998). Equality, employment, and budgetary restraint: the trilemma
     of the service economy. *World Politics*, Vol.50, No.4, pp. 507-46.

Jaeger, U.; Zellner, K.; Kromeyer-Hauschild, K.; Lüdde, R.; Eisele, R. & Hebebrand, J. (2001).
     Body height, body weight and body mass index of German military recruits.
     Historical retrospect and current status. *Anthropologischer Anzeiger*, Vol.59, No.3,
     pp. 251-73.

Komlos, J. & Lauderdale, B.E. (2007). The mystery trend in American heights in the 20th
     century. *Annals of Human Biology*, Vol.34, No.2, pp. 206-15.

Landon, B.E.; Normand, S.L.; Blumenthal, D. & Daley, J. (2003). Physician clinical
     performance assessment: prospects and barriers, *The Journal of the American Medical
     Association*, Vol.290, No.9, pp. 1183-9.

Machluf, Y.; Pirogovsky, A.; Palma, E.; Yona, A.; Navon, A.; Shohat, T.; Ytzhak, A.; Tal, O.;
     Ash, N.; Nachmann, M. & Chaiter, Y. (2011). Coordinated computerized systems
     aimed at management, control, and quality assurance of medical processes and

informatics. *International Journal of Health Care Quality Assurance*, Accepted for publication.

Mandel, D.; Amital, H.; Zimlichman, E.; Wartenfeld, R.; Benyamini, L.; Shochat, T.; Mimouni, F.B. & Kreiss, Y. (2004). Quality assessment program in primary care clinics: a tool for quality improvement. *International Journal for Quality in Health Care*, Vol.16, No.2, pp. 175-80.

Mandel, D.; Zimlichman, E.; Ash, N., Mimouni, F.B.; Ezra, Y. & Kreiss, Y. (2003). Quality assessment of primary health care in a military setting, *Milltary Medicine*, Vol.168, No.11, pp. 890-2.

Munro, R.A. (2009). *Lean six Sigma for the healthcare practice: A pocket guide*, Amer Society for Quality Press, ISBN 0873897609, Milwaukee, Wisconsin, USA.

Navon, A.; Machluf, Y.; Cohen, A.; Pirogovsky, A.; Palma, E.; Tal, O.; Frenkel-Nir, Y.; Ash, N. & Chaiter, Y. (2011). Quality Assurance of Administrative Aspects of Medical Processes within the Framework of Medical Committees. *Journal of Israeli Military Medicine*, Accepted for publication.

Ovretveit, J. (1992). *Health service quality: An introduction to quality methods for health services*, Blackwell Scientific Publications, ISBN 0632032790, Oxford, England.

Ransom, E.R.; Joshi, M.S.; Nash, D.B. & Ransom, S.B. (2008). *The Healthcare Quality Book: Vision, strategy, and tools* (2nd edition), Health Administration Press, ISBN 1567933017, Chicago, USA.

The State Controller and Ombudsman Office. (2002). *Annual Report 53a*, Israel.

Zalmanovitch, Y. & Vashdi, D.R. (2010). Trade-offs are unavailable. *British Medical Journal*, Vol.340, pp. c1259.

5

Dose Optimization for the Quality Control Tests of X-Ray Equipment

Mana Sezdi
Istanbul University
Turkey

1. Introduction

Radiation is a major risk in diagnostic and therapeutic medical imaging. The problem is caused from incorrect use of radiography equipment and from the radiation exposure to patients much more than required. Exposure of different dose values for the same clinical examination, is an enough reason to draw attention to this issue.

International Commission on Radiation Protection (ICRP), the International Atomic Energy Agency (IAEA) and other various independent institutions have been making publications in relation to ionizing radiation protection for more than fifty years. Report 60 of the ICRP and the Basic Safety Standards that was published in the IAEA report have three basic principles related to the radiation protection (ICRP, 1991; IAEA, 1996).

The most important issue in these principles is the optimization of radiation. In the mentioned policy, the lowest dose is aimed by considering the country's economic and social factors for acceptable applications. Personnel already receive low dose with protection systems in the working areas. However, the patient doses must be taken under control based on the principle of optimization as much as possible.

There are two important points when performing a radiological procedure:

• To obtain the best possible image for a clear diagnosis of the disease,
• To apply the lowest dose for protecting the patient while getting the best image.

The second point indicates that the patient's radiation dose level must be kept at the lowest possible dose. In other words, it indicates dose optimization. The dose optimization meaning "the minimum radiation dose of the optimum image quality", is achieved by applying quality control procedures, calibration and dosimetric measurements.

In the Radiology Quality Control systems, the biggest problem is dose control and dose optimization. Neither patient nor users knows how much dose is exposed because there is no any system in the x-ray device for measuring or showing dose during exposure.

Since there is no dose adjustment on the equipment, the systems are operated by using the usual parameters; kVp and mAs. Because dose can not be adjusted, the patient may receive more dose than the aimed dose.

For dose optimization, all exposures should be kept at the minimum dose level in according to the ALARA principle (ALARA-as low as reasonably achievable). The aim of the optimization is not to download the risks of irradiation to zero. It is to reduce them to an acceptable level. This can be possible only by examining all parameters that affect the X-ray, by investigating the relationship between dose and these parameters, on the basis of this relationship, by performing the necessary regulations.

In all x-ray equipment, the operator can control the quantity and the quality of the radiation with kVp and mAs controls. If the equipment is not properly controlled, it will not be possible to control the radiation output. For this reason, optimization consists of not only improving of image quality and low dose but also establishing quality assurance and quality control programmes to ensure a proper performance of the x-ray equipment.

As frequently documented in the scientific literature, patient dose and image quality are basic aspects of any quality control (QC) tests in diagnostic radiology. Image quality must be adequate for diagnosis and it must be obtained with low doses.

The following QC tests are performed for both patient dose and image quality evaluation;

- kVp Accuracy and Repeatability
- Dose-kVp Linearity Test
- Dose-mAs Linearity Test
- X-ray Tube Output-kVp Relation
- HVL (Half Value Layer)
- Image Quality (Beam alignment, collimation alignment, contrast and resolution)

The quality control tests' methods, as well as the criteria for scoring the results, are in full agreement with those specified in the American Association of Physicists in Medicine (AAPM) Report No.4 and IEC 61223-3-1 (AAPM, 1981; IEC 61223-3-1, 1999).

There are a number of recent studies about dose optimization. Some of them are the surveys about image quality and patient dose in radiographic examinations in the authors' countries (Bouzarjomehri, 2004; Ciraj et al., 2005; Ramanandraibe, 2009; Papadimitriou, 2001; Shahbazi-Gahrouei, 2006). Some investigators focused only patient dose optimization (Brix et al., 2005; Vano & Fernandez, 2007; Seibert, 2004; Williams & Catling, 1998), whereas the others examined both the patient dose and image quality in radiographic devices (Aldrich et al., 2006; Schaefer-Prokop et al., 2008; Geijer, 2002). There are also studies that give reference values for clinical x-ray examinations by measuring phantom dose (Gray et al., 2005). But there is no any study focused to the dose optimization during quality control tests of x-ray devices. Dose optimization is very important because of the quality and quantity of quality control tests of x-ray equipments.

The aim of this study is to provide optimal x-ray parameters that may be used for quality control tests in order to make quality control activities more efficient and can be controlled. The staff know how the quality control tests are performed, but they don't know which parameters' values give which qualified image. They have problems during evaluation of test results, although there are some recommendations in the standards (AAPM, 1981; IEC 61223-3-1, 1999). They need proven parameter values for comparison. In this study, it was examined during quality control tests which parameters give a high quality image and how much dose is measured when these parameters were applied.

This study was performed by investigating the effects of X-ray parameters' changes on dose and by modeling of dose related to these parameters. After the modeling, in according to the related parameters, the dose level can be controlled, and in different x-ray units the dose levels that are obtained by applying the same parameter setting, can be compared.

Thus, in addition to obtain optimal parameters, controlling of the accuracy of the measured dose values may be possible by calculating the dose value during quality control tests.

2. Parameters of x-ray

In radiography, dose and image quality are dependent on radiographic parameters. This study is concerned with the quantification of these parameters and an assessment of their effect on patient dose and image quality. The focus of this study is on the relationship between dose, image quality and other radiographic parameters.

2.1 Absorbed dose
Absorbed dose is the quantity that expresses the radiation concentration delivered to a point, such as the entrance surface of patient's body. Absorbed dose in air is recognized as air kerma and it is a measure of the amount of radiation energy, in the unit of joules (J), actually deposited in or absorbed in a unit mass (kg) of air. Therefore, the quantity, kerma, is expressed in the units of J/kg which is also the radiation unit, the gray (G) (Sprawls, 1987; Hendee et al., 1984).
In this study, the word of "dose" will be used instead of air kerma (absorbed dose in air).

2.2 kVp
The high energy of the x-ray spectrum is determined by the kilovoltage applied to the x-ray tube. The maximum photon energy is numerically equal to the maximum applied potential in kilovolts. The maximum photon energy is determined by the voltage during the exposure time. This value is generally referred as the kilovolt peak (kVp) and is one of the adjustable factors of x-ray equipment (Sprawls, 1987).

2.3 mAs
The x-ray cathode is heated electrically by a current from a separate low voltage power supply. The output of this supply is controlled by the mA selector on the x-ray unit. Additionally, the duration of the x-ray exposure is controlled by the time selector. mAs is described by multiplying of these two values (mA x second) (Hendee et al., 1984).

2.4 Half Value Layer (HVL)
Half value layer describes both the penetrating ability of specific radiations and the penetration through specific objects. HVL is the thickness of material that reduces the intensity of an x-ray beam by half, and is expressed in unit of distance (mm) (Sprawls, 1987).

2.5 Image quality
The purpose of the radiographic image is to provide information about the medical condition of the patient. A quality image is one that provides all the information required for diagnosis of the patient's condition (Hendee et al., 1984).
Image quality is not a single factor but is described with beam alignment, collimation alignment, contrast and resolution. Contrast means differences in the form of gray scales or light intensities, whereas the resolution is a measure of its ability to differentiate between two objects a small distance apart; such that they appear distinct from one another.
An image is acceptable as qualified only if it has high resolution and high contrast.

3. Material and method

The radiographic measurements were performed in ten stationary X-ray units in five hospitals. The X-ray units including: Siemens, Philips, Toshiba, General Electric and Shimadzu were participated in this study. The reason for chosing these x-ray units is that their age is between 5 and 7 years old and the machines have 3 phase generators, thus their HVL value is kept in a narrow range, such as between 3 and 3,2mmAl.

Dosimax Plus A (Wellhöfer, Scanditronix, IBA, Germany) dosimeter was used to measure radiation dose. Dosimax Plus A dosimeter is a universal basic device and is designed according to IEC 61674 for acceptance tests and for quality checks at radiographic X-ray units. In Dosimax Plus A, dose measurements are performed by using solid state detectors (RQA). The dose range is from 200nGy to 9999mGy (Iba Dosimetry, 2008). It was calibrated by the Iba Laboratory of Germany and found to be capable of performing within recommended level of precision and accuracy.

Dose measurement applications has been included in recent recommendations (AAPM, 1981; IEC 61223-3-1, 1999). The measurement procedures that were realized in this study, are explained below step by step.

Before starting dose measurements, kVp accuracy tests were performed for 10 units and it was seen that they have acceptable accuracy in according to the standards (AAPM, 2002).

3.1 Measurement procedure of X-ray dose variation with kVp

The dosimeter was positioned in central beam axis such that the X-ray tube focal spot-dedector distance (FDD) was 100cm for the measurements. The radiation field size was set to cover the dosimeter in order to avoid the possible scatter radiation to the dosimeter.

In order to investigate the effect of kVp to the dose, the unit was set at 20mAs and 50kVp value. An X-ray exposure was made and the dosimeter reading was recorded. This step was repeated at same constant mAs and different kVp settings (50, 70, 80 and 100kVp) and dosimeter reading was determined. Similar X-ray dose measurements were also determined for 40 and 50mAs settings for each kVp value (50, 70, 80 and 100kVp). All measurements were repeated for 60cm (FDD). The measured dose values were plotted against the corresponding kVp for each X-ray unit separately.

3.2 Measurement procedure of X-ray dose variation with mAs

The dosimeter was positioned at 100cm (FDD) from the focal spot of the X-ray tube.

In order to determine the effect of mAs to the dose, the exposures were performed with constant kVp (50kVp), but with gradually increasing mAs (10, 20, 40 and 50mAs). Similar X-ray dose measurements were also determined for 70 and 100kVp settings for each mAs value (10, 20, 40 and 50mAs). All measurements were repeated for distance of 60cm. The measurement results for each X-ray unit were plotted against the corresponding mAs.

3.3 Measurement procedure of X-ray tube output variation with kVp

The X-ray tube output was determined as the ratio of dose reading to the mAs setting. The values of X-ray tube output were plotted against kVp by using dose values obtained from two measurement procedures (Section 3.1 and 3.2).

3.4 Measurement procedure for Half Value Layer (HVL)

For dose measurements, filtration was realized by using aluminum (Al) filters with 1mm and 0,5mm.

During the measurements, mAs and kVp were stable (20mAs, 50kVp) and the distance was determined as 100cm. Initially, the dose measurement without the filter was generalized. After this, the dose measurement was repeated by using filter with different thickness. Each filter thickness was obtained by adding 1mmAl and 0,5mmAl. The dose measurements were taken in the conditions; without filter, 1mmAl, 2mmAl, 3mmAl and 3,5mmAl.

3.5 Observing of image quality
Test tool ETR1 (Iba Dosimetry, 2008) was used for image quality tests. The ETR1 is a multi-purpose test tool. With a single exposure on X-ray film made by using this tool, all criterias (alignments, contrast and resolution) can be checked for quality control of image.
Before exposure, a cassette with x-ray film was placed on the patient table. The distance between the film and the focal spot was set to 100cm. The test tool was placed over the cassette and the collimator was adjusted to ensure that the light beam covers exactly the inner pattern of the test tool. An exposure was performed with 50kVp and 20mAs. The exposure was repeated for each setting value adjusted for dose measurements mentioned in Section 3.1 and 3.2 (50kVp-40mAs, 50mAs; 70kVp-20mAs, 40mAs, 50mAs; etc...).
After developing the film, the image on the film was compared with the real test tool image. Beam alignment, collimation alignment, contrast and resolution factors were determined and recorded.

4. Results

During the quality control of x-ray equipment, it is essential to know the effects of x-ray parameters to the image quality. The x-ray parameters' effects were measured by using quality control test procedures and they were analysed graphically.
In result, the optimized dose in which parameters' value gave the high quality image was determined.

4.1 Assessment of X-ray dose variation with kVp
The measured doses by changing kVp are given in Table 1. During measurements, mAs was firstly kept stable (20mAs) and kVp was changed as 50, 70, 80 and 100kVp to investigate the effects of kVp to the dose at stable mAs.
After this, the same measurement procedure was applied to other mAs values (40 and 50mAs). All measurements were performed at distance of 100 and 60cm.
Graphical representations of the relationship between dose and kVp value for constant mAs (20, 40 and 50mAs) at 100cm and 60cm are given in Fig. 1, 2, 3 and Fig. 4, 5, 6, respectively.

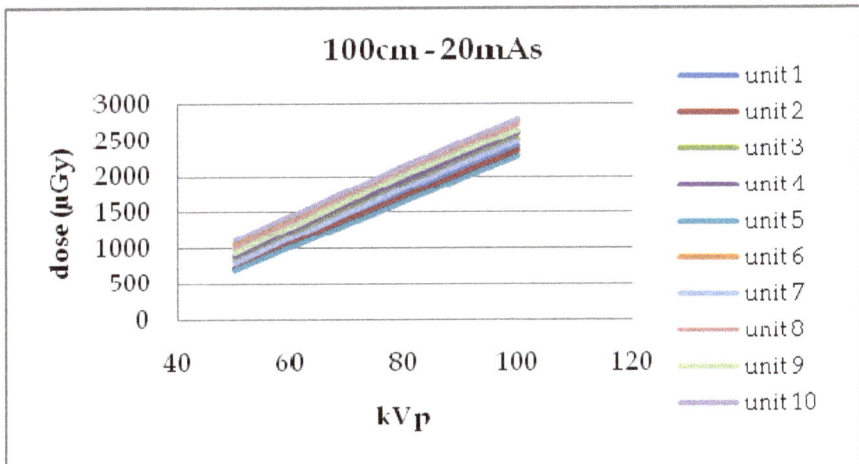

Fig. 1. Measured dose values of 10 x-ray units versus kVp for 20mAs at distance of 100cm.

	Unit	50 kVp	70 kVp	80 kVp	100 kVp	50 kVp	70 kVp	80 kVp	100 kVp	
					Dose (µGy)					
100cm - 20mAs	1	736,0	1439	1766	2420	1950	4230	5450	8050	**60cm - 20mAs**
	2	710,7	1388	1707	2357	1883	3955	5243	7588	
	3	824,3	1538	1866	2538	2327	4640	5905	8608	
	4	894,4	1605	1935	2601	2478	4915	6181	8909	
	5	690,4	1324	1642	2282	1554	3717	4981	7268	
	6	1048	1734	2101	2757	3310	5758	7006	9498	
	7	792,1	1481	1826	2483	2077	4353	5627	8277	
	8	988,4	1704	2070	2726	2999	5477	6722	9300	
	9	934,7	1657	2017	2652	2796	5157	6445	9062	
	10	1101	1784	2136	2792	3484	5872	7195	9741	
100cm - 40mAs	1	1482	2938	3622	5056	4190	8260	10800	15600	**60cm - 40mAs**
	2	1493	2863	3580	4912	3774	7983	10511	15123	
	3	1587	3129	3797	5276	4511	8588	11383	15515	
	4	1600	3136	3872	5335	4684	8981	11515	16089	
	5	1400	2738	3460	4861	3464	7613	10099	14656	
	6	1923	3461	4207	5716	5468	10097	12415	17247	
	7	1528	3082	3702	5185	4271	8716	10984	15797	
	8	1858	3335	4104	5561	5249	9773	12037	16986	
	9	1692	3290	3946	5405	4870	9408	11788	16487	
	10	2053	3581	4320	5855	5805	10618	12869	17610	
100cm - 50mAs	1	1790	3612	4590	6380	5350	12548	15800	22700	**60cm - 50mAs**
	2	1722	3443	4384	6194	4991	12168	15553	22313	
	3	1933	3678	4609	6608	6144	13528	15973	23548	
	4	1985	3697	4481	6652	6588	13249	16348	23704	
	5	1604	3345	4185	6008	4585	11712	15033	21972	
	6	2245	4020	4965	6995	8101	14631	17750	24808	
	7	1853	3665	4575	6518	5574	12999	15799	23255	
	8	2198	3813	4759	6823	7623	14240	17234	24484	
	9	2091	3799	4691	6712	7064	13751	16857	23892	
	10	2391	4213	5167	7194	8724	15027	18203	25283	

Table 1. Measured doses (µGy) for constant mAs but increasing kVp at different distances.

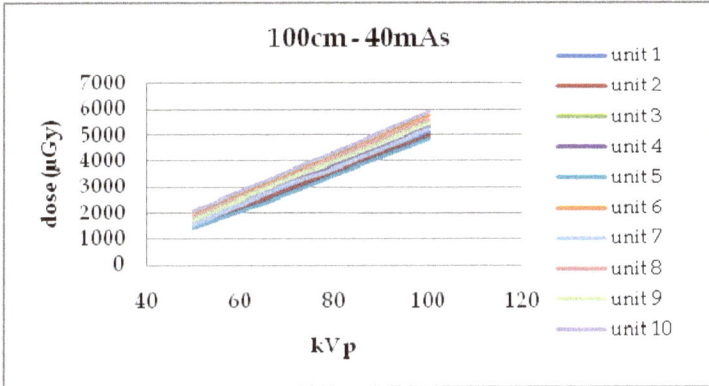

Fig. 2. Measured dose values of 10 x-ray units versus kVp for 40mAs at distance of 100cm.

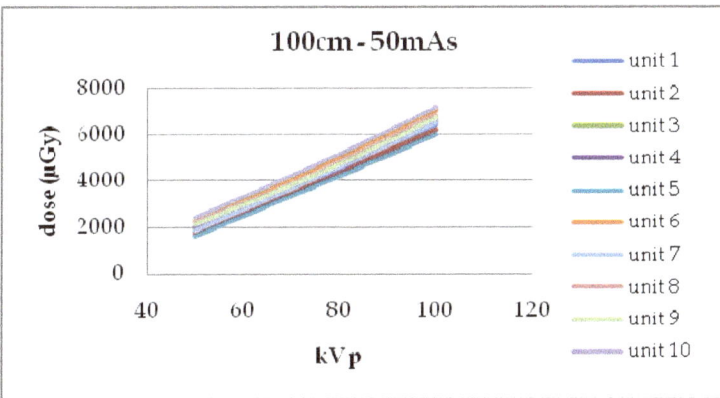

Fig. 3. Measured dose values of 10 x-ray units versus kVp for 50mAs at distance of 100cm.

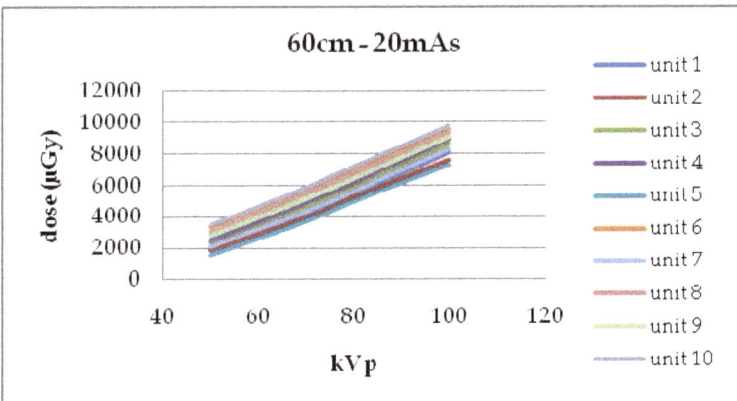

Fig. 4. Measured dose values of 10 x-ray units versus kVp for 20mAs at distance of 60cm.

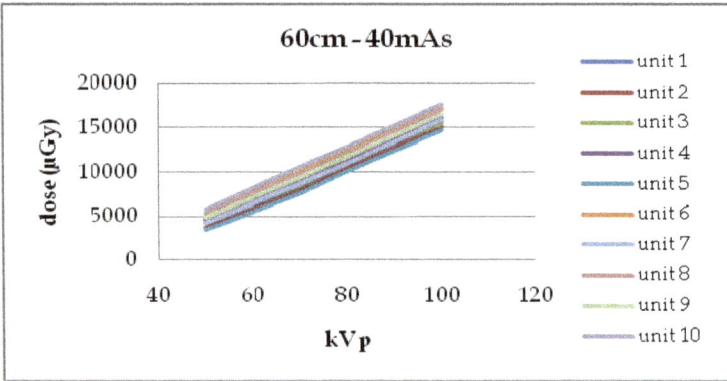

Fig. 5. Measured dose values of 10 x-ray units versus kVp for 40mAs at distance of 60cm.

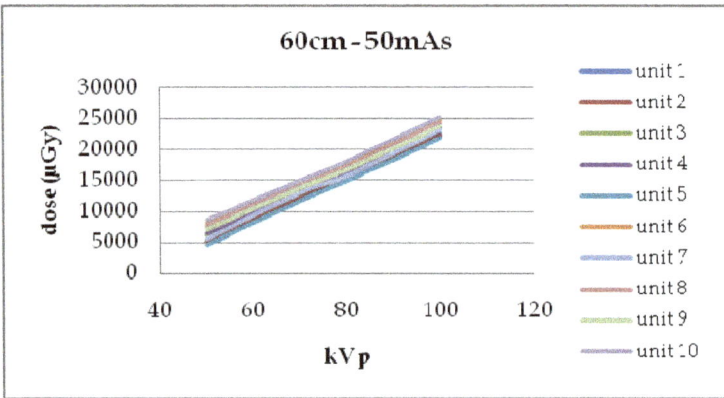

Fig. 6. Measured dose values of 10 x-ray units versus kVp for 50mAs at distance of 60cm.

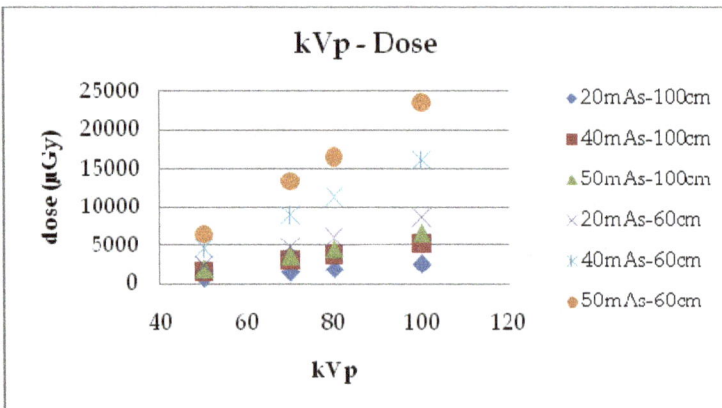

Fig. 7. Mean dose values of 10 x-ray units versus kVp for different mAs and distance setting.

The dose values obtained from 10 x-ray units were analysed statistically and the mean dose values for each setting parameter were defined with standard deviation in Table 2.
For different distance and mAs settings, the mean dose values were plotted against kVp (Fig. 7). Hence, the small differences that are caused from unit changes were eliminated, and the effect of kVp to dose variation was focused.

Dose	100cm - 20mAs				60cm - 20mAs			
(μGy)	50kVp	70kVp	80kVp	100kVp	50kVp	70kVp	80kVp	100kVp
Mean	872,0	1565,4	1906,5	2560,7	2485,8	4807,5	6075,5	8630,2
std	144,3	155,9	172,4	174,8	645,3	753,0	760,0	823,6
Dose	100cm - 40mAs				60cm - 40mAs			
(μGy)	50kVp	70kVp	80kVp	100kVp	50kVp	70kVp	80kVp	100kVp
Mean	1661,6	3155,3	3861,1	5316,0	4628,7	9003,7	11440,1	16110,9
std	215,4	266,5	283,4	329,3	742,1	962,9	866,6	957,6
Dose	100cm - 50mAs				60cm - 50mAs			
(μGy)	50kVp	70kVp	80kVp	100kVp	50kVp	70kVp	80kVp	100kVp
Mean	1981,3	3728,4	4640,7	6608,4	6474,2	13385,3	16455,0	23595,8
std	249,7	254,1	280,0	356,4	1387,9	1069,0	1024,1	1075,5

Table 2. The statistic analysis of dose from 10 units for different kVp at constant mAs.

4.2 Assessment of X-ray dose variation with mAs

The obtained dose values at constant kVp by increasing mAs can be seen in Table 3. The dose measurements were performed at 50, 70 and 100kVp with changing mAs (10, 20, 40 and 50mAs) in distance of 100 and 60cm.
Graphical representations of the relationship between dose and mAs for constant kVp (50, 70 and 100kVp) at 100 and 60cm are shown in Fig. 8, 9, 10 and Fig. 11, 12, 13, respectively.

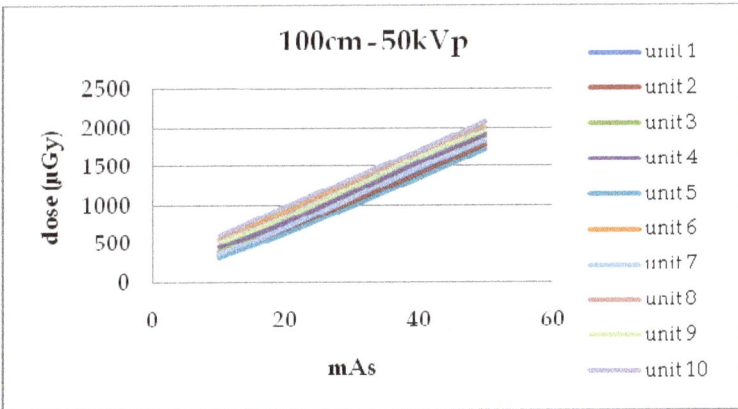

Fig. 8. Measured dose values of 10 x-ray units versus mAs for 50kVp at distance of 100cm.

	Dose (µGy)								
	Unit	10 mAs	20 mAs	40 mAs	50 mAs	10 mAs	20 mAs	40 mAs	50 mAs	
100cm - 50kVp	1	378,0	736,0	1432	1810	1050	2089	4188	5120	**60cm - 50kVp**
	2	339,8	695,4	1397	1773	932	1940	4056	5075	
	3	402,2	796,4	1527	1886	1202	2231	4276	5245	
	4	464,0	786,5	1554	1911	1247	2333	4329	5394	
	5	321,1	632,5	1344	1717	826	1846	3975	5026	
	6	565,8	934,5	1688	2068	1494	2532	4680	5715	
	7	370,2	712,0	1483	1834	1123	2152	4245	5206	
	8	530,9	879,0	1641	2006	1342	2402	4516	5657	
	9	510,4	844,6	1610	1952	1272	2342	4447	5551	
	10	606,9	987,6	1711	2091	1597	2614	4801	5862	
100cm - 70kVp	1	714,4	1469	2871	3652	2310	4180	8140	10077	**60cm - 70kVp**
	2	641,8	1372	2787	3528	2187	4040	7966	9846	
	3	801,0	1618	3055	3874	2604	4401	8428	10480	
	4	874,9	1687	3180	3902	2799	4675	8672	10585	
	5	574,9	1292	2683	3414	1993	3828	7692	9613	
	6	1001	1881	3401	4157	3317	5186	9123	10959	
	7	754,2	1558	2955	3773	2580	4309	8359	10228	
	8	987,4	1764	3387	4084	3221	4956	8972	10745	
	9	956,6	1718	3292	3952	2989	4936	8832	10645	
	10	1075	1926	3515	4219	3503	5387	9280	11210	
100cm - 100kVp	1	1404	2808	5616	7020	3250	7288	15435	19426	**60cm - 100kVp**
	2	1198	2623	5387	6819	3006	6870	14811	18926	
	3	1641	2927	5751	7255	3870	8097	15872	20058	
	4	1690	2922	5964	7285	4278	8296	16080	20415	
	5	1045	2444	5308	6622	2421	6342	14354	18465	
	6	1854	3182	6215	7558	4940	8955	17157	21594	
	7	1582	2885	5703	7131	3537	7601	15343	19500	
	8	1797	3060	6070	7429	4509	8735	17102	21306	
	9	1717	3055	6070	7407	4448	8584	16399	20958	
	10	1889	3343	6266	7638	5290	9310	17697	21944	

Table 3. Measured doses (µGy) for constant kVp but increasing mAs at different distances.

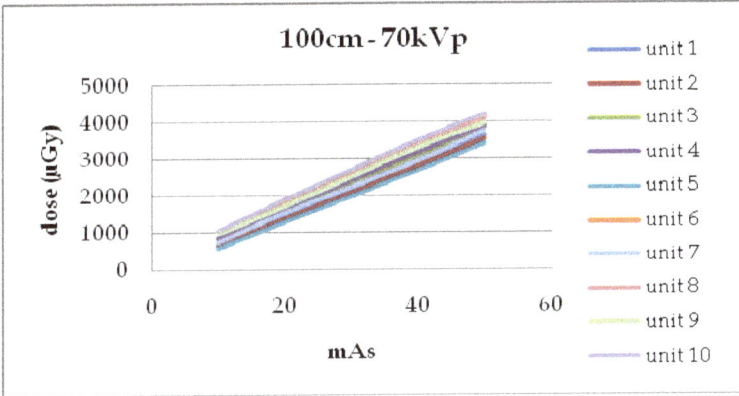

Fig. 9. Measured dose values of 10 x-ray units versus mAs for 70kVp at distance of 100cm.

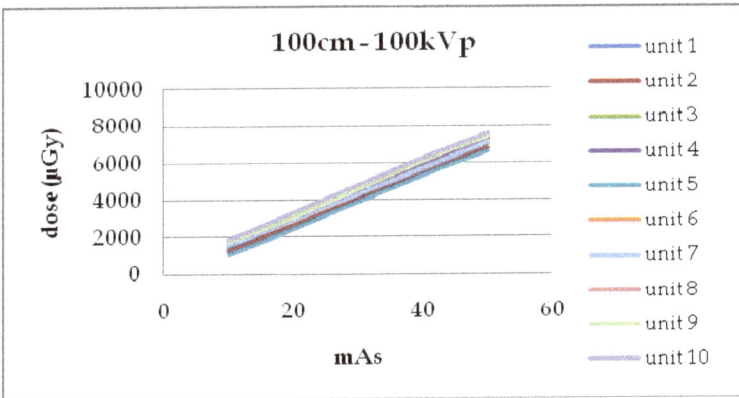

Fig. 10. Measured dose values of 10 x-ray units versus mAs for 100kVp at distance of 100cm.

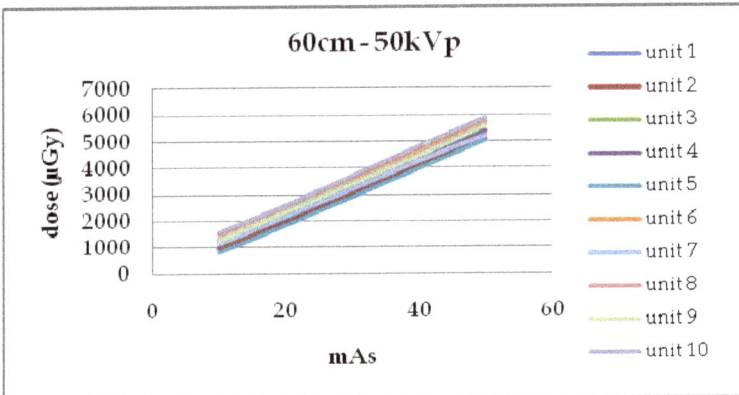

Fig. 11. Measured dose values of 10 x-ray units versus mAs for 50kVp at distance of 60cm.

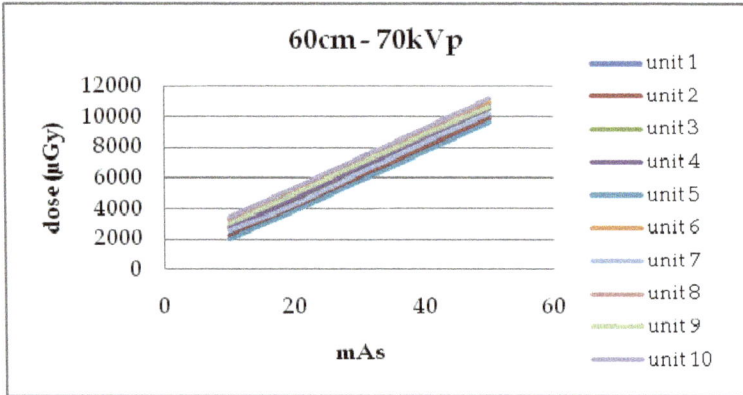

Fig. 12. Measured dose values of 10 x-ray units versus mAs for 70kVp at distance of 60cm.

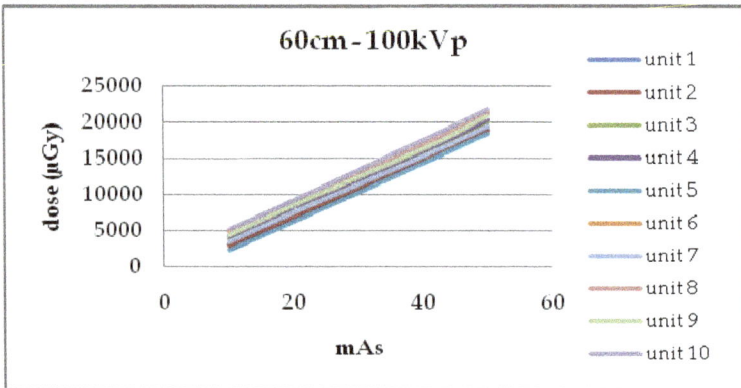

Fig. 13. Measured dose values of 10 x-ray units versus mAs for 100kVp at distance of 60cm.

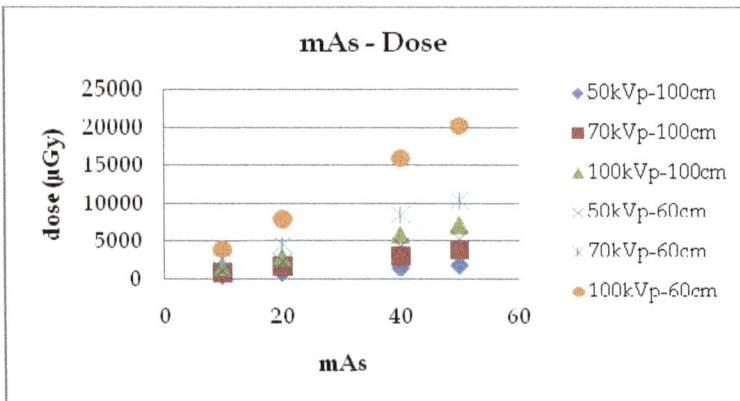

Fig. 14. Mean dose values of 10 x-ray units versus mAs for different kVp and distances.

The dose values obtained from 10 x-ray units were analysed statistically and the mean dose values for each setting parameter were defined with standard deviation in Table 4. For different distance and kVp settings, the mean dose values were plotted against mAs (Fig. 14). Hence, the small differences that are caused from unit changes were eliminated, and the effect of mAs to dose variation was focused.

Dose (µGy)	100cm - 50kVp				60cm - 50kVp			
	10mAs	20mAs	40mAs	50mAs	10mAs	20mAs	40mAs	50mAs
Mean	448,9	800,5	1538,6	1904,9	1208,4	2248,1	4351,4	5385,1
std	100,5	111,8	125,0	125,0	238,3	246,1	262,3	295,3
Dose (µGy)	100cm - 70kVp				60cm - 70kVp			
	10mAs	20mAs	40mAs	50mAs	10mAs	20mAs	40mAs	50mAs
Mean	838,2	1628,5	3112,6	3855,6	2750,4	4589,7	8546,2	10438,6
std	167,4	208,6	285,7	266,1	526,1	518,1	519,6	498,2
Dose (µGy)	100cm - 100kVp				60cm - 100kVp			
	10mAs	20mAs	40mAs	50mAs	10mAs	20mAs	40mAs	50mAs
Mean	1581,7	2924,9	5835,1	7216,4	3954,8	8008,0	16025,0	20259,2
std	281,9	261,2	334,7	322,8	903,1	960,3	1079,2	1180,2

Table 4. The statistic analysis of dose from 10 units for different mAs at constant kVp.

4.3 Assessment of X-ray tube output variation with kVp

In order to investigate the relationship between the x-ray tube output and kVp, firstly the x-ray tube outputs for 10 x-ray units, were calculated by dividing the measured dose values to the mAs values. It was seen that there is a dose distribution because of the measured different doses of each x-ray unit. Therefore, the mean of the x-ray tube output values for each mAs values were used for plotting of the x-ray tube output against the kVp.

Additionally, the graphics show that there is a different distribution that are caused from different distances although all distributions were similar for each mAs value.

For each different distance, the mean of the calculated tube output for different mAs were plotted with equations.

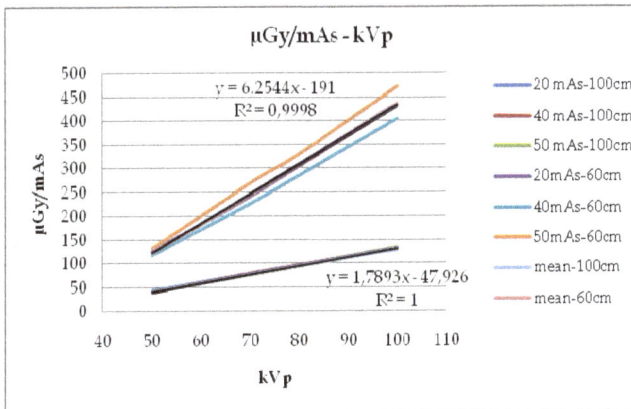

Fig. 15. X-ray tube output changes with kVp (dose values from procedure in Section 3.1).

In Figure 15, the dose values that were obtained by applying the procedure mentioned in Section 3.1 and the procedure settings (mAs, kVp), were plotted, whereas in Figure 16 the values obtained from procedure in section 3.2, were used.

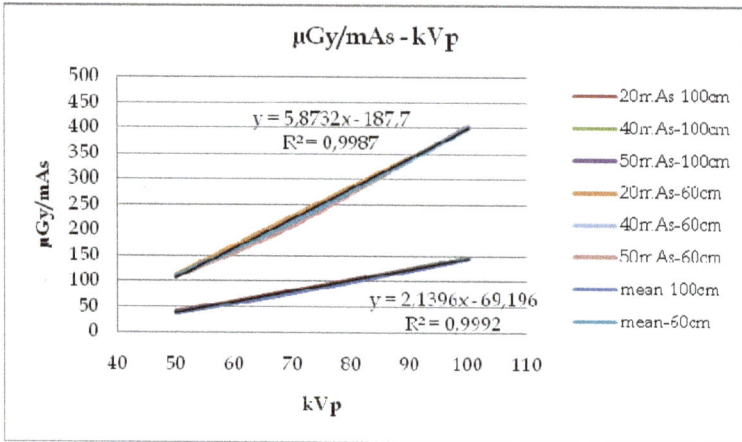

Fig. 16. X-ray tube output changes with kVp (dose values from procedure in Section 3.2).

The tube output which is derived from direct measurement can be expressed in equations obtained from Figure 15 and Figure 16, because kVp is related to tube output directly.
For distance of 100cm, tube output can be written separately in different two equations that were obtained from graphics in Fig. 15 and Fig. 16.

$$\text{Tube output } (\mu Gy/mAs) = 1{,}7893 kVp - 47{,}926 \qquad R^2 = 1 \tag{1}$$

$$\text{Tube output } (\mu Gy/mAs) = 2{,}1396 kVp - 69{,}196 \qquad R^2 = 0{,}9992 \tag{2}$$

Dose can be determined from tube output, and mAs can be placed in Equations 1 and 2.

$$\text{Dose } (\mu Gy) = (1{,}7893 kVp - 47{,}926) \times mAs \tag{3}$$

$$\text{Dose } (\mu Gy) = (2{,}1396 kVp - 69{,}196) \times mAs \tag{4}$$

For distance of 60 cm, again tube output can be written separately in different two equations obtained from Figure 15 and Figure 16, respectively.

$$\text{Tube output } (\mu Gy/mAs) = 6{,}2544 kVp - 191{,}0 \qquad R^2 = 0{,}9998 \tag{5}$$

$$\text{Tube output } (\mu Gy/mAs) = 5{,}8732 kVp - 187{,}7 \qquad R^2 = 0{,}9987 \tag{6}$$

When mAs is placed in Equations 5 and 6, the following Equations 7 and 8 are derived.

$$\text{Dose } (\mu Gy) = (6{,}2544 kVp - 191{,}0) \times mAs \tag{7}$$

$$\text{Dose } (\mu Gy) = (5{,}8732 kVp - 187{,}7) \times mAs \tag{8}$$

To test the validity of these equations, an external set of dose values obtained from measurements (10mAs-50kVp, 10mAs-70kVp and 10mAs-100kVp for 100cm and 60cm) was selected. kVp and mAs values were plugged into the Equations 3, 4, and Equations 7, 8 to predict the dose values for applied kVp and mAs values of measurement procedures (Section 3.1 and 3.2). The results were then compared with the measured dose values, as shown in Table 5.

It is seen from Table 5, the predicted dose values are within the measured dose value with standard deviations for each measurement procedure (in different distance, firstly constant mAs with increasing kVp, afterly constant kVp with increasing mAs). The predicted dose values obtained from equations that shows the relationship between x-ray tube output and kVp during measurement procedure (mAs is increased with constant kVp), are approximately similar with the predicted dose values that were derived from measurement procedure of constant mAs and increasing kVp. Relatively it can be said that dose measurements are not affected from the application style of parameters (kVp and mAs). Not only keeping mAs as constant and increasing kVp, but also keeping kVp as constant and increasing mAs doesn't affect the measured dose values for the same kVp and mAs. For example, the dose value obtained from measurement of 50kVp-40mAs are approximately similar during application of both constant 50kVp with increasing mAs and constant 40mAs with increasing kVp. This result showed that taking into account the results that were obtained from only one measurement procedure is sufficient. Especially, Equation 3 and Equation 7 can be preferred for dose estimation because of their best R^2.

Distance 100cm	Mean dose from direct measurement (µGy)	Dose calculated from Equation 3 (µGy)	Dose calculated from Equation 4 (µGy)
10mAs-50kVp	448,9 ± 100,5	415,39	377,84
10mAs-70kVp	838,2 ± 167,4	773,25	805,76
10mAs-100kVp	1581,7 ± 281,9	1310,04	1447,64
Distance 60cm	Mean dose from direct measurement (µGy)	Dose calculated from Equation 7 (µGy)	Dose calculated from Equation 8 (µGy)
10mAs-50kVp	1208,4 ± 238,3	1217,2	1059,60
10mAs-70kVp	2750,4 ± 526,1	2468,08	2234,24
10mAs-100kVp	3954,8 ± 903,1	4344,4	3996,20

Table 5. Measured and calculated dose values.

4.4 Assessment of Half Value Layer (HVL)

Testing of half value layer is performed by measuring dose values with different Al thickness and it verifies that half value layer is sufficient to reduce patient exposure to low energy radiation. The obtained dose measurement results of each x-ray unit in this study for stable mAs and kVp are given in Table 6.

The dose measurement results were plotted against the aluminum (Al) thickness (Fig. 17). Dose (µGy) equations were obtained as a function of Al thickness and from these equations, the Al thickness in which the dose decreased to its half value was calculated (Table 7).

kVp = 50, mAs = 20, Distance = 100cm					
Unit	Dose (μGy)				
	0mmAl	1mmAl	2mmAl	3mmAl	4mmAl
1	736,0	520,9	451,8	374,0	342,8
2	710,7	519,9	446,8	370,1	337,7
3	824,3	647,5	522,0	431,6	388,9
4	894,4	643,1	554,2	462,3	415,7
5	690,4	502,1	431,8	359,5	322,0
6	1047,8	767,3	654,0	524,1	493,8
7	792,1	598,8	485,7	401,8	360,0
8	988,0	702,1	588,9	499,6	454,9
9	934,7	704,2	577,1	474,8	425,2
10	1101,0	782,2	656,0	556,6	506,8

Table 6. Dose measurements for different aluminum thickness (mm).

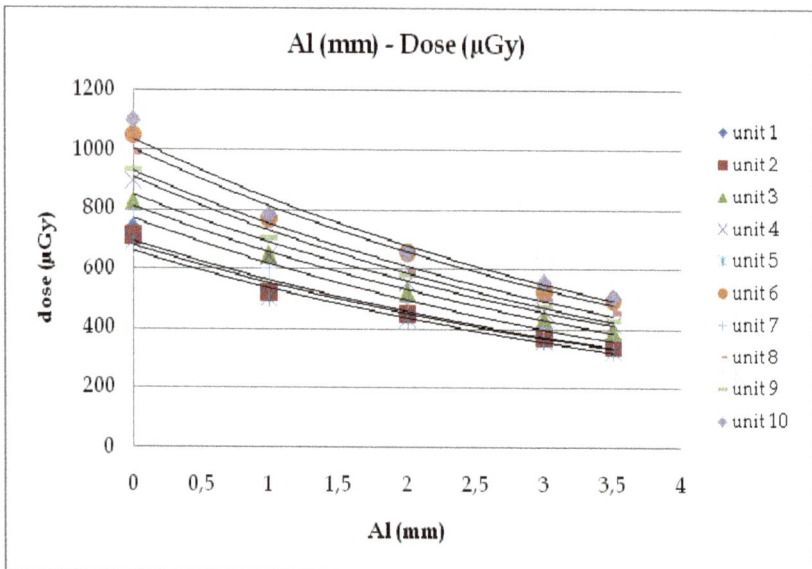

Fig. 17. Al(mm)-Dose(μGy) graphic for each X-ray unit.

Unit	Dose (µGy) = f(Al (mm))	Calculated HVL (mm)
1	$y = 693{,}18\ e^{-0{,}208x}$ $R^2 = 0{,}9714$	3,0
2	$y = 678{,}13\ e^{-0{,}204x}$ $R^2 = 0{,}9819$	3,1
3	$y = 811{,}49\ e^{-0{,}212x}$ $R^2 = 0{,}9979$	3,2
4	$y = 848{,}79\ e^{-0{,}208x}$ $R^2 = 0{,}9775$	3,1
5	$y = 658{,}65\ e^{-0{,}207x}$ $R^2 = 0{,}9812$	3,1
6	$y = 1002{,}5\ e^{-0{,}211x}$ $R^2 = 0{,}9830$	3,1
7	$y = 769{,}50\ e^{-0{,}220x}$ $R^2 = 0{,}9940$	3,0
8	$y = 929{,}33\ e^{-0{,}212x}$ $R^2 = 0{,}9723$	3,0
9	$y = 907{,}66\ e^{-0{,}219x}$ $R^2 = 0{,}9939$	3,0
10	$y = 1035{,}5\ e^{-0{,}212x}$ $R^2 = 0{,}9722$	3,0

Table 7. Dose=f(Al) equations and calculated HVL values.

As it is seen from the table, the observed x-ray units' HVL values change from 3,0 to 3,2 mmAl. Because it is required that the HVL of an acceptable x-ray unit with 3 phase generator must exceed 2,9mm, the observed 10 x-ray units were appropriate to the international standards (AAPM, 1981).

4.5 Image quality

Image quality tests were performed by controlling of beam alignment, collimation alignment, contrast and resolution of image.

As a result of the beam aligment and collimation alignment tests, it was seen that beam alignment and collimation alignment are only related to the quality of x-ray tube, are not dependent to the x-ray parameters, such as kVp, mAs and dose. For this reason, the test results that were obtained from only one measurement setting (50kVp, 20mAs, 100cm), are sufficient to obtain information about alignments of each x-ray units (Table 8).

Beam alignment test gives the deviation of the centre from the middle of the exposed film to the middle of the test tool (point "a" in Figure 18). The test's results that were given in Table 8 show that the beam misalignments were less than 10mm for 10 x-ray units.

In the collimation alignment test, the vertical misalignment was defined as the sum of the deviation of the top and bottom edges, horizontal as the sum of the deviation of the right and left edges (point "b" in Figure 18). In according to the international standards, the misalignment must each be less than 25mm (AAPM, 1981). As it is seen from Table 8, all misalignmenst for 10 units are appropriate to the standards.

For the measurement of the resolution, parallel lead strips separated by a distance equal to the width of the strips, that are placed on the test tool (point "c" in Figure 18) were used. The common practice is to describe the line width and separation distance in terms of line pairs (lp) per unit distance (millimeters) (Lp/mm). One line pair consists of one lead strip and adjacent separation space. The number of line pairs per millimeter is actually an expression of spatial frequency. As the lines get smaller and closer together, the spatial frequency increases (Sprawls, 1987). The test pattern contains areas with different spatial frequencies. To evaluate an imaging system, the visible line group is recorded as line pairs per mm. In according to the international standards, resolution below 0,8Lp/mm is not acceptable (AAPM, 1981). The obtained test results in this study, are shown in Table 9.

Evaluating the contrast was performed by looking at the copper step wedge from the test pattern that are placed on the test tool (point "d" in Figure 18). The visible copper step wedges were recorded in order to describe the resolution quality (Table 10). In according to the international standards, all copper steps have to be clearly visible (AAPM, 1981).

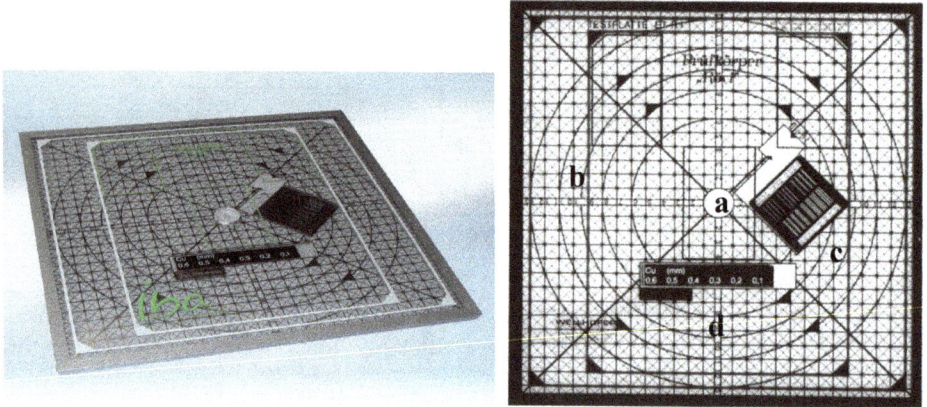

Fig. 18. ETR1 test tool used for image quality test.

Unit	Beam Alignment (mm)		Collimation Alignment (mm)			
1	< 10mm	OK	Top: 3mm Right: 1mm	Bottom: 2mm Left: 1mm	Total: 5mm Total: 2mm	OK
2	< 10mm	OK	Top: 5mm Right: 3mm	Bottom: 3mm Left: 1mm	Total: 8mm Total: 4mm	OK
3	< 10mm	OK	Top: 1mm Right: 2mm	Bottom: 1mm Left: 1mm	Total: 2mm Total: 3mm	OK
4	< 10mm	OK	Top: 3mm Right: 2mm	Bottom: 2mm Left: 3mm	Total: 5mm Total: 5mm	OK
5	< 10mm	OK	Top: 2mm Right: 2mm	Bottom: 1mm Left: 2mm	Total: 3mm Total: 4mm	OK
6	< 10mm	OK	Top: 6mm Right: 3mm	Bottom: 4mm Left: 3mm	Total: 10mm Total: 6mm	OK
7	< 10mm	OK	Top: 2mm Right: 3mm	Bottom: 2mm Left: 1mm	Total: 4mm Total: 4mm	OK
8	< 10mm	OK	Top: 4mm Right: 1mm	Bottom: 3mm Left: 1mm	Total: 7mm Total: 2mm	OK
9	< 10mm	OK	Top: 5mm Right: 2mm	Bottom: 4mm Left: 2mm	Total: 9mm Total: 4mm	OK
10	< 10mm	OK	Top: 7mm Right: 4mm	Bottom: 5mm Left: 3mm	Total: 12mm Total: 7mm	OK

Table 8. Beam alignment and collimation alignment test results.

As it is seen from Table 8, both beam alignments and collimation alignments of 10 x-ray units are appropriate to the international standards and there is no unwanted effect on the image quality.

Although beam alignment and collimation alignment are not dependent to the kVp and mAs value, resolution and contrast are directly related to these parameters. On Table 9, it is seen that resolution increases with increasing parameter setting values, especially with kVp. From Table 10, it can be said that contrast is good on the values of 70kVp, especially on 70kVp-40mAs for 10 x-ray units. At this value of parameters, all copper steps on the test tool can be seen definitely. While the values on the Table 10 decreases from 0,6 to 0,1, the contrast also decreases and the seeable points on the film loss step by step.

5. Discussion

In this study, the x-ray units with ages between 5 and 7 years old were selected to prevent the wide distribution of measured dose because the x-ray tubes don't produce the same exposure and the output decreases with age of x-ray unit.

Again, in this study, three phase generators were preferred because they produces more radiation exposure per unit mAs. This characteristic is essential for modeling of dose.

A difference in tube output among tubes is often caused by variations in the filtration. For this reason, this study were performed on the x-ray units with HVL values that changes approximately from 3,0 to 3,1mmAl. It also prevented the wide distribution of measured dose. The obtained HVL values in this study are acceptable in according to the international standards (AAPM, 1981).

It is known that dose is more sensitive to the kVp changes than mAs changes. Exposure errors can occur if the actual kVp generated by the x-ray generator is different from the adjusting kVp value. Before dose measurements, kVp accuracy testing were performed correctly and it was seen that the kVp during exposure was the close within the acceptable deviation to the selected kVp value.

All dose measurements were performed at different distance of 100cm and 60cm. With this application, the distance effects on dose were investigated and it was used for dose modeling because of the inverse-square effect.

For dose measurements, two different measurement procedures were used. In the first procedure, mAs value was kept constant and kVp values were changed to investigate the dose variation with kVp. In the second procedure, kVp value was kept constant and mAs values were changed to investigate the dose variation with mAs. Thus, the effects of kVp and mAs were examined separately.

Because the x-ray units were selected in according to the criterias mentioned above, the measured dose values didn't show wide distribution for each measurement setup in all 10 x-ray units. In this condition, the mean of the dose values of 10 x-ray units for each measurement setup was used to show the tube output variations with kVp. Plotting of tube output to kVp (Figure 15 and 16) were performed by using dose values obtained from two different measurement procedures. By this way, it was seen that the tube output variations related to kVp were approximately similar at different mAs value. Hence, the mean variations were used for modeling of dose.

Modeling was realized twice for dose values at different distances of 100cm and 60cm, because the different variations were seen between measurement values obtained different distances. By using equations (Equation 3 and Equation 7) in the models related to the

	Resolution (Lp/mm)								
	Unit	50kVp	70kVp	80kVp	100kVp	50kVp	70kVp	80kVp	100kVp	
100cm - 20mAs	1	2,2	3,1	3,4	4,0	2,2	3,4	3,4	4,0	**60cm - 20mAs**
	2	2,2	3,4	3,7	4,3	2,2	3,1	3,4	4,0	
	3	2,5	3,1	3,7	4,0	2,0	2,8	3,0	3,7	
	4	2,5	3,4	3,4	4,0	2,2	3,1	3,4	4,3	
	5	2,5	3,4	3,4	4,3	2,5	3,4	3,4	4,3	
	6	2,0	2,8	3,7	4,3	2,2	3,1	3,4	4,0	
	7	2,5	3,1	3,7	4,3	2,2	3,1	3,4	4,3	
	8	2,0	2,8	3,4	4,0	2,0	2,8	3,1	4,0	
	9	2,0	2,8	3,4	4,0	2,2	3,1	3,4	4,3	
	10	2,2	3,1	3,7	4,3	2,2	3,1	3,7	4,3	
100cm - 40mAs	1	2,2	3,4	3,7	4,3	2,5	3,4	3,7	4,3	**60cm - 40mAs**
	2	2,2	3,4	3,7	4,3	2,0	3,1	3,4	4,0	
	3	2,5	3,4	4,0	4,3	2,5	3,4	3,7	4,3	
	4	2,5	3,1	3,4	4,0	2,2	3,1	3,4	4,0	
	5	2,2	3,4	3,7	4,3	2,5	3,4	3,7	4,0	
	6	2,8	3,4	3,7	4,3	2,8	3,7	4,0	4,3	
	7	2,8	3,7	4,0	4,3	2,5	3,4	3,7	4,0	
	8	2,2	3,4	3,7	4,0	2,5	3,4	3,7	4,0	
	9	2,5	3,4	3,7	4,3	2,2	3,1	3,7	4,3	
	10	2,2	3,1	4,0	4,3	2,2	3,4	4,0	4,3	
100cm - 50mAs	1	2,5	3,1	4,0	4,3	2,2	3,1	3,7	4,3	**60cm - 50mAs**
	2	2,8	3,4	4,0	4,6	2,5	3,1	3,7	4,3	
	3	2,8	3,4	4,0	4,6	2,8	3,4	4,0	4,6	
	4	2,5	3,1	3,7	4,3	2,2	3,1	3,7	4,3	
	5	2,5	2,8	3,7	4,3	2,2	3,1	3,7	4,6	
	6	2,8	3,1	4,0	4,6	2,0	2,8	3,7	4,3	
	7	3,1	3,4	4,3	4,6	2,8	3,1	4,0	4,6	
	8	2,8	3,1	4,0	4,3	2,0	2,8	3,7	4,0	
	9	2,8	3,1	4,0	4,6	2,5	3,1	4,0	4,6	
	10	3,1	3,7	4,3	4,6	3,1	3,4	4,3	4,3	

Table 9. Resolution test results for 10 x-ray units

		Contrast (mmCu)								
	Unit	50kVp	70kVp	80kVp	100kVp	50kVp	70kVp	80kVp	100kVp	
100cm - 20mAs	1	0,3	0,5	0,4	0,1	0,4	0,5	0,3	0,1	60cm - 20mAs
	2	0,4	0,4	0,3	0,2	0,3	0,5	0,3	0,1	
	3	0,3	0,5	0,4	0,1	0,4	0,4	0,4	0,2	
	4	0,4	0,5	0,3	0,1	0,4	0,4	0,4	0,1	
	5	0,4	0,4	0,3	0,3	0,5	0,5	0,3	0,2	
	6	0,3	0,5	0,3	0,2	0,4	0,5	0,4	0,1	
	7	0,4	0,5	0,4	0,3	0,3	0,6	0,4	0,3	
	8	0,5	0,6	0,4	0,1	0,4	0,5	0,3	0,1	
	9	0,4	0,5	0,3	0,2	0,5	0,6	0,3	0,1	
	10	0,3	0,4	0,4	0,2	0,4	0,4	0,4	0,2	
100cm - 40mAs	1	0,5	0,6	0,5	0,3	0,4	0,6	0,5	0,2	60cm - 40mAs
	2	0,5	0,6	0,4	0,2	0,5	0,6	0,4	0,2	
	3	0,4	0,6	0,4	0,2	0,3	0,6	0,5	0,3	
	4	0,4	0,6	0,5	0,3	0,5	0,6	0,5	0,2	
	5	0,5	0,6	0,5	0,3	0,4	0,6	0,4	0,2	
	6	0,4	0,6	0,5	0,2	0,4	0,6	0,5	0,3	
	7	0,5	0,6	0,4	0,1	0,5	0,6	0,4	0,1	
	8	0,5	0,6	0,4	0,2	0,5	0,6	0,5	0,2	
	9	0,4	0,6	0,5	0,3	0,3	0,6	0,5	0,3	
	10	0,4	0,6	0,4	0,2	0,5	0,6	0,4	0,2	
100cm - 50mAs	1	0,4	0,6	0,5	0,3	0,5	0,5	0,4	0,3	60cm - 50mAs
	2	0,5	0,6	0,5	0,2	0,5	0,6	0,5	0,2	
	3	0,5	0,5	0,4	0,3	0,4	0,5	0,4	0,3	
	4	0,4	0,6	0,5	0,3	0,4	0,6	0,5	0,3	
	5	0,4	0,6	0,5	0,4	0,4	0,6	0,5	0,4	
	6	0,5	0,6	0,5	0,3	0,5	0,6	0,4	0,3	
	7	0,4	0,6	0,5	0,2	0,5	0,6	0,5	0,2	
	8	0,5	0,5	0,4	0,2	0,4	0,6	0,4	0,2	
	9	0,5	0,6	0,5	0,3	0,4	0,6	0,5	0,3	
	10	0,5	0,6	0,5	0,3	0,5	0,6	0,5	0,3	

Table 10. Contrast test results for 10 x-ray units.

distances, the dose was calculated for the parameter settings that are different from the parameter settings used for dose modeling. After estimation, the measured and calculated dose values were compared. And, it was seen that the dose estimation was very successful.

For observing of image quality, a film was exposed during each dose measurement, after this, it was developed. Contrast and resolution tests were performed on these films. From Table 10, it can be said that contrast decreased with increasing kVp. It was seen that the best contrast is possible at the values of 70kVps, especially at 70kVp-40mAs. Although the other mAs values with constant 70kVp show good contrast, the best contrast with low dose is determined at 70kVp-40mAs.

In the resolution tests, from Table 9, it can be said that resolution increased related to increasing kVp. Because of this, the resolution is good in kVp values of 100kVp with different mAs.

But, in this study, because our aim is to obtain high image quality (both good contrast and good resolution), the optimum parameter values were selected as recommendation. The parameter setting values of 70kVp-40mAs can be accepted as the recommended technical parameters to obtain high quality image and low dose. If it is wanted to increase the number of recommended parameters, all mAs changes with constant 70kVp (20, 40, 50mAs) can be used as quality control test parameters.

If a radiographic staff adjusts these recommended parameters in an x-ray device, he/she will know which characteristics will appear on the image and how much dose will be measured. Hence, by this way, the staff can control and evaluate his/her tests' results during quality control tests of x-ray units.

6. Conclusion

The technical x-ray parameters are very important to reduce the dose and to obtain the image with good quality. The dose reduction can be obtained by adequate changes of physical parameters without lose of image quality. The optimal radiation dose for optimal image quality can be achieved by understanding of the parameters that affect radiation dose and image quality. The dose optimization process also consists of quality control programs to test radiographic devices periodically. In this study, it was studied in which parameters' values were appropriate to obtain high quality image and to reduce dose, in other words, dose optimization, during quality control tests of x-ray units.

This study shows that optimization of technical factors may lead to a substantial dose reduction. If the optimized parameters are applied to X-ray equipment during quality control tests, it is possible to determine how much good image quality will be obtained with this optimized parameters and how much dose will be measured when this qualified image is developed.

The results show the importance of radiographic staff training about the recommended parameters that are applied to the x-ray units for a qualified quality control system. It is essential to provide relevant education and training to staff in the radiology departments.

It can be sure that with such a study the questions on many professional staff's mind will be answered, and the dose and the image characteristics will be parameters that are controlled and managed.

7. Acknowledgment

I would like to thank the co-operation of radiographers at all of the radiological departments participating in this study.

8. References

Aldrich, J., Duran, E., Dunlop, P., & Mayo, J. (2006). Optimization of dose and image quality for computed radiography and digital radiography. *Journal of Digital Imaging,* Vol.19, No.2, (June 2006), pp. 126-131

AAPM (1981). Basic quality control in diagnostic radiology. *AAPM Report No. 4,* (1981)

AAPM (2002). Quality control in diagnostic radiology. *AAPM Report No. 74,* (2002)

Bouzarjomehri, F. (2004). Patient dose in routine X-ray examinations in Yazd state. *Iran. J. Radiat. Res.,* Vol.1, No.4, (2004), pp. 199-204

Brix, G., Nekolla, E., & Griebel, J. (2005). Radiation exposure of patients from diagnostic and interventional X-ray procedures. *Radiologe,* Vol.45, No.4, (April 2005), pp. 340-349

Ciraj, O., Markovic, S., & Kosutic, D. (2005). First results on patient dose measurements from conventional diagnostic radiology procedures in Serbia and Montenegro. *Radiation Protection Dosimetry,* Vol.113, No.3, (March 2005), pp. 330-335

Geijer, H. (2002). Radiation dose and image quality in diagnostic radiology. Optimization of the dose-image quality relationship with clinical experience from scoliosis radiography, coronary intervention and a flat-panel digital detector. *Acta Radiol. Suppl.,* Vol.43, (March 2002), pp. 1-43

Gray, J., Archer, B., Butler, P., Hobbs, B., Mettler, F., Pizzutiello, R., Schueler, B., Strauss, K., Suleiman, O., & Yaffe, M. (2005). Reference values for diagnostic radiology: Application and impact. *Radiology,* Vol.235, No.2, (May 2005), pp. 354-358

Hendee, W., Chaney, E., & Rossi, R. (1984). *Radiologic Physics, Equipment and Quality Control,* Year Book Medical Publishers, Chicago, USA

IAEA (1996). *International Basic Safety Standards for Protection against Ionizing Radiation and for the Safety of Radiation Sources.* IAEA Safety Series 15, ISBN 92-0-104295-7, Vienna, Austria

Iba Dosimetry (2008). Dosimax Plus A User Manual, 21 March 2011, Available from: www.iba-dosimetry.com

ICRP (1991). 1990 Recommendations of the international commission on radiological protection. ICRP Publication 60, *Annals of the ICRP,* Vol.21, No.1-3, (1991)

IEC 61223-3-1:1999 (1999). *Evaluation and routine testing in medical imaging departments. Acceptance tests. Imaging performance of X-ray equipment for radiographic and radiscopic systems,* BSI, ISBN 0-580-32753-1, London, England

Papadimitriou, D., Perris, A., Molfetas, M., Panagiotakis, A., Manetou, A., Tsourouflis, G., Vassileva, J., Chronopoulos, P., Karapanagiotou, O., & Kottou, S. (2001). Patient dose, image quality and radiographic techniques for common X-ray examinations in two Greek hospitals and comparison with European guidelines. *Radiation Protection Dosimetry,* Vol.95, No.1, (2001), pp. 43-48

Ramanandraibe, M., Andriambololona, R., Rakotoson, E., Tsapaki, V., & Gfirtner, H. (2009). Survey of image quality and patient dose in simple radiographic examinations in Madagascar: Initial results, *Proceedings of HEP-MAD 09,* Antananarivo, Madagascar, August 21-28, (2009)

Schaefer-Prokop, C., Neitzel, U., Venema, H., Uffmann, M., & Prokop, M. (2008). Digital chest radiography: an update on modern technology, dose containment and control of image quality. *Eur. Radiol.*, Vol.18, (April 2008), pp. 1818-1830

Seibert, A. (2004). Tradeoffs between image quality and dose. *Pediatr. Radiol.*, Vol.34, No.3, (2004), pp. 183-195

Shahbazi-Gahrouei, D. (2006). Entrance surface dose measurements for routine X-ray examinations in Chaharmahal and Bakhtiari hospitals. *Iran. J. Radiat. Res.*, Vol.4, No.1, (2006), pp. 29-33

Sprawls, P. (1987). *Physical Principles of Medical Imaging*, Aspen, ISBN 0-87189-644-3, Maryland, USA

Vano, E., & Fernandez Soto, J. (2007). Patient dose management in digital radiography. *Biomedical Imaging and Intervention Journal*, Vol.3, No.2, (2007)

Williams, J., & Catling, M. (1998). An investigation of X-ray equipment factors influencing patient dose in radiography. *The British Journal of Radiology*, Vol.71, (November 1998), pp. 1192-1198

6

The Significance of Board-Certified Registered Breast Specialist of the Japanese Breast Cancer Society in Regional Medical Activities

Noriyuki Tohnosu[1,2], Jun Hasegawa[2,3], Yosio Masuda[2,4], Taku Kato[5],
Satoru Ishii[6] and Kanae Iwata[7]
[1]Department of Surgery,Funabashi Municipal Medical Center
[2]Breast cancer screening committee of Funabashi Municipal Medical Association,
[3]Funabashi Futawa Hospital,
[4]Masuda Clinic of breast and thyroid diseases,
[5]Laboratory Section of Cytology, Funabashi Municipal Medical Center,
[6]Radiotechnical Department, Funabashi Municipal Medical Center
[7]Department of Pharmacy, Funabashi Municipal Medical Center
Funabashi, Chiba,
Japan

1. Introduction

Although the mortality of breast cancer patients in the Western countries has declined due to high screening rate, the number of Japanese breast cancer patients has seen a sharp rise and the most common cause of death of middle-aged women. Since one of every sixteen Japanese women have been diagnosed with breast cancer and more than 10,000 patients die from breast cancer every year, it is a goal to reduce the mortality through detection and treatment in its early stages. Board certified breast specialists of the Japanese Breast Cancer Society have been established in 1997 to contribute for the benefit of welfare of the nation and meet the social needs. In addition, The Ministry of Health, Welfare and Labor authorized the advertisement of specialists via the home page (http://www.jbcs.gr.jp/) in October, 2004 (Sonoo, 2005). As our institution has been designated as the region-based affiliated hospital for cancer treatment since January, 2007, the significance of the breast specialists was surveyed.

2. Breast specialists, board certified institutions and its affiliated institutions

2.1.Breast specialists

It is the minimum requirement for breast specialists to be authorized experts or qualified doctors in any one of the fields of surgery, oncological internal medicine, radiology and gynecology. The standards for qualifying breast specialists are different depending on each field. The standards for surgeons is: (1) It is required to be specialists in surgery and authorized breast doctors whose titles can possibly be acquired 4 years after graduation. (2) It is necessary to deal with breast diseases for over 7 years and experience the treatment and/or diagnosis of more than 100 breast cancer patients. (3) It is mandatory to be engaged in clinical works at the certified institutions. (4) Academic achievements on breast diseases (publications or presentations) have to exceed the compulsory score. (5) Passing the written and oral examinations is needed (Table 1). The qualification has to be renewed every 5 year.

1.Concerning surgeons, it is required to be specialists in surgery and authorized breast doctors whose titles can possibly be acquired 4 years after graduation.
2.It is necessary to deal with breast diseases for over 7 years and experience the treatment and/or diagnosis of more than 100 breast cancer patients.
3. It is mandatory to be engaged in clinical works at the certified institutions.
4. Academic achievements on breast diseases (publications or presentations) have to exceed the compulsory score.
5. Passing the written and oral examinations is needed.

Table 1. Standards of board-certified registered breast specialist.

The breast specialists have been registered from seven regional blocs in Japan: Hokkaido, Tohoku, Kanto, Kinki, Chubu, Chugoku-Shikoku and Kyusyu-Okinawa. The present number of the nationwide breast specialists has been still as small as 837 and 303 in Kanto bloc or eastern part of Japan, 40 in Chiba prefecture of Kanto bloc and 2 in Funabashi city of Chiba prefecture, respectively (Fig.1.)

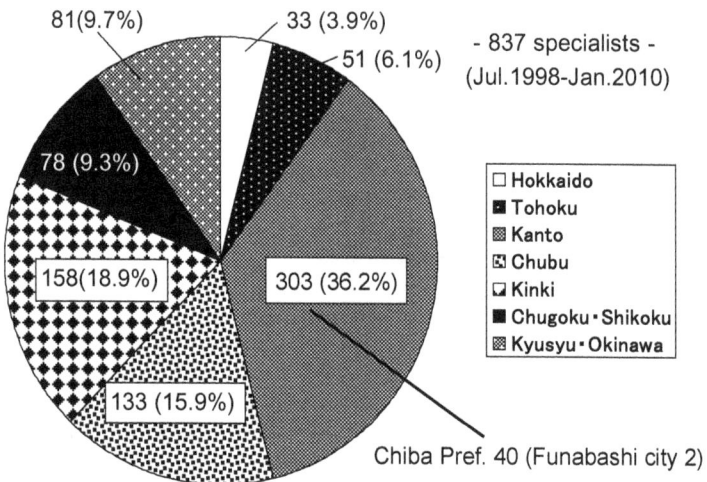

Fig. 1. Distribution of the numbers of board-certified registered breast specialists in Japan.

2.2 Board certified institutions and its affiliated institutions

In1998, the Japanese Breast Cancer Society has designated certified institutions and its affiliated institutions in the seven blocs of Hokkaido, Tohoku, Kanto, Chubu, Kinki, Chugoku-Shikoku and Kyusyu-Okinawa throughout Japan and our hospital has acted as the certified institution since then. Certified institutions have to meet the following standards (Tab. 2).

The Significance of Board-Certified Registered Breast Specialist of the Japanese Breast Cancer
Society in Regional Medical Activities

103

1. It is required to have beds for surgical treatment or diagnosis and/or non surgical treatment of more than 20 breast cancer patients in a year.
2. Board -certified registered breast specialists have to regularly work and adequately instruct.
3. Laboratories and libraries are well-equipped.
4. Records of anamnesis are well-written and preserved in ample care.
5. Autopsy room is equipped.
6. Instructive events on breast diseases are regularly held.
7. Publications or presentations on breast diseases have to be continued.
8. It is compulsory that board-certified registered breast specialists belonged to certified institutions instruct at the certified affiliated institutions and report its contents.

Table 2. Standards of board certified institutions.

It is impossible for doctors who aim at breast specialists to be qualified although they even practice hard in the non-certified institutions where non-certified but skillful surgeons treat many breast cancer patients. Therefore, considering the present small number of breast specialists in Japan, it is necessary for non-certified institutions to be affiliated with the regional certified institutions so that the doctors could be qualified. If the number of breast specialists is large in the future, the affiliated institutions may possibly be abolished (Sonoo, 2005,2008). The standards of the affiliated institutions are shown in Table3. The number of board-certified institutions and its affiliated institutions are 357 and 410, respectively, Kanto bloc occupying most in the former and Chubu bloc occupying most in the latter (Fig. 2).

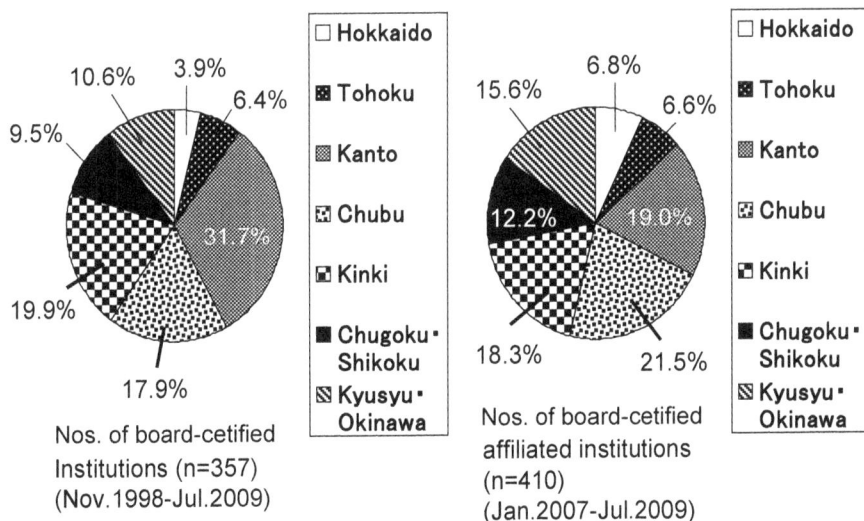

Fig. 2. Distribution of the number of board-certified institutions and its affiliated institutions in Japan.

1. It is required to have beds for surgical treatment or diagnosis and/or non surgical treatment of more than 20 breast cancer patients in a year.
2. Board -certified registered breast specialists have to regularly work and adequately instruct.
3. Laboratories and libraries are well-equipped.
4. Records of anamnesis are well-written and preserved in ample care.
5. Autopsy room is equipped.
6. Instructive events on breast diseases are regularly held.
7. Publications or presentations on breast diseases have to be continued.
8. It is compulsory that board-certified registered breast specialists belonged to certified institutions instruct at the certified affiliated institutions and report its contents.

Table 3. Standards of board-certified affiliated institution.

3. Medical activities as region-based affiliated hospital for cancer treatment

Japanese government has issued an Act of Strategy for Cancer in June, 2006 to treat the Japanese major cancers of lung, stomach, liver, colorectum and breast with equal high medical quality throughout the nation. For that purpose, affiliated hospitals for cancer treatment have been designated in each region and our institution has been the hospital since January, 2007.

The present number of the hospitals is 13 in Chiba Prefecture and the qualification has to be renewed every 5 year. The main works for breast specialists in our region are raising the percentage of examinees, reducing the percentage of detailed examinations, maintaining quality control in breast cancer screening, promoting close cooperation with the board certified affiliated hospital and community hospitals or clinics, optimal team management for breast cancer, an education of trainee doctors and providing citizens with information on breast cancer including extension lectures.

3.1 Breast cancer screening

As for breast cancer screening in Funabashi city, a mammography has been applied to women aged 40 and over since 2004. The percentage of examinees has annually risen from 8.9% in 2004 to 17.0% in 2009, whereas the percentage of detailed examinations has been almost constantly 6−7% exept in 2008 due to the effect of good quality control, the detection rate of breast cancer being 0.28% (Fig.3).

Although there were only four qualified doctors to read mammography when mammographic screening started, there have been 10 qualified doctors at present to form five teams in which a pair of two doctors reads alternately every week. On mammographic technics and knowledge, a qualified technologist at our hospital who plays a leading role in Japan has called a monthly meeting to educate technologists involved in screening mammography in Funabashi city. Breast cancer screening using stereotactic guided Mammotome has begun in our hospital by request of the community hospitals and clinics in March, 2011.

3.2 Coordination with the affiliated hospital

As far as coordination with the affiliated hospital is concerned, a conference has been held monthly to compare mammography with pathology of the postoperative cases of both

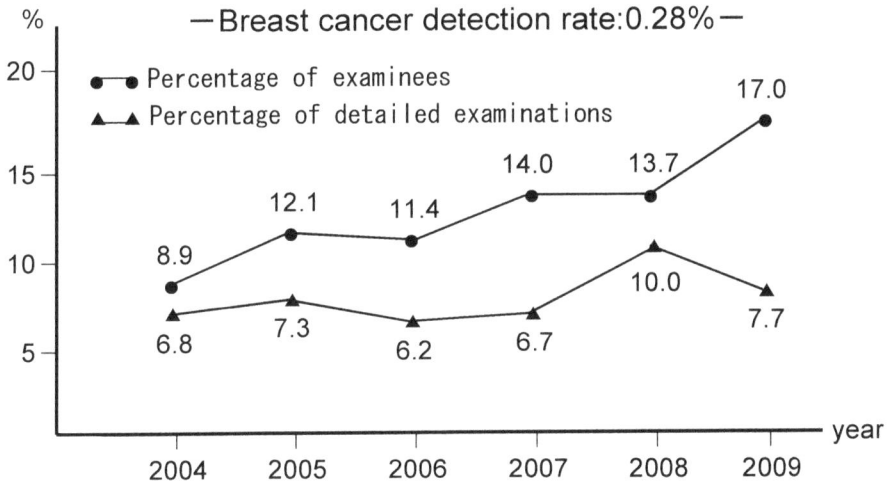

Fig. 3. Change in percentage of breast cancer examinees and detailed examinations in Funabashi city.

hospitals. Doctors including gynecologists, cytologists and technologist attend to share knowledge and information on breast cancer. In addition, breast cancer patients who want to undergo surgery at an earlier date are referred to the affiliated hospital. Those women who are indicated to screening are referred to the hospital also.

3.3 Team management
In our hospital activities on breast cancer patients, patient and the family-centered team management has been carried out and comedical staffs play an important role in each field. Apart from nurses, a pharmacist not only immediately reacts to advise doctors on optimal use of drugs whenever asked but also kindly responds to the patients soon after adverse events particularly of chemotherapy are seen. She is well informed about recent drug news both domestic and abroad on breast cancer and often has presentations even in the Japanese Breast Cancer Society. An experienced cytologist who plays a leading part in Japan is specifically of great help in the diagnosis of breast diseases both on the outpatient and intraoperative basis to contribute much for the benefit of not only our hospital but also the community hospitals or clinics which ask consultations of microscopic specimen.

Education of trainee doctors contain mammography reading, academic presentation and surgery of breast cancer. On mammographic reading ability, a third-year trainee doctor in our hospital was the first successful candidate in Chiba prefecture in the nationwide examination several years ago and it was also quite rare in Japan. Six trainee doctors of 2 general surgeons, 2 lung surgeons and 2 gynecologists have been accredited up until now.

Since present team management basically requires certified nurses (breast care nurse) in particular, it is an urgent task to have the staffs in our team although the nationwide examination is relatively difficult to pass. As for the accredited pharmacist for cancer drug, The Japanese Society of Pharmatheutical Health Care and Sciences has also adopted a board-certified system since 2006 and the standards of certified pharmacist for cancer drugs are as follows (Table 4).

1. It is required to have a career as a pharmacist for over five years and have to be a member of the Society for over five years when applied for the examination.
2. It is necessary to attend the Society or symposium of the Society more than two times.
3. Academic presentation on medical pharmacy has to be carried out as a coauthor over three times in meetings and one of them must be given as an author.
4. Publishing more than three papers on medical pharmacy is needed.

Table 4. Standars of board certified pharmacist for cancer drugs.

Strengthening a supportive system of cancer consultation is so important also that we started the team formed of a nurse who serves exclusively and three social workers in November, 2008. Concerning best supportive care, a specialist of former lung surgeon has gone into action in a newly built ward with 20 beds since April, 2009.

As the region-based affiliated hospital for cancer treatment, we will have to achieve the goal to have common clinical path of breast cancer among the community hospitals or clinics in our city and carry out breast cancer screening using ultrasound especially for the examinees aged 30-40 years.

3.4 Clinical path

A clinical path is useful for low-risk postoperative breast cancer patients in shortening time to visit and wait for receiving the same standard treatment without concentrating only in the region-based affiliated hospital . It is also beneficial for the region-based affiliated hospital to have more ample time on treating more seriously ill patients with metastases or recurrence. When the patients are operated on at the region-based hospitals and found to be at low risk (i.e. node negative, positive hormone receptors for prescription of oral medicines alone and no chemotherapy or no therapy), they are treated there with or without hormonal medicines for 1- 6 months, then are referred to community hospitals or clinics if they consent the use of path. They are checked up only with prescription and blood collection and then return to the region-based affiliated hospitals every 6-12 month for detailed examinations like mammography, ultrasound and CT or bone scan if required. If some emergencies, recurrence and/or metastases happen, the region-based affiliated hospitals have to respond immediately not only for the patients but also the community physicians (Fig. 4).

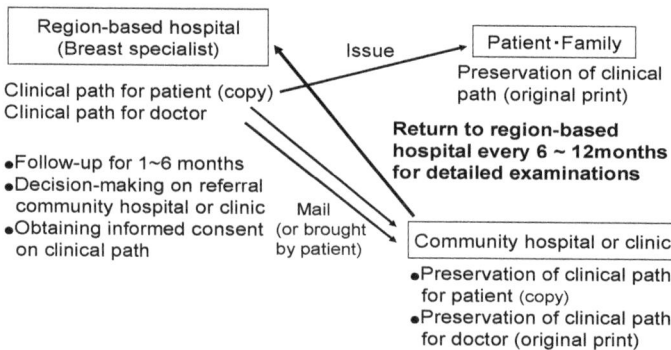

Fig. 4. Flowchart of clinical path for postoperative breast cancer patients.

The Significance of Board-Certified Registered Breast Specialist of the Japanese Breast Cancer
Society in Regional Medical Activities

107

3.5 Extension lectures

Together with the affiliated hospital, community hospitals and clinics, we have held four-time extension lectures for citizens on breast cancer each with certain theme at every other year since 2006 (Tab. 5.). Specifically, a lecture on liaison clinical path was given for the first time in October 2010, and the majority of the audience including men was in their 40s and 50s, younger than the age group at the previous lectures.

Year	Sponsored organization	Attended institutions
Sep. 2006	A private enterprise	Our hospital and three community hospitals
Mar. 2008	Funabashi city	Our hospital
Jun. 2008	A private enterprise	Our hospital, two ommunity hospitals and a breast clinic
Oct. 2010 *	A private enterprise	" "

Table 5. Extension lectures ever held. ＊A theme 'clinical path ' was included.

4. Discussion

Quite different from Western countries, Japanese specialist system has been privately established to improve and maintain the quality of the members in each society. In 1962, specialist system for anesthetists started for the first time in the Japanese Anestheology Society followed by those for radiologists and brain surgeons in 1964. Since around mid-1970, the Japanese Society of Internal Medicine has taken the iniative not only to let the board-certified doctors and specialists be socially accepted but also to enable them to advertise in April, 2004 (Sakai, 2005). Health, Welfare and Labor Ministry has also authorized the advertisement of breast specialists through the home page in October, 2004. Although the field of breast cancer is related to not only surgery but also radiology, gynecology and oncological internal medicine, surgeons occupy 95% of breast specialists in Japan, quite unlike Western countries. Therefore, Japanese breast specialists are busy working even for rapidly-advancing chemotherapy, however, they can possibly acquire a broad range of knowledge and experience through medical activities in team management. Since there are few emergency patients, even woman doctors including surgeons and gynecologists could work by taking advantage of being female while they rear children. As the number of breast specialists is still small in Japan, they will turn out to be important human resources for Japanese breast cancer patients who have continued to rise in number (Sonoo, 2007, 2008).

Before applying to the examination for breast specialist, the following curriculum has to be finished; 1) to master general knowledge on breast diseases, clinical judgement, the ability to solve problems without regard to each expertise, 2) to master basic knowledge on anatomy, physiology, hormonal environment, epidemiology, pathology, biology of the mammary gland and breast cancer screening and to be able to clinically respond, 3) to master basic treatment technics on diagnosis by imaging modalities, aspiration cytology, core needle biopsy, biopsy using Mammotome, surgical biopsy , sentinel lymphnode biopsy and treatment with surgery, radiation, anticancer drugs, hormonal medicines, best supportive care and postoperative rehabilitation, medical ethics including informed consent, second opinion and clinical trials. 4) to master special treatment technics on breast diseases required for each expertise. 5) to actively attend conference or academic meetings, research or treat via evidence based medicine and give academic presentations on case reports or clinical study. 6) to understand the importance of medical administration involving risk management, medical cost, team management, etc. for carrying out practical medical activities (Sawai, 2006).

The number of board certified institutions and their affiliated institutions has seen a gradual rise and the largest number is centered on Kanto bloc or eastern part of Japan as well as breast specialists. There have been only two breast specialists in Funabashi city with a population of 600,000, therefore it is essential to increase the number of breast specialists for treating the steadily increasing breast cancer patients.

Under a guidance of The Ministry of Health, Welfare and Labor, nationwide mammographic breast cancer screening for women aged 50 and over has been introduced at intervals of two years since April, 2000 followed by the screening for women aged 40-49 since 2003 (Ohuchi, 2007). However, the percentage of examinees aged 40 and over has still been as low as 20.3% in Japan and the percentage of mammographic screening at the age of 50-69 was 23.8% compared to 60-90% in Western countries in 2006 (OECD Health Data 2009). According to the Japanese government statistics in fiscal year 2008, the nationwide average percentage of examinees is as low as 1.5% at mammography screening alone and 7.6% at combined mammography screening with palpation. The Ministry of Health, Welfare and Labor has started to distribute free coupons to raise the percentage up to 50% since October, 2009. The average rate of using coupons has been approximately 30% at the age of 40-69 (Japanese government bulletin, 2009). There have been some reports showing the effect of distributing free coupons to improve the percentage of examinees (Komoto & Ishiyama, 2010). Similarly, the percentage of examinees was raised from 81% to 88% in some age groups after the introduction of free of charge in Finland (Kauhava, 2008). Whereas the percentage was improved from 18.8% to 40.7% in Sendai, 2011's quake and tsunami-hit Miyagi prefecture, northern part of Japan through not using coupons but making various efforts of increase in consultations with universities or institutions to exchange views, women's cancer screening promotion project and distribution of application form to all houses (Satake, 2011). The percentage of detailed examinations and the detection rate of breast cancer screening in 2008 in Japan is 8.6% and 0.32%, respectively (Japanese government report, 2010). Our series have shown almost same as the nationwide data.

In order to maintain a quality control not only on facility but also personnel qualification of mammography reading ability for doctors and mammographic skills for technologists, The Central Committee on Quality Control of Mammographic Screening was launched as a non-profit organization in November, 1997 and has offered nationwide training

The Significance of Board-Certified Registered Breast Specialist of the Japanese Breast Cancer
Society in Regional Medical Activities

109

seminars frequently including examinations since November, 2000 (Tsunoda, 2008 & Endo, 2009).

The examination of reading mammography includes 100 cases half with two views and candidates have to fill out the marksheets for judging categories within a limited time of 100 minutes. The results of examinations are classified into the ranks of A, B, C , D and A+B are certified. Rank A is accepted as the ability of reading and teaching of screening mammography. The possibility of obtaining rank A in a review of rigorous testing is less than 10 percent of the total candidates. The Committee has not only accredited the doctors and technologists involved in breast cancer screening program but also given the members examinations every five year for the renewal of accreditation. Futhermore, the lecturer's and staff member's meeting has also been held once a year for keeping the knowledge and information about mammography screening system (Tsunoda, 2008). Although there has been legally-authorized Mammography Quality Standard ACT (MQSA) of the United States, the Japanese committee is considered the pioneer system in the world to evaluate the ability of individuals who engage in mammographic screening (Ohuchi, 2007).

Although ultrasound screening has been strongly recommended for examinees with dense breast particularly in their 40s or younger (Hashimoto & Ban 2010), there has been no worldwide evidence to prove whether ultrasound screening would have a potential to reduce the mortality of breast cancer for women aged 40-49. Therefore, in order to clarify sensitivity, specificity and detection rate as primary endpoint and cumulative advanced breast cancer incidence rate as secondary endpoint, 5-year Japan Strategic Anti-Cancer Randomized Trial (J-START) (http://jsrtfall.umin.jp/) has been conducted between the two groups of mammography combined with ultrasound and mammography alone in the respective number of 100,000 examinees in their 40s since 2006 with the initiative of The Ministry of Health, Welfare and Labor and the outcome will be shown in 2012. If effectiveness of ultrasound is confirmed, ultrasound screening would start in Japan for the first time in the world and we will have to prepare for the screening to especially secure a certain number of well-trained breast sonographers in Funabashi city also.

As far as clinical path in Japan is concerned, The Ministry of Health, Welfare and Labor authorized the payment for using paths for neck of femur fracture for the first time in 2006. Then, the Ministry issued Strategic Anti-Cancer Promotion Project in 2007 to oblige the use of clinical paths of the five major cancers of stomach, colorectum, liver, lung and breast. According to the Ministry's survey in 2010, the number of the clinical paths used and patients enrolled has been larger in a comparison between before-2008 and 2009. Specifically, breast cancer patients increased even five-fold, largest in number compared to the other cancer patients. However, questionnaires to 410 hospitals by Health, Welfare and Labor Ministry in 2010 revealed that those patients including their families who understand paths have occupied only 30% even in the hospitals using paths. In other words, hospital staffs seem unlikely to give ample explanation or enlighten on paths considering that there actually have been many other path-using community hospitals throughout Japan. Likewise, the number of Chiba prefecture-based common paths used has been still as small as four, similar to lung cancer (1) and liver cancer (4) from April 2010 to January 2011. According to the other questionnaire by a certain group formed of breast cancer patients and their families, there have been the following anxieties for patients; 1) Whether the physicians of the community hospitals or clinics can treat enough or not, as well as the breast specialists in the region-based hospitals. 2) When clinical path is suddenly mentioned under treatment, some patients feel like losing a relationship of mutual trust. 3) Without close communication between the breast specialists and the community physicians, patients may possibly think to forcedly be sent out.

Region-based affiliated hospitals for cancer treatment have now been required to more closely cooperate to promote coordination with the community hospitals or clinics using clinical paths for the purpose of improving and maintaining treatment quality (Ando, 2004 & Aogi, 2008).

Since The Ministry of Health, Welfare and Labor revised the score of payment for treatment in fiscal year 2010, the region-based affiliated hospitals and community hospitals or clinics can get a certain additional score if the patients requiring a 10-year follw-up are treated using the common clinical path. Apart from the low-risk breast cancer patients, a path for postoperative chemotherapy using trastuzumab has been applied in Kitakyusyu city, Fukuoka prefecture, southern part of Japan (Ohno, 2008). For success of the path, Ohno et al stress the importance and benefit of close communication with the community physicians by holding regular meetings three times a year before starting its use.

We have had regular meetings since 2009 like Ohno et al to start Funabashi city-based clinical path in April, 2011. In addition to the meetings, it is required for success to fully inform the patients both before and after admission for surgery and citizens through regular extension lectures like ours, bulletins or website.

5. Conclusion

It is essential to have a more number of breast specialists in order to treat the steadily increasing Japanese breast cancer patients. To raise the percentage of examinees in breast cancer screening, the efforts must be continued including various campaigns by non-profit organization or consecutive distribution of free coupons by the government. Mammographic reading ability for doctors and mammographic skills for technologists have to be renewed every five year as before to maintain quality control. Evidence-based ultrasound breast cancer screening may possibly start in Japan for women aged 30-40. Clinical path has to be popular for low-risk breast cancer patients via adequate information before and/or after admission for surgery, regular extension lectures for citizens and bulletins or website.

6. Acknowledgement

I wish to thank Dr. Hasegawa for collecting continuous data on breast cancer screening in Funabashi city, Dr. Masuda for referring many early breast cancer patients, cyotologist Mr. Kato for diagnosing quickly on the outpatient and intraoperative basis, pharmacist Ms. Iwata for offering quick assistance on drug information and radiotechnologist Mr. Ishii for maintaining quality control on mammography as well as continuing technical education to other technologists involved in municipal breast cancer screening.

7. References

Sonoo, H. (2005). Certification system of breast diseases of the Japanese Breast Cancer Society (in Japanese with English abstract). *Jpn J Breast Cancer*, vol.20, No.1, 2005, pp. 59-63

The Significance of Board-Certified Registered Breast Specialist of the Japanese Breast Cancer
Society in Regional Medical Activities

111

Sakai, O. (2005). For the establishment of specialist system (in Japanese). *Treatment*, vol87, No.2, 2005, pp.400-403

Sonoo, H. (2007). Breast surgery; progress, perspective and future directions (in Japanese). *Surgery*, vol.69, No.4, pp. 396-401

Sonoo, H. (2008). Board certified breast specialists of the Japanese breast cancer society (in Japanese). *Surgical Treatment*, vol.98, No.3, 2008, pp. 260-266

Sawai, S. & Nakajima, H. (2006). Breast specialist (in Japanese). *All about easy-to-understand breast cancer*, Nagai Book Co. 2006, pp. 515-519.

Ohuchi, N. (2007). Current status of breast cancer screening (in Japanese). *Nippon Rinsho*, vol 65, Suppl 6, 2007, pp. 213-219

http://www.gankenshin50.go.jp/. OECD Health Data 2009 (in Japanese).

http://www.mhlw.go.jp/stf/shingi. Promotion of the Status of the 2009 women's cancer screening in 1990 article (in Japanese), pp. 5

Komoto, S. et al. (2010). The effect of free shipping coupons and breast cancer screening handbook (in Japanese). *Proceeding of J. Jpn Assoc. Breast Cancer Screen*, vol.19, No.3, 2010, pp. 282

Ishiyama, K. (2010). Current status of breast cancer screening in Akita city in 2009 — Status after introduction of free coupons (in Japanese). *Proceeding of J Jpn. Assoc. Breast Cancer Screen*, vol.19, No.3, 2010, pp. 357

Kauhava, L. et al. (2008). Results on mammography screening in Turku, Finland. *Proceeding of J. Jpn.Assoc. Breast Cancer Screen*, vol7, No.3, 2008, pp.341

Satake H. et al (2011) Effort toward examination rate 50% of breast cancer screening with special reference to activity of Sendai city (in Japanese). *J. Jpn. Assoc. Breast Cancer Screen*. Vol.20, No.2, 2011, pp.102-105

http://www.mmjp.or.jp/kawakami-clinic/data. *The Ministry of Health Care Community Health Project Report FY* 2009 (in Japanese), 2010, pp. 1

Tsunoda-Shimizu H. et al. (2008). Quality control of breast cancer screening and future problems (in Japanese with English abstract). *Jpn J Breast Cancer*, vol.23, No.3, 2008, pp.191-196

Endo, T. (2009). Progress of quality control in breast cancer screening and future problems (in Japanese). *J Jpn. Assoc. Breast Cancer Screen*, vol.18, No.2, 2009, pp.107-114

Hashimoto,H. (2010). Present and future of ultrasound breast cancer screening (in Japanese). *Inner Vision* , vol.25,No.8, 2010, pp.27-29

Ban, K. et al. (2010). Consideration of appororiate scrutiny of concurrent ultrasound screening with mammography screening (in Japanese). *J Jpn. Assoc. Breast Cancer Screen*, vol.25, No.6, 2010, pp.649-656

http://jsrtfall.umin.jp/. Present status and perspective of J-START (in Japanese). *Proceeding of an instructive lecture*, 2010, pp.1-2

Ando, T. et al. (2004). Breast cancer surgery and clinical path (in Japanese). *Surgical Treatment* , vol.90, No.5, pp.937-943

Aogi, K. et al. (2008). Breast cancer surgery — The present status of development of clinical path (in Japanese). *Surgical Treatment* , vol.99, No.1, pp.48-56

Ohno S. et al. (2008) Present status and problems of introduction of regional medicine clinical path of breast cancer (in Japanese). *J New Rem. & Clin,* vol.57, No.12 , pp.12-21

7

Infectious Aetiology of Cancer: Developing World Perspective

Shahid Pervez

Professor, Section of Histopathology,
Department of Pathology & Microbiology,
Aga Khan University Hospital Karachi,
Pakistan

1. Introduction

Infection attributable cancers contribute over 1/4th of all cancers in the developing countries (26.3%) compared to the developed countries (7.7%), (Parkin, 2006). Overwhelming majority are related to viral infections. In contrast to other carcinogens where it is usually a *'hit and run'* kind of situation, with infectious agents particularly viruses one may precisely demonstrate and prove its presence and integration within host neoplastic cells. Oncogenic DNA viral genome incorporates itself directly into host cells DNA while oncogenic RNA viral genome is transcribed into host cell DNA by reverse transcriptase. Neoplastic transformation usually follows. Oncogenic mechanisms include acting as promoter, transforming protooncogenes into oncogenes. Credit goes to Dr Peyton Rous, a noble laureate pathologist who demonstrated that it was possible to transmit tumours from one animal to other like transmission of an infection.

Human tumours with proven or proposed viral aetiology include 'Human papillomavirus (HPV)', Epstein-Barr Virus (EBV), Hepatitis B and C viruses, RNA retroviruses like 'Human T-lymphotropic virus (HTLV1)', 'Human Herpes Virus-8 (HHV-8). Bacteria with proven carcinogenic potential include 'Helicobacter pylori'. Among fungi aflatoxins produced by 'Aspergillus flavus' are potent carcinogens. Among parasites 'Schistosoma' and 'Clonorchis sinensis' are implicated in the causation of cancer.

2. Human papillomavirus (HPV)

HPV is a small epitheliotropic, non enveloped DNA virus belonging to papovaviridae family. Its genome comprises 7000-8000 base pairs of double-stranded closed-circular DNA. At least 70 genetically distinct types of HPV have been identified in humans. According to their oncogenic potential HPV is classified in a high oncogenic risk group (i.e., HPV16, 18, 31, 33, 35, 39, 45, 51, 52, 56, 58, 65, 66) and low oncogenic risk group (i.e., HPV6, 11, 42, 43, 44). High risk HPV association with cervical and anogenital cancers is established beyond doubt. HPV16 and 18 are declared as human carcinogens by 'international Agency for Research on Cancer (IARC)'. HPV association with other cancers in particular with *'oral cancer'* is also being investigated with evidence of significant association.

2.1 HPV & cervical carcinoma

Cervical carcinoma is one of the most common malignancies in women worldwide. However with effective preventive measures like *'cervical screening'* programs in developed countries more and more cases are picked at an early stage where complete cure is possible. A significant recent breakthrough has come in the form of *'HPV vaccine'* against high risk HPV16 & 18. This is gaining momentum in developed countries with high risk and burden of disease. It is administered at age 11-13, three shots are given intramuscularly. In contrast in the developing countries data is patchy or non-existent. In most countries no cervical screening programs are in place. In developing muslim countries situation is even worse. For instance in Pakistan, a populous muslim country of about 170 million inhabitants there is no cervical screening program and the only source of cervical smears are the sporadic smears obtained at the time of consultation in obstetrics & gynecology clinics. The problem is further compounded by the social taboos on matters of sexual practices and sexually transmitted infections (STI). These socio-cultural prohibitions create a substantial barrier to such investigations.

Recently a study was carried out with the help of IARC in women of Karachi, Pakistan (largest port city of Pakistan with an estimated population of 15 million from diverse ethnic backgrounds), (Raza et al, 2010). A sum of 899 married women aged 15-59 years living in a densely populated suburb of Karachi consented to participate. HPV prevalence was found to be 2.8%. Cervical abnormalities were diagnosed in 2.4% of whom 27.3% were HPV positive. HPV16 was detected as the most common type among women with both normal (0.5%) and abnormal (9.1%) cytology. This study also included 91 invasive cervical carcinomas (ICC) from two major university hospitals of Karachi, Pakistan. HPV16 was also the predominant HPV type (75.8%) in ICC followed by HPV18 (6.6%). This study led to the suggestion of very low burden of HPV infection in general female population, considerably lower than neighboring India (17%, Franceschi et al, 2005), China (15-18%, Dai et al, 2006, Wu et al, 2007) and Nepal (9%, Sherpa et al, 2010).

In another study from Karachi, Pakistan (Khan et al, 2007) women visiting two major tertiary care hospitals in Karachi, diagnosed with ICC, sixty (60) paraffin-embedded biopsies were analysed for HPV subtypes by PCR. Out of the 60 samples only one was negative for HPV, the rest were positive excluding two samples where subtype could not be determined. Fifty six (56) were HPV16 positive and only one was HPV18 positive.

2.1.1 Conclusion

- In most developing countries particularly developing muslim countries data regarding sexually transmitted infections (STI) is either non-existent or sparse.
- A comprehensive well planned study utilizing samples from asymptomatic married women of Karachi, Pakistan revealed very low incidence of HPV (2.8%).
- Samples from both asymptomatic women as well as from invasive cervical cancer (ICC) showed overwhelming predominance of HPV16.
- Results from Karachi, Pakistan underscore the importance of cervical screening programs & HPV vaccination in resource constrained economies.

3. HPV and oral cancer

Oral cancer (OC) / Oral squamous cell carcinoma (OSCC) excluding salivary gland cancers ranks 6th overall in the world in both sexes with much higher incidence in the developing countries. In Karachi it ranked 2nd with an identical risk in both genders (Bhurgri et al, 2003). However if combined with pharynx and larynx cancers which have the same histologic type (squamous cell carcinoma) & risk factors it ranks number 1. Major risk factors in the developed world include *'smoking'* and *'alcohol;* however in the developing world though smoking is a common major risk factor, role of alcohol drinking is possibly a minor risk factor particularly in developing muslim countries. In subcontinent (Pakistan, India & Bangladesh) alternate chewing habits like betel quid and areca nut are major risk factors. Areca nut is now declared by WHO as a bonafide carcinogen. People using paan (betel leaf) are about 8- 9 times more likely to develop oral cancers as compared to non-users (Merchant et al, 2000). Smokeless tobacco, including *'gudka'* and *'niswar'* is an extremely addictive substance with a high rate of use in younger age groups, as well is contributing toward endemic rise of oral cancers in Pakistan (Ali et al, 2009 & Nair et al, 2004). (Figure 1) This habit commonly leads to a pre-malignant condition *'Submucosal fibrosis'* which commonly transforms into OSCC. Poor oral hygiene is another contributory factor in this population.

A significant proportion of OSCC patients however deny exposure to conventional and well known risk factors. This has led to search of other risk factors and associations including microbes (Scully et al, 1985). The striking commonality between oral cavity and cervical cavity paved the way to look for epitheliotropic viruses like HPV. Although these two areas are anatomically different, the squamous epithelium found in both areas has several similarities. For instance the squamous epithelium of ecto-cervix and oral cavity including pharynx and larynx are composed of squamous epithelium with a thin layer of keratin or no keratin. In both areas the epithelium is subject to microtrauma of various types as well as to bacteria and varying chemical irritants. Most common malignancy at both anatomic sites is also SCC with varied differentiation. (Figure 2) These factors may directly expose to HPV infections of cells resulting in malignant transformation. Furthermore the HPV subtypes isolated from lesions of squamous epithelium of cervix are similar to the type found in both normal epithelium and various lesions of the oral cavity, pharynx and larynx. These include HPV subtypes16, 18, 31 & 33.

The reported prevalence of HPV in OSCC varies widely in various studies depending on the population and ethnicity studied and/or sensitivity of the methods used and viral DNA sequence targeted. HPV in particular HPV-16, like in cervix is implicated in the aetiology of OSCC (Gillison, 2004; Miller & Johnstone, 2001) About 40 – 60% of patients with tumours of oropharynx are reported to be positive for HPV infection (Gillison, 2004; Kreimer et al, 2005). HPV-positive tumours are distinct from HPV-negative tumours in their biological characteristics and clinical behaviour. Data from retrospective analyses as well as a prospective clinical trial demonstrated that HPV-positive oropharyngeal tumours are more sensitive to chemotherapy and radiation treatment and have a markedly improved prognosis and favourable clinical outcome compared with HPV-negative tumours (Fakhry et al, 2008; Settle et al, 2009)

Recently, another significant observation has emerged in terms of HPV status of oropharyngeal tumours and racial disparities. Black Americans are known to have a higher incidence of and mortality from head and neck squamous cell carcinoma (HNSCC) than the whites and present with more advanced disease at a younger age (Goodwin et al, 2008; Morse & Kerr 2006; Ryerson et al, 2008; Shiboski et al, 2007). The greatest survival difference between blacks and whites was detected specifically in oropharyngeal cancers, but there was no racial difference between the overall survival rates of patients with non-oropharyngeal tumours (Settle et al, 2009). Most importantly, the recently published prospective analysis demonstrated that a marked difference exists between black and white Americans in terms of HPV infection. HPV positivity was about 9-fold higher in white (34%) than in black (4%) patients, directly correlating HPV infection with significant survival disparities between the two populations (Settle et al, 2009). Clearly, the HPV status of patients with OSCC would be an important determinant for prognosis and treatment options in the future.

Recently, in a retrospective study of 140 patients with primary OSCC and a long-term follow up, Ali et al reported from Karachi, Pakistan, 68% of cases to be positive for HPV (Ali et al, 2008). Approximately 90% of these cases were infected with HPV16, (Figure 3 & 4) the predominant subtype in the US population as well. HPV infection was detected entirely in tumours of the cheek and tongue in the oral cavity; this was consistent with the occurrence frequency in the Karachi population for oral cancers which is as follows: 55.9% for cheek, 28.4% for tongue, 6.8% for palate, 4.4% for gum, 3.1% for lip and 1.4% for floor of the mouth (Bhurgri et al, 2003). Furthermore, though HPV positive patients had comparatively prolonged overall survival when compared with HPV negative patients but the difference was not statistically significant (P=0.97) (Figure 5). Betel quid chewers were comparatively more prone to HPV positivity (OR=2.34; 95 CI= 1.1-4.31). These findings are in contrast with the results from US studies where the ratio of oropharyngeal tumours with respect to other sites was 2:1 and the HPV-positive tumours were consistently associated with a better clinical outcome in terms of both overall and disease-free survival (Fakhry et al, 2008; Settle et al, 2009). The reason(s) for these different findings are not clear.

3.1 Oncogenic HPV pathways
The chief oncoproteins of HPV16 are encoded by the genes E6 and E7. The E6 protein targets the tumour suppressor gene p53 for degradation. In fact, degradation of p53 in HPV positive cells is fully dependent on the presence of E6 (Ali et al, 2010, Figure 6). The E7 oncoprotein is involved in suppression of retinoblastoma protein (pRb) function. Reduced pRb expression is common in HPV-positive tonsillar cancer.

3.2 Mode of transmission
Two questions immediately come to mind, first how HPV gets there and second why patients with HPV association will have better survival. In response to question 1, haematogenous spread from genital tract is proposed besides atypical sexual habits. In response to question 2 one possible explanation is that HPV infection may lead to genome instability, paradoxically making tumour cells more susceptible to radiotherapy.

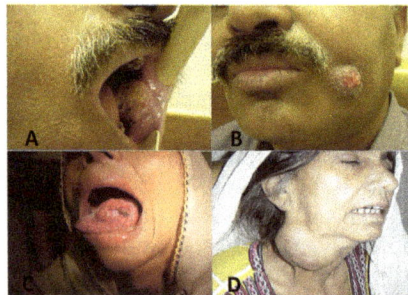

Fig. 1. Clinical presentations of patients with oral squamous cell carcinoma (OSCC) in Pakistani patients (Photographs were taken with patient's consent).

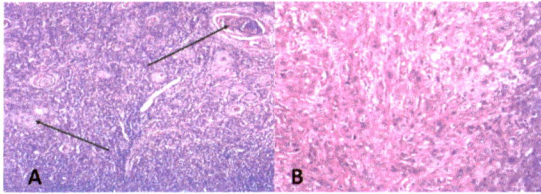

Fig. 2. Photomicrograph of H & E stained (A) well differentiated oral squamous cell carcinoma showing diffuse sheets of squamous cell with prominent keratinization and keratin pearl formations, Magnification X 10. (B) poorly differentiated oral Squamous cell carcinoma, Magnification X 20.

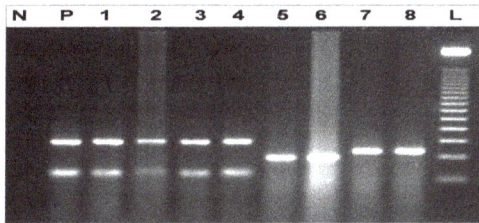

Fig. 3. PCR amplification of HPV general, HPV type 16 and HPV type 18 in OSCC samples. The products were electophoresed on 2% agarose gel and stained with ethidium bromide. Lane N: negative control, lane P: positive control, lanes 1-4 HPV (general primer) positive tumour samples, lanes 5-6 HPV 16 positive tumour samples, lanes 7-8 HPV 18 positive tumour samples, Lane L: molecular size marker (50-bp ladder marker).

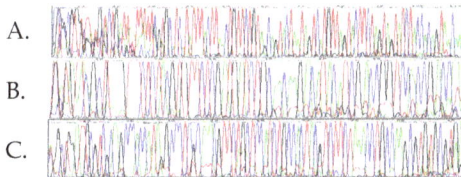

Fig. 4. Result of sequence analysis of PCR products, (A), Gene Sequencing HPV General. (B), Gene Sequencing HPV Type 16. (C), Gene Sequencing HPV Type 18.

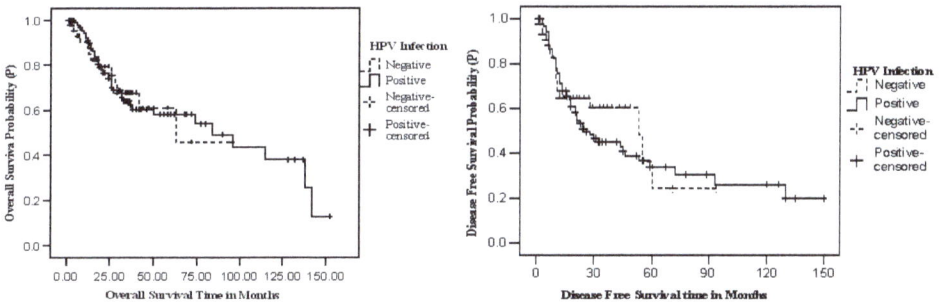

Fig. 5. Kaplan-Meier curves of overall survival (OS) of (A) human papillomavirus (HPV) positive patients as compared with HPV-negative patients. (B) Disease Free Survival of human papillomavirus (HPV) positive patients as compared with HPV-negative patients.

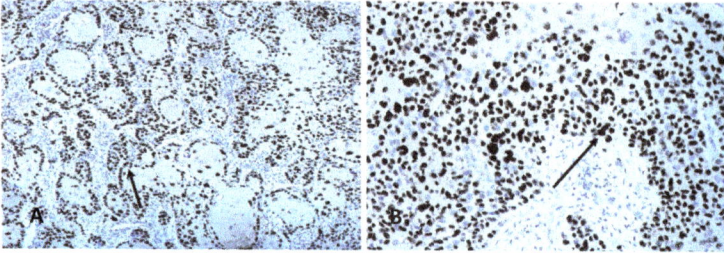

Fig. 6. Photomicrograph of a well-differentiated OSCC demonstrating diffuse strong nuclear TP53 staining. The arrows indicate positive dark brown intranuclear staining (magnification x 10 (A) & 20 (B)).

3.3 Conclusions
- Incidence of Oral Cancer in subcontinent (Pakistan, India & Bangladesh) is one of the highest in the world.
- Alternate chewing habits alongwith cigarette smoking are major risk factors in this part of the world.
- High risk HPV association was seen in 68% of the cases of OC in a high risk population of Karachi (Pakistan) with 90% containing HPV 16.
- Survival advantage was seen in OC patients with HPV association albeit not coming to statistical significance as seen in American whites.

4. Epstein-Barr virus (EBV)

EBV was initially discovered from cell cultures of a high grade B-cell lymphoma 'Endemic (African) Burkitt Lymphoma (BL)', which is highly prevalent in paraequatorial Africa and New Guinea. The disease affects children and adolescents and has strong association with malaria. Endemic BL commonly involves extra-nodal sites particularly jaw. In rest of the world 'sporadic form of BL' is seen having a weaker association with EBV and commonly affecting gastro-intestinal tract (GIT) particularly small intestine.

EBV associated other lymphomas include 'classic Hodgkin lymphoma (cHL)' particularly 'mixed cellularity' type (~60%), 'B-cell lymphoma in immunosuppressed', 'mature T-cell lymphoproliferative disorders' in particular 'Angioimmunoblastic T-cell lymphoma (AILT)', 'Angiocentric (Nasal) T-cell lymphoma'. Non-lymphoid associations include 'Nasopharyngeal carcinoma'.

4.1 EBV & mature T-cell non-Hodgkin lymphoma (T-NHL)
EBV association with certain subsets of T-NHL is now well established. In a study conducted by us in Pakistan (Noorali et al, 2003), mature T-NHL comprised 22.2% of total mature NHLs. These cases were characterised on the basis of morphology, immunohistochemistry and T-cell receptor (TCR) gene rearrangement studies. This study demonstrated frequent presence of EBV in mature T-NHL cases (55.4%) by 'PCR' (Figure 7) and 'in-situ hybridization (ISH)'. While analysing various subsets of mature T-NHL 'Peripheral T-cell lymphoma (PTCL) - unspecified' (n=88) showed 51.2% EBV positive cases. EBV can be differentiated according to size polymorphism depending on the number of

internal repeats in the Bam HI, E, K, N and Z regions. We also studied the extent of polymorphism in EBV genome by *'single stranded conformation polymorphism (SSCP)'* technique for *'Bam HI E, K, N and Z regions'*. Hypervariability in Bam HI, K and N regions was noticeably higher compared to *E or Z* regions. All in all no association was established between EBV variants differentiated on the basis sequence heterogeneity in *Bam HI, K, N, E and Z regions* in different subsets of T-NHL.

Mode of infection of T-cells by EBV is complex and poorly understood. Nazaruk et al, 1998 proposed that initially virus infects the B-cells and remains in the latent phase but under immunosuppressive conditions IL10 is secreted by EBV specific CD8+ T-cells activating B-cells. Subsequently reactivation of EBV lytic cycle occurs that may contribute to the development of EBV-associated T-cell lymphoproliferative disorders.

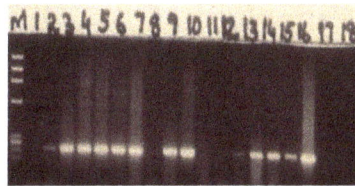

Fig. 7. Ethidium Bromide stained agarose gel showing PCR products of EBV-DNA amplified with primers specific for gp200 region.

4.2 EBV & angioimmunoblastic T-cell lymphoma (AILT)

AILT is an uncommon form of mature T-NHL characterised by systemic disease that occurs predominantly in middle-aged and elderly patients. The clinico-pathologic syndrome is characterised by fever, night sweats, weight loss, generalised lymphadenopathy, hepatomegaly and splenomegaly. Histologic examination of lymph nodes typically shows effacement of lymph nodes architecture, a polymorphous infiltrate including immunoblasts, lymphocytes, plasma cells, eosinophils, epithelioid histiocytes and a prominent arborizing postcapillary vasculature (Figure 8). In a study conducted by us a total of 13 well characterised cases of AILT based on morphology, IHC and TcR gene rearrangement studies were analysed for EBV by PCR and ISH (EBER). Association of EBV was seen in 11 out of 13 cases (84.6%) by PCR. By ISH (EBER), EBV was detected in 8 out of 9 cases (88.8%) cases. (Figure 8) So all in all strongest correlation of EBV was seen in this type of T-NHL. (Noorali et al, 2005).

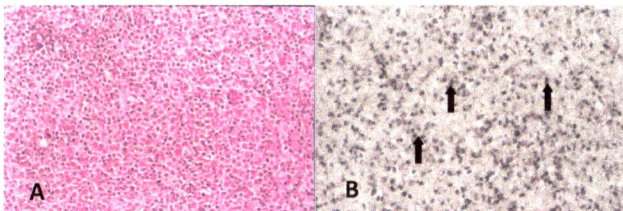

Fig. 8. *In situ hybridization* photomicrograph of a lymph node showing the localisation of EBV in the nuclei of neoplastic lymphocytes indicated by blackish signal (↑) (B), H& E of the same (A).

4.3 EBV & Mycosis fungoides (MF)

MF is an indolent T-cell lymphoma of skin. In a study conducted by us a total of 14 well characterised cases of MF were analysed for EBV by PCR and ISH (EBER). EBV was identified in 3 out of 6 cases (50%) by PCR but all these were negative on ISH (EBER). This discrepancy is most likely caused by low copy number of infected cells in tissue sections not amplified as in PCR based studies (Noorali et al, 2002).

4.4 EBV & anaplastic large cell lymphoma (ALCL)

ALCL is a peculiar type of T-NHL. In a recent study by us (Syed et al, 2011) ALCL was turned out to be the most common T-NHL in the archives of the largest referral centre of Pakistan. This variant of T-NHL however has the weakest association with EBV (Noorali et al 2004).

5. HTLV-1 & T-NHL

HTLV1 is a RNA oncogenic virus which is associated with *'adult T-cell leukemia /lymphoma'* and is endemic in southern Japan and Caribbean basin. Like HIV which causes AIDS, HTLV1 also shows tropism for CD4+ T cells, hence this subset is the main victim for neoplastic transformation. In our local studies HTLV1 association was absent in mature T-lymphoproliferative disorders. This is in line with relatively low burden of HIV-AIDS in Pakistan so far (Noorali et al, 2004). (Figure 9)

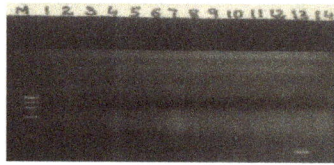

Fig. 9. Agarose gel showing samples of mature T-NHL negative for HTLV-1 DNA by PCR.

6. Role of EBV detection by PCR, ISH & IHC in diagnostic pathology

The ability to amplify specific regions of DNA from paraffin-embedded tissue by PCR has a profound impact on diagnostic pathology. For routine histopathological diagnosis of various lymphoproliferative disorders EBV-ISH (EBV-encoded nuclear RNA -1(EBER-1) and IHC by using an antibody to 'Latent Membrane Protein-1 (LMP-1) are frequently used in diagnostic dilemmas. For instance in the differential diagnosis of cHL and ALCL, EBER or LMP-1 positivity in neoplastic cells will strongly favour cHL as EBV association with ALCL is very *weak*. (Figure 10)

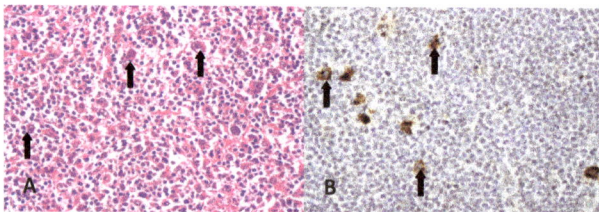

Fig. 10. Photomicrograph of a case of Hodgkin lymphoma (mixed cellularity, H&E, A ↑) stained with an antibody to LMP-1, note cytoplasmic staining of large Hodgkin cells (↑ B).

6.1 Conclusions

- EBV association with 'classic Hodgkin lymphoma' is well established and is strongest with 'mixed cellularity variant'
- Endemic (African) 'Burkitt lymphoma (BL)' involving jaw is strongly associated with EBV while association with 'sporadic BL' is relatively weak.
- EBV association with mature T-NHL is variable
- Strongest association of EBV is seen in AILT subtype while weakest association with ALCL subtypes.
- EBV immunohistochemistry using an antibody for LMP-1 and 'in situ hybridization (EBER)' are commonly used in routine diagnostic pathology
- HTLV-1 association with 'adult T-cell leukemia/lymphoma' endemic in Japan & Caribbean is insignificant in our experience in Pakistan.

7. EBV & nasopharyngeal carcinoma

Nasopharyngeal carcinomas are particularly common in some parts of Africa and southern China. In former they constitute most frequent childhood cancer while in the later adults are mostly affected. Association of EBV with nasopharyngeal carcinoma is well established. In fact this association is literally 100%. EBV associated protein LMP-1 is expressed in most cases. (Figure 11)

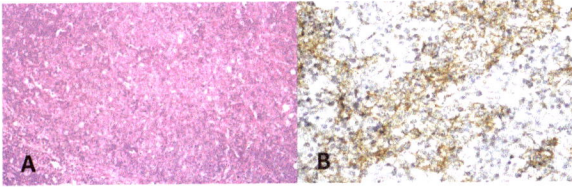

Fig. 11. Photomicrograph of a case of Nasopharyngeal carcinoma (H& E, A) stained with an antibody to LMP-1 (B), note cytoplasmic staining of neoplastic cells.

7.1 Conclusions

- EBV association in 'nasopharyngeal carcinoma' is literally 100%

8. HHV8 & Kaposi sarcoma

Relatively recently in 1994 'Human Herpesvirus-8 (HHV-8) was identified in an AIDS patient with cutaneous 'Kaposi sarcoma (KS)'. Later it was found that over 95% KS are associated with HHV-8. This virus is largely transmitted sexually. An antibody against HHV-8 shows positive reactivity in about 100% of cases and is a useful tool to confirm the diagnosis. Although 'Kaposi sarcoma' is uncommon in our practice in Pakistan, it is highly prevalent in developing world with high AIDS incidence. (Figure 12) Four forms are recognized based primarily on population demographics and risk factors. These include a) 'Chronic KS', also called European KS b) 'Lymphoadenopathic KS' also called African or endemic KS c) 'Transplant associated KS' and d) 'AIDs-associated KS'.

8.1 Conclusions

- HHV-8 association with KS is near 100%

- KS incidence in countries like Pakistan with relatively low burden of HIV carriers & AIDS is very low

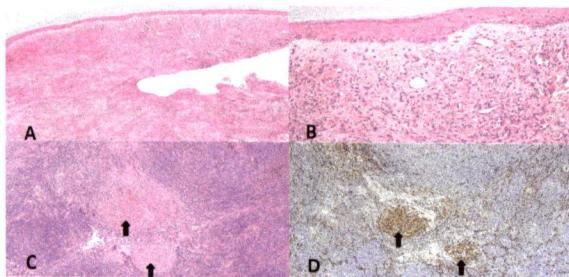

Fig. 12. Photomicrograph of Kaposi sarcoma resected from skin (H&E, A & B). Figure C shows lymph node metastases of the same (↑) highlighted on immunohistochemistry with CD31 (↑D).

9. *Helicobacter pylori* & gastric MALT lymphoma

'MALT lymphomas' were first described in 1983 by Peter Isaacson and Dennis Wright. They noted that primary low grade gastric B cell lymphomas recapitulate the histology of *'Mucosa Associated Lymphoid Tissue (MALT)'* exemplified by the Peyer patches and coined the term *'MALT lymphoma'*. These lymphomas are currently recognized as *'Extranodal marginal zone B cell lymphomas of MALT type'* according to the *'WHO Classification for Tumours of Haematopoietic and Lymphoid Tissues'* (Issaacson et al, 2008) (Figure 13). The stomach is the most reported and best studied site of *'MALT Lymphomas'*. An intimate relationship has been reported between the presence of *'Helicobacter pylori (HP)'* in the stomach and the development of *'MALT Lymphoma'* (Figure 13). In fact the pathogenesis of gastric *'MALT Lymphoma'* is believed to be caused by repeated antigenic stimulation of the immune system in the stomach by HP. The role of HP in the pathogenesis of *'gastric MALTomas'* can be illustrated by the fact that 75% of the patients who have gastric MALToma undergo remission if treated with antibiotics to eradicate HP (Ono et al, 2008). About half the people in the world have HP colonized in their gastrointestinal tract. Of these most remain asymptomatic. Despite the fact that, a high prevalence of HP is reported from Pakistan (Pervez et al, 2011), the prevalence of *'gastric MALTomas'* is very low in our experience. Seroprevalence of HP infection in the Pakistani population has been reported as high as 58%. This correlates with the *'Asian enigma'* described by various authors where less developed Asian countries like Pakistan, India, Bangladesh and Thailand have lower rates of gastric carcinoma compared to well developed countries like Japan and China, despite a higher prevalence of HP infection in the population. HP has been established to have a role in the aetiology of gastric carcinoma and its paradoxical high prevalence in areas with few cases of gastric carcinoma has long puzzled researchers. Available evidences do not support difference in HP strains as the sole explanation for this enigma.

9.1 Conclusions
- Helicobacter Pylori (HP) association with gastric adenocarcinoma & MALT lymphoma is well established.

- In Pakistan though prevalence of HP is very high, associated gastric adenocarcinoma & MALT lymphoma is low.

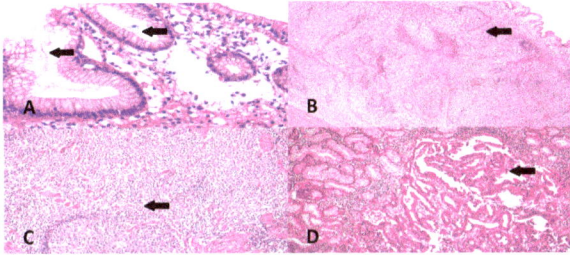

Fig. 13. Photomicrograph of gastric biopsy showing abundant 'Helicobacter pylori' organisms on epithelial surface (←A), Figure B & C shows gastric MALT lymphoma arising from the marginal zone of lymphoid follicle ←. Figure D shows well differentiated adenocarcinoma of stomach (intestinal type ←).

10. Immunoproliferative small intestinal disease (IPSID) & *Campylobacter jejuni*

Immunoproliferative small intestinal disease (IPSID) is a special variant of, *'Extranodal marginal zone B cell lymphoma'*, which affects the small intestine. In early to mid 1960s it was referred to as *'Mediterranean lymphomas'*, during late 1960s the term *'a-heavy chain disease'* was also used for patients with similar clinico-pathological presentations. Later it was realized that both *'Mediterranean lymphomas'* and *'a-heavy chain disease'* represented a spectrum of the same disease which presents in different stages i.e., benign, intermediate and overtly malignant (stage A, B & C) and the disease was named IPSID (Fine & Stone 2000). IPSID is predominantly found in patients of *'Mediterranean origin'*; however a few cases of IPSID are also diagnosed in the subcontinent (Pervez et al, 2011). IPSID involves the production of truncated alpha heavy chains which may appear in the serum and other body fluids. It can be treated with broad spectrum antibiotics at its early stages.

It is postulated that IPSID occurs in patients with repeated intestinal infections. Recent studies suggest association with *Campylobacter jejuni* (Lecuit et al, 2004). It is postulated that this results in continuous chronic antigenic stimulation of IgA secreting lymphoid tissue common in small intestine with a resultant clonal proliferation of IgA secreting lymphoid cells. Subsequently most cases lose the ability to synthesize light chain. In early stages it may be very difficult to differentiate IPSID, from chronic inflammatory process by the reporting pathologists. In such circumstances it may be impossible to diagnose without the help of clonal studies for IgH chain gene rearrangement (Figure 14). The other close mimicry includes *'Coeliac disease'* as both IPSID and *'Coeliac disease'* are characterised by lympho-plasmacytic infiltrate and villous atrophy. In these cases demographics are important; also gluten free diet will lead to improvement of *'Coeliac disease'* cases. Intra-intraepithelial lymphocytosis with surface epithelial damage shall also favour Coeliac disease. As some cases of IPSID particularly if untreated may transform into aggressive lymphomas like *'Diffuse large B-cell lymphoma'* (DLBCL), recognition of subtle features and follow-up is of paramount importance, particularly in endemic regions.

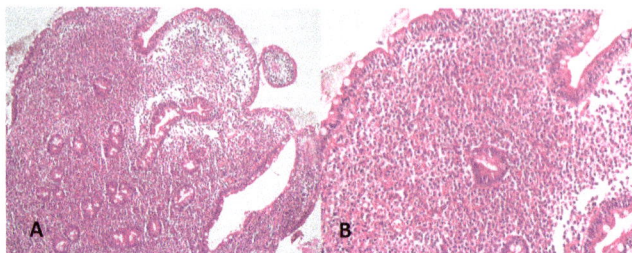

Fig. 14. Photomicrograph of duodenal biopsy from an IPSID patient diagnosed at stage A. Note flattening of mucosa with loss of villous architecture. Lamina propria shows diffuse sheets of plasma cells (H&E A & B).

10.1 Conclusions
- IPSID is a special variant of *'Extranodal marginal zone B-cell lymhoma'* predominantly found in Mediterranean region with sporadic cases in sub-continent.
- Recent studies suggest association with *'Compylobacter jejuni'*.
- In early stages treatment with broad spectrum antibiotics like tetracycline is curative.

11. Hepatitis B virus (HBV), Hepatitis C virus (HCV) & Hepatocellular carcinoma (HCC)

Developing countries bear major burden of 'Hepatitis B & C' for the obvious reasons i.e., insufficient or no screening of transfused blood, multiple use of contaminated needles, drug abuse and overall poor safety standards (Jafri et al, 2006). Pakistan for instance carries a very high burden of hepatitis B & C. There are estimated 7-9 million carriers of hepatitis B with a carrier rate of 3-5% (Ali et al, 2011). Genotype D (63.71%) is the most prevalent genotype in Pakistani population (Ali et al, 2011). The overall anti-HCV prevalence rate is 14-15% in general population of Pakistan (Idrees et al, 2009). Though hepatitis C is a major culprit for the reasons including increased potential to cause *'chronic liver disease'* and *'no vaccination'*; hepatitis B is still highly prevalent as well. A large proportion of population is still not vaccinated for hepatitis B, though now it is included in EPI (Extended Program of Immunization) program by the government and all newborns do get it.

In a recent study from Pakistan out of 161 subjects with HCC, chronic HCV infection was identified as a major risk factor (63.44% of tested HCC patients) for the development of HCC (Idrees et a, 2009). The time from HCV infection to the clinical appearance of cancer ranged from 10-50 years. In this population with HCC among various genotypes of HCV, genotype 3a was predominant (40.96%), followed by 3b in 15.66%, 1a in 9.63% and 1b in 2.40%.

On the face of such a high burden of Hepatitis B & C, hepatocellular carcinoma (HCC) is one of the common malignancies in our practice arising in a background of liver cirrhosis (Figure 15). Besides several other environmental factors are also playing their role in the causation of HCC. In Karachi, a port city of about 15 million inhabitants with hot and humid climate, it is reported that in wholesale markets selling food commodities without proper packing and preservation, a very high content of *'aspergillus flavus'* is isolated which is a known cause of HCC. Unfortunately HCC is a bad cancer and in our experience life expectancy at the time of diagnosis is not more than six months.

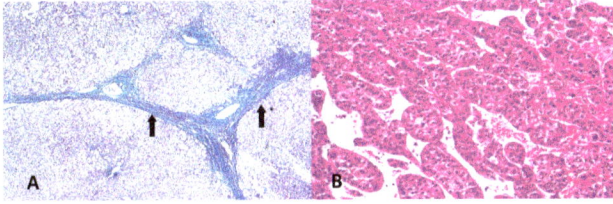

Fig. 15. Photomicrograph of a liver biopsy in a patient infected with Hepatitis C. Note fibrous band dividing the liver parenchyma into varying size nodules (↑A, Trichrome). Figure B shows a well differentiated hepatocellular carcinoma (HCC) arising in this patient (H&E).

11.1 Conclusions
- Hepatitis B & C are highly prevalent in the developing countries like Pakistan
- In Pakistan Hepatitis B carrier rate is 3-5% while anti-HCV prevalence rate is up to 15%.
- HCC is s common cancer in Pakistan mostly arising in a background of liver cirrhosis secondary to Hepatitis B & C.
- Over 60% of HCC are associated with HCV in Pakistan.

12. Acknowledgement

Dr Samina Noorali & Dr Syed Adnan Ali who completed their PhD under my supervision and participated in experimental work & original work on EBV and HPV in T-NHL & Oral cancer respectively included in this chapter are duly acknowledged. Ms. Shamsha Punjwani is acknowledged for her help in formatting this manuscript.

13. References

Ali M, Idress M, Ali L, Hussain A & Rehman I U, Saleem S, Afzal S & Butt S. (2011). Hepatitis B virus in Pakistan: A systematic review of prevalence, risk factors, awareness status and genotypes. *Virology journal*, 8:102-110.

Ali NS, Khuwaja AK, Ali T & Hameed R. (2009). Smokeless tobacco use among adult patients who visited family practice clinics in Karachi, Pakistan. *J Oral Pathol Med.* May;38(5):416-421.

Ali SM., Awan MS., Ghaffar S,., Slahuddin I., Khan S., Mehraj V & Pervez S. (2008). Human Papillomavirus infection in oral squamous cell carcinomas: Correlation with histologic variables and survival outcome in a high risk population, *Oral Surgery*, 1, 96-105

Ali SM, Awan MS, Ghaffar S, Azam SI & Pervez S. (2010). TP53 protein overexpression in oral squamous cell carcinomas (OSCC);correlation with histologic variables and survival oucome in Pakistani patients. *Oral Surgery*, 3, 83-95.

Bhurgri Y, Bhurgri A, Hussainy AS, Usman A, Faridi N, Malik J, Zaidi ZA, Muzaffar S, Kayani N, Pervez S & Hasan SH.(2003). Cancer of oral cavity and pharynx in Karachi-identification of potential risk factors. *Asia Pacific J Cancer Prevention*, Apr-Jun, 4:125-130.

Dai M, Bao YP, Li N, Clifford GM, Vacarrela S, Snijders PJ, Huang RD, Sun LX, Meijer CJ, Qiao YL & Franceschi S (2006). Human papillomavirus infection in Shanxi province, People's republic of China: a population based study, *Br J Cancer, July,* 95:96-101.

Fakhry C, Westra WH, Li S, Cmelak A, Ridge JA, Pinto H, Forastiere A & Gillison ML. (2008). Improved survival of patients with human papillomavirus-positive head and neck squamous cell carcinoma in a prospective clinical trial. *J Natl Cancer Inst,* Feb, 100(4): 261-269.

Fine KD & Stone MJ. (1999). Alpha-heavy chain disease, Mediterranean lymphoma and immunoproliferative small intestinal disease:a review of clinicopathological features, pathogenesis and differential diagnoses. *Am J Gastroenterology,* May, 94:1139-1152.

Franceschi S, Rajkumar R, Snijders PJF, Arslan A, Mahe C, Plummer M, Sankaranarayanan R, Cherian J, Meijer CJ & Weiderpass E. (2005). Papillomavirus infection in rural women in southern India. *Br J Cancer,* Feb, 92:601-606

Gillison ML. Human papillomavirus-associated head and neck cancer is a distinct epidemiologic, clinical, and molecular entity. (2004), *Semin Oncol,* Dec 31(6): 744-754.

Goodwin WJ, Thomas GR, Parker DF, Joseph D, Levis S, Franzmann E, Anello C & Hu JJ. (2008). Unequal burden of head and neck cancer in the United States. *Head & Neck,* March, 30(3): 358-371.

Idrees M, Rafique S, Rehman I, Akbar H, Yousaf MZ, Butt S, Awan Z, Manzoor S, Akram M, Aftab M, Khubaib B & Riazuddin S. (2009). Hepatitis C virus genotype 3a infection and hepatocellular carcinoma: Pakistani experience. *World J Gastroenterology,* Oct,15(40):5080-5085.

Issaacson PG, Chott A, Nakamura S, Hermelink M, Harris Nl & Swerdlow SH, editors. (2008). WHO classification of tumours of haematopoeitic and lymphoid tissues. Extranodal marginal zone lymphoma of mucosa-associated lymphoid tissue (MALT lymphoma), 4th edition, France IARC; 214-219.

Jafri W, Jafri N, Yakoob J, Islam M, Farhan S, Tirmizi A, Jafar T, Akhtar S, Hamid S, Shah HA & Nizami SQ. (2006). Hepatitis B and C: prevalence and risk factors associated with seropositivity among children in Karachi, Pakistan. *BMC infectious diseases,June,* 6:101-110.

Khan S, Jaffer NN, Khan MN, Rai MA, Shafiq M, Ali A, Pervez S, Khan N, Aziz A, Ali SH. (2007). Human papillomavirus subtype 16 is common in Pakistani women with cervical carcinoma. *Int J Infect Diseases,* July, 11(4):313-317.

Kreimer AR, Clifford GM, Boyle P & Franceschi S. (2005). Human papillomavirus types in head and neck squamous cell carcinomas worldwide: a systematic review. Cancer Epidemiol Biomarkers Prev. Feb 14(2): 467-475.

Lecuit M, Abachin E, Martin A, Poyart G, Suarez F, Bengoufa D, Feuillard J, Lavergne A, Gordon JI, Berche P, Guillevin L & Lortholary O. (2004). Immunoproliferative small intestinal disease associated with Campylobacter jejuni. *N Engl J Med,* Jan, 350 (3):239-248.

Merchant A, Husain SS, Hosain M, Fikree F, Pitiphat W, Siddiqui AR, Hayder SJ, Haider SM, Ikram M, Chuang SK & Saeed SA.. (2000). Paan without tobacco: an independent risk factor for oral cancer. *Int J Cancer, April,* 86 (1): 128-131.

Miller CS & Johnstone BM. (2001). Human papillomavirus as a risk factor for oral squamous cell carcinoma: a meta-analysis, 1982-1997. (2001). *Oral Surg Oral Med Oral Pathol Oral Radiol Endod*. June 91 (6): 622-635.

Morse DE & Kerr AR (2006). Disparities in oral and pharyngeal cancer incidence, mortality and survival among black and white Americans. *J Am Dent Assoc*, April, 137(4): 203-212.

Nair U, Bartsch J & Nair J. (2004). Alert for an epidemic of oral cancer due to use of the betel quid substitutes gutkha and pan masala: a review of agents and causative mechanisms. *Mutagenesis*, July, 19 (4) 251-262.

Nazaruk RA, Rochford R, Hobbs MV & Cannon MJ. (1998). Functional diversity of the CD8+ T-cell response to Epstein-Barr virus (EBV): implications for the pathogenesis of EBV-associated lymphoproliferative disorders. *Blood*, May, 91(10): 3875-3883.

Noorali S, Pervez S, Nasir MI & Smith JL. (2005) Characterization of Angioimmunobalstic T-cell lymphomas (AILT): and its association with Epstein-Barr virus (EBV) in Pakistani patients. *JCPSP*, 15(7):404-408.

Noorali S, Pervez S, Yaqoob N, Moatter T, Nasir MI, Hodges E and Smith JL. (2004). Prevalence and Characterization of Anaplastic Large Cell Lymphoma and Its Association with Epstein-BarrVirus in Pakistani Patients. *Pathology Research & Practice*; 200(10): 669-679.

Noorali S., Pervez S., Moatter T., Soomro IN., Kazmi SU., Nasir MI & Smith JL.(2003). Characterization of T-cell Non-Hodgkin's Lymphoma and its Association with Epstein - Barr virus in Pakistani patients. *Leukemia & Lymphoma*, 44(5), 807-813.

Noorali S., Yaqoob N., Nasir MI., Moatter T & Pervez S. (2002). Prevalence of Mycosis Fungoides and its association with EBV and HTLV1 in Pakistanian patients. *Pathol Oncol Res*, 8(3): 194-199.

Ono S, Kato M, Ono Y, Itoh T, Kubota K, Nakagawa M, Shimizu Y, Asaka M.. (2008). Characteristics of magnified endoscopic images of gastric extranodal marginal zone B-cell lymphoma of the mucosa associated lymphoid tissue including changes after treatment. *Gastrointestinal Endos*, Oct, 68(4): 624-631.

Parkin DM. (2006). The global health burden of infection-associated cancers in the year 2002. *Int J Cancer*. Jun 15;118(12):3030-3044,

Pervez S, Ali N, Aaqil H, Mumtaz K Ullah SS & Akhtar N. (2011). Gastric MALT lymphoma; a rarity, *JCPSP*, Mar, 21(3):171-172.

Pervez S, Mumtaz K, Ullah SS, Akhtar N, Ali N & Aaqil H. (2011) Immunoproliferative small intestinal disease. JCPSP, Jan; 21(1):57-58.

Raza SA, Franceschi S, Pallardy S, Malik FR, Avan BI, Zafar A, Ali SH, Pervez S, Serajuddaula S, Snijders PJ, van Kemenade FJ, Meijer CJ, Shershah S & Clifford GM. (2010). Human papillomavirus infection in women with and without cervical cancer in Karachi, Pakistan. *Br J Cancer*, May, 102(11):1657-1660.

Ryerson AB, Peters ES, Coughlin SS,. Chen VW, Gillison ML, Reichman ME, Wu X, Chaturvedi AK, Kawaoka K. (2008). Burden of potentially human papillomavirus-associated cancers of the oropharynx and oral cavity in the US, 1998-2003. *Cancer*, Nov, 113 (10 suppl): 2901-2909.

Scully S, Prime S & Maitland NJ. (1985). Papillomavirus: their possible role in oral disease. *Oral Surg Oral Med Oral Pathol*, Aug, 60(2):166-174.

Settle K, Posner MR, Schumaker LM, Tan M. Suntharalingam M. Goloubeva O, Strome SE, Haddad RI, Patel SS, Cambell EV 3rd, Sarlis N, Lorch J & Cullen KJ. (2009). Racial survival disparity in head and neck cancer results from low prevalence of human papillomavirus infection in black oropharyngeal cancer patients. *Cancer Prev Res* (Philadel) Sept, 2(9) 776-781.

Settle K, Taylor R, Wolf J, Kwok Y, Cullen K, Carter K, Ord R, Zimrin A, Strome S, Suntharalingam M. (2009). Race impacts outcome in stage III/IV squamous cell carcinomas of the head and neck after concurrent chemoradiation therapy. *Cancer*, April, 115 (8) 1744-1752.

Sherpa ATL, Clifford G, Vaccarella S, Shrestha S, Nygard N, Karki BS, Snijders PJ, Meijer CJ, & Franceschi S.. (2010). Human papillomavirus infection in women with and without cervical cancer in Nepal. *Cancer Causes Control, March*, 21 (3):313-330.

Shiboski CH, Schmidt BL & Jordan RC. (2007). Racial disparity in stage at diagnosis and survival among adults with oral cancer in the US. Community. *Dent Oral Epidemiol*, June, 35(3):233-240.

Syed S, Khalil S & Pervez S. (2011). Anaplastic Large Cell Lymphoma: The most common T-cell lymphoma in Pakistan, Asia Pacific J Cancer Prev, 12(3):685-689.

Wu RF, Dai M, Qiao YL, Clifford GM, Liu ZH, Arslan A, Li N, Shi JF, Snijders PJ, Meijer CJ &, Franceschi S. (2007). Human papillomavirus infection in women in Shenzhen city, People's republic of China, a population typical of recent Chinese urbanisation. *Int J Cancer*, Sept, 121(6):1306-1311.

8

Blood Irradiation

Sezer Saglam[1], Aydin Cakir[2] and Seyfettin Kuter[3]

[1]Istanbul University, Oncology Institute,
Department of Medical Oncology, Fatih, Istanbul
[2]Istanbul University, Oncology Institute,
Department of Medical Pyhsics, Fatih, Istanbul
[3]Istanbul University, Oncology Institute,
Department of Medical Pyhsics, Fatih, Istanbul,
Turkey

1. Introduction

Transfusion-associated graft-versus-host disease (TA-GVHD) is a possible complication of blood transfusion that occurs when viable donor T-lymphocytes proliferate and engraft in immunodeficient patients after transfusion. Presently, the only method accepted to prevent TA-GVHD is the irradiation of blood and its components before transfusion (Moroff and Luban 1997)). Ionizing irradiation eliminates the functional and proliferative capacities of T-lymphocytes leaving other blood components, especially erythrocytes, granulocytes and platelets, functional and viable. This is possible because T-lymphocytes are more radiosensitive than other blood components (Masterson and Febo, 1992).To carry out the irradiation of blood specially designed commercial irradiators exist, usually localized in blood banks, and dedicated exclusively to this task.

Blood and blood components may be treated with ionizing radiation, such as gamma rays from ^{137}Cs or ^{60}Co sources, and from self-contained X-ray (bremsstrahlung) units and medical linear X-ray (bremsstrahlung) and electron accelerators used primarily for radiotherapy. However, teletherapy machines, such as linear accelerators or ^{60}Co units already available at the hospital, may also be used for the same purpose (Moroff 1997), improving the cost/benefit ratio of the process.

Blood irradiation specifications include a lower limit of absorbed dose, and may include an upper limit or central target dose. For a given application, any of these values may be prescribed by regulations that have been established on the basis of available scientific data.

The absorbed dose range for blood irradiation is typically 15 Gy to 50 Gy. In some jurisdictions, the absorbed dose range for blood irradiation is 25 Gy to 50 Gy. The energy range is typically from approximately 40 keV to 5 MeV for photons, and up to 10 MeV for electrons.

For each blood irradiator, an absorbed-dose rate at a reference position within the canister is measured by the manufacturer as part of acceptance testing using a reference-standard dosimetry system. That reference-standard measurement is used to calculate the timer setting required to deliver the specified absorbed dose to the center of the canister with blood and blood components, or other reference position. Either relative or absolute absorbed dose measurements are performed within the blood or blood-equivalent volume for determining the absorbed-dose distribution. Accurate radiation dosimetry at a reference position which could be the position of the maximum absorbed dose (Dmax) or minimum absorbed dose (Dmin) offers a quantitative, independent method to monitor the radiation process.

Dosimetry is part of a measurement quality assurance program that is applied to ensure that the radiation process meets predetermined specifications.

2. Blood irradiators

The basic operating principles and configurations of a free-standing irradiator with either ^{137}Cs source or a linear accelerator are shown schematically in Figure 1. With a freestanding ^{137}Cs irradiator, the blood components are contained within a metal canister that is a rotating turntable. Continuous rotation allows for the γ rays, originating from one to four closely positioned pencil sources, to penetrate all portions of the blood component. The number of sources and their placement depend on the instrument and model. The speed of rotation of the turntable also depends on the make or model of the instrument. A lead shield encloses the irradiation chamber. Free standing irradiators employing ^{60}Co as the source of γ rays are comparable except that the canister containing the blood component does not rotate during the irradiation process; rather, tubes of ^{60}Co are placed in a circular array around the entire canister within the lead chamber. When free standing irradiators are used, the rays are attenuated as they pass through air and blood but at different rates. The magnitude of attenuation is greater with ^{137}Cs than with ^{60}Co.

Linear accelerators generate a beam of X-rays over a field of given dimension. Routinely, the field is projected on a table-top structure. The blood component is placed (flat) between two sheets of biocompatible plastic several centimeters thick.The plastic on the top of the blood component (ie,nearer to the radiation source) generates electronic equilibrium of the secondary electrons at the point where they pass through the component container.The plastic sheet on the bottom of the blood component provides for irradiation back-scattering that helps to ensure the homogenous delivery of the x-rays. The blood component is usually left stationary when the entire x-ray dose is being delivered. Alternatively it may be flipped over when one half of the dose has been delivered; this process involves turning off and restarting the linear accelerator during the irradiation procedure. Although it seems as if the practice of flipping is not required,further data are needed.

3. Blood components

The risk of GVHD for patients, all components that might contain viable T lymphocytes should be irradiated. These include units of whole blood and cellular components (red cells, platelets, granulocytes),whether prepared from whole blood or by apheresis. All types of red cells should be irradiated, whether they are suspended in citrated plasma or in an additive solution. There are recent data supporting the retention of the quality of irradiated red cells after freezing and thawing.If frozen thawed units are intended for GVHD-susceptible individuals and have not been previously irradiated,they should be irradiated because it is known that such components contain viable T lymphocytes.Filtered red cell products should also be irradiated.Extensive leucoreduction through filtration may decrease the potential for GVHD and serve as an alternative to irradiation in the future when questions about the minimum level of viable T lymphocytes that can lead to GVHD are resolved. There are reports of TA-GVHD in patients who received leucodepleted (filtered) red cells; however,the extent of leucoreduction of the components was not uniformly quantified in such reports. In addition, investigators have suggested that the number of T lymphocytes present in a product that causes GVHD may depend on the extent of patient immunocompetence at the time of transfusion . It is likely that the greater the degree of immunosuppression, the fewer the viable T lymphocytes that will be required to produce GVHD in susceptible patients. In a recent review, it was suggested that cytotoxic T lymphocytes, or interleukin-2-secreting precursors of helper T lymphocytes,may be more predictive of GVHD than the number of proliferating T cells alone. Accordingly,this suggests that until further data are available to confirm adequate removal of these T-cell subtypes by leucoreduction, irradiation should be used for blood products destined for patients at risk for GVHD. Irradiated red cells undergo an enhanced efflux of potassium during storage at 1^0 to 6^0 C.Comparable levels of potassium leakage occur with or without prestorage leucoreduction. Washing units of red cells before transfusion to reduce the supernatant potassium load does not seem to be warranted for most red cell transfusions because posfinfusion dilution prevents the increase in plasma potassium. On the other hand, when irradiated red cells are used for neonatal exchange transfusion or the equivalent of a whole blood exchange is anticipated, red cell washing should be considered to prevent the possible adverse effects caused by hyperkalemia associated with irradiation and storage. Blood components given to recipients, whether immunocompromised or immunocompetent, that contain lymphocytes that are homozygous for an HLA haplotype that is shared with the recipient, pose a specific risk for TA-GVHD. This circumstance occurs when first and second degree relatives serve as directed donors s-ll and when HLA matched platelet components donated by related or unrelated individuals are being transfused. Irradiation of blood components has been recommended in these situations.

Platelet components that have low levels of leucocytes because of the apheresis process and/or leucofiltration should also be irradiated if intended for transfusion to susceptible patients. This is because the minimum number of T lymphocytes that induces TA-GVHD has not yet been delineated.

Fresh frozen plasma does not need to be irradiated routinely because it is generally accepted that the freezing and thawing processes destroy the T lymphocytes that are present in such plasma.

During the past 2 years, there have been two brief articles suggesting that immunocompetent progenitor cells may be present in frozen-thawed plasma; the authors therefore suggested that frozen-thawed plasma may need to be irradiated. Further studies are needed to validate these findings and to assess whether the number of immunocompetent cells, that may be present in thawed fresh frozen plasma, is sufficient to induce GVHD. In rare instances, when nonfrozen plasma (termed *fresh plasma*) is transfused, it should be irradiated because of the presence of a sizable number of viable lymphocytes, approximately 1×10^7 cells in a component prepared from a unit of whole blood.

4. Quality assurance guidelines

Various dosimetry techniques have been used to measure the dose to blood products. These include thermoluminescent dosimeters (TLD); alanine, ferrous sulphate, red perspex, metaloxide semiconductor field effect transistors (MOSFETs) and chloroform /dithoizone/parafin mixture (Hillyer *et al* 1993). Recently radiochromic film was shown to be an adequate dosimeter for blood irradiation (Butson *et al* 1999). The most prevalent method relies on TLDs and tries to ascertain the causes and variations in delivered *in vitro* dose across an 'active' treatment volume in a dedicated blood box for standard x-ray beams.

Most dosimeters have significant energy dependence at photon and electron energies less than 100 keV, so great care must be exercised when measuring absorbed dose in that energy range.

This practice outlines irradiator installation qualification, operational qualification, performance qualification, and routine product processing dosimetric procedures to be followed in the irradiation of blood and blood components by the blood-banking community. If followed, these procedures will help to ensure that the products processed with ionizing radiation from gamma, X-rays (bremsstrahlung), or electron sources receive absorbed doses within a predetermined range.

One must document that the instrument being used for irradiation is operating appropriately and confirm that blood components had been irradiated.To assure that the irradiation process is being conducted correctly, specific procedures are recommended for free-standing irradiators and linear accelerators, which are summarized in Tables 1 and 2. The procedures to be used with free-standing irradiators are an update to the guidelines provided several years ago by Anderson. Included are current recommendations from the FDA.

Measure	Frequency
Isotop decay factor	Annually for ^{137}Cs ;montly for ^{60}Co
Dose map	Annualy for ^{137}Cs; annually for ^{60}Co
Radiation leakage	daily
Timer accuracy	montly
Turntable	daily

Table 1. Recommended Quality Assurance Measures to be Used with Free-Standing Gamma Irradiators.

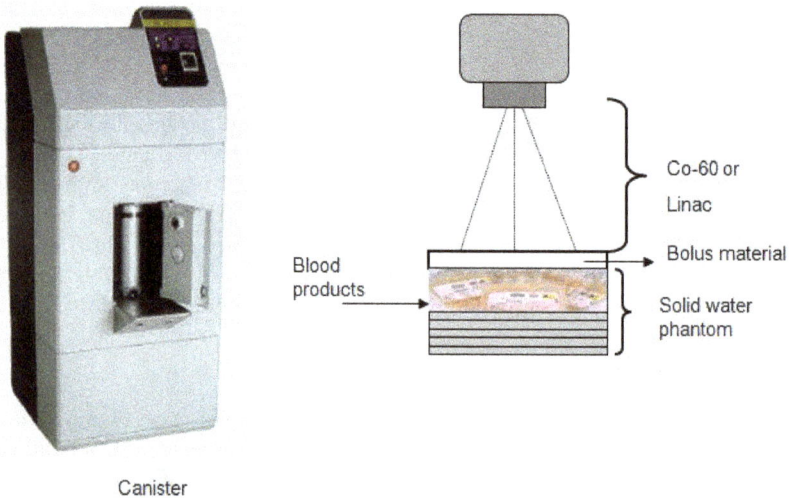

Fig. 1. With a freestanding ^{137}Cs irradiator, the blood components are contained within a metal canister that is positioned on a rotating turntable.

Dose mapping measures the delivery of radiation within a simulated blood component or over an area in which a blood component is placed. This applies to an irradiation field when a linear accelerator is used or to the canister of a free-standing irradiator. Dose mapping is the primary means of ensuring that the irradiation process is being conducted correctly. It documents that the intended dose of irradiation is being delivered at a specific location (such as the central midplane of a canister), and it describes how the delivered irradiation dose varies within a simulated component or over a given area. This allows conclusions to be drawn about the maximum and minimum doses being delivered. Dose mapping should be performed with sensitive dosimetry techniques. A number of commercially available systems have been developed in recent years. Other quality assurance measures that need to be done include the routine confirmation that the turntable is operating correctly (for ^{137}Cs rradiators), measurements to ensure that the timing device is accurate, and the periodic lengthening of the irradiation time to correct for source decay. With linear accelerators, it is necessary to measure the characteristics of the x-ray beam to ensure consistency of delivery. Confirming that a blood component has, in actuality, been irradiated is also an important part of a quality assurance program. At least one commercial firm has developed an indicator label for this purpose.

5. Dose mapping with free-standing irradiators

For free-standing irradiators, a dose-mapping procedure will measure the delivered dose throughout the circular canister in which the blood component is placed. To establish a two-dimensional map, a dosimetry system is placed in a canister that is completely filled with a blood/tissue-compatible phantom composed of water or an appropriate plastic such as polystyrene. The dosimetry material is placed within the phantom in a predetermined way. This approach provides data that describe the minimum levels of irradiation that would be absorbed by a blood component placed in the canister and recognizes that maximum attenuation will occur when the canister is completely filled with a blood-compatible material. Relevantly, it was shown recently that the absorbed dose at the central midplane of a canister (ie, at the center point) decreased by approximately 25% (from 3100 to 2500 cGy) in a ^{137}Cs irradiator (JL Shepherd and Associates, San Francisco, CA) when the loading of the canister was changed from 0% (air) to 100% (with blood components). An irradiationsensitive film dosimetry system (International Speciality Products) that will be described later in this report was used for this purpose. A linear relationship was observed between the amount of fill and the measured central dose. With 1 and 2 units of blood components, the central dose relative to air was 0.98 and 0.93. The minimum and maximum levels were influenced in the same manner as the central dose on decreasing the proportion of the canister that contained air. Other studies have shown that the extent of variability in the dose delivered to the interior of simulated blood units (water or saline in plastic blood storage containers) depended on the model of the ^{137}Cs free-standing irradiator. An immobilized grid of thennoluminescent dosimeters in a plastic sheet were placed within the simulated blood units to measure dose delivery. See the section on dosimetry systems in use. It was also shown that a spacer into the bottom of the canister increases the minimum level of radiation within the simulated blood units as expected from the results of full-canister dose mapping involving a phantom. The extent of variability with ^{137}Cs irradiations is influenced by a number of factors, including the number of sources, turntable speed, and the presence of a spacer at the bottom of the canister. These studies underscore the need for consistency in loading the canister. Attenuation of the irradiation dose delivered is a function of physical density, electronic density, and atomic number with three major processes: photoelectric, Compton, and pair production. In practical terms, attenuation is caused when the irradiation enters a liquid, such as water or blood. The extent of attenuation depends on a number of factors, including the dimensions of the canister. In a fully filled canister, as is used for dose mapping, the attenuation will increase as the irradiation transverses to the center point. The dose map that is generated describes the dose distribution. As depicted in the theoretical dose map shown in Figure 2, the edges of the canister are exposed to a greater dose of irradiation compared with the center line because the attenuation is less in the periphery. The attenuation with ^{60}Co is less than that seen with ^{137}Cs.

When an irradiator is purchased, the distributor will provide a central dose level that is determined in a blood-compatible environment. In the 1970s and 1980s manufacturers provided a central dose that was determined in air, resulting in the use of timer settings that provided for a dose level that was somewhat less than what was expected. Subsequent to the issuing of the FDA guidelines in July 1993 and the use of dose mapping, it has been necessary to readjust irradiation times with some instruments because the attenuation effect had not been considered previously.

A theoretical two-dimensional dose map describing the irradiation dose distribution through a fully filled canister of a free-standing ^{137}Cs irradiator is shown in Figure 2. To obtain this dose map, dosimeters would have been positioned in the central axis and the edge of circular canister from the top to the bottom of the canister. The y dimension of the map depicts the top to bottom axis of the canister, whereas the x dimension depicts the cross-sectional axis. For the theoretical situation described in Figure 2, the central midplane dose is 2560 cGy, slightly above the minimum standard of 2500 cGy, and the minimum dose is 1750 cGy. In this irradiation dose map, the minimum dose is at the central bottom of the canister, a common finding in actual practice.

The dose map can also be used to assess whether the turntable of a ^{137}Cs irradiator is rotating in an appropriate manner. The occurrence of comparable readings at the two edges of the two-dimensional map, as depicted in the theoretical dose map, indicates that the canister is rotating evenly in front of the ^{137}Cs source. If the turntable were not rotating, the dose levels at the edge of the map closest to the source would be much higher than that found on the opposite edge, ie, the side located distant to the source. According to the 1993 recommendations from the FDA, dose mapping should be performed routinely on an annual basis and after a major repair, especially one involving the sample handling apparatus such as the turntable.

6. Dosimetry systems

The delivered irradiation dose can be measured by a variety of dosimetry systems. In recent years, several commercial interests have developed complete systems for use with free-standing irradiators; each system consists of a phantom that fills the canister and a sensitive dosimetry system. Three main types of dosimetry measurement systems are available (Table 3).

These dosimeters are referred to as routine dosimeters. They are calibrated against standard systems, usually at national reference laboratories such as the National Institute of Standards and Technology in the United States. The routine dosimeter measurement systems were initially developed for use with ^{137}Cs irradiators because this is the predominant irradiation source for blood. More recently, they have been developed also for use with ^{60}Co irradiators. Thermolumeniscent dosimeters (TLD chips) are one type of routine dosimeter. TLD chips are small plastic chips with millimeter dimensions having a crystal lattice that absorbs ionizing radiation. Specialized equipment is used to release and measure the energy absorbed by the TLD chip at the time of the test irradiation. In one commercially available system, chips are placed at nine different locations within a polystyrene phantom that fits into the canister of the IBL 437C irradiator (CIS US, Inc,Bedford, MA). The timer setting used routinely for an instrument is used in the test procedure.

Method	Measurement type
Thermoluminescent Dosimetry	Emission of light
Radiochromic(GafChromic) film	Optic density
Mosfet(metal –oxide field effect transistors)	Voltage detection
Alanine/ESR	ESR signal-Magnetic field

Table 3. Dosimetric systems in different clinics.

There are two systems that use radiochromic film. On exposure to irradiation, the film darkens, resulting in an increase in optical density. The optical density, determined at various locations on the film, is linearly proportional to the absorbed irradiation dose. Standard films that are irradiated at a given dose level with a calibrated source at a national reference laboratory provide the means to assess the absolute level of absorbed irradiation.This type of dosimeter is basically an x-ray film comparable with that used in clinical practice. With this device, the map that is developed identifies the absorbed irradiation dose that is measured at a large number of locations. In one system, a film contained in a thin water-tight casement is placed into the canister (International Specialty Products,Wayne, NJ), This approach is being used with a variety of irradiators. The canister is filled completely with water before the irradiation procedure.This system provides a direct readout of the dose that is delivered throughout the canister. The timer setting used routinely is employed for the test procedure. In a second system, a film having different radiation sensitive characteristics is embedded between two halves of a circular-fitting polystyrene plastic phantom (Nordion Internation, Canada, Ontario). Irradiation of specialized films is performed with a number of timer settings, each being larger than that used routinely. The map produced is normalized for a central midplane dose of 2500 cGy. The time to produce the 2500 cGy will have been predetermined with a different dosimeter system, the Fricke system, in which absorbed radiation causes a change in the state of a iron salt that can be assessed spectrophotometrically. Another approach to irradiation dose mapping employs a solid-state electronic dosimeter that is technically referred to as a *metal-oxide silicon field effect transistor* (MOSFET). A board contains a number of small transistors in an arrangement that provides data for a dose map. This board is placed between two halves of a circular polystyrene phantom that fits into the canister. This dosimeter absorbs and stores the radiation dose imparted to it electronically . The radiation causes the formation of holes in the metal-oxide layer that becomes trapped within the transistor. The magnitude of the holes is evaluated by measuring the voltage across the transistor with a voltmeter. The voltages measured are converted to absorbed dose. With each dosimetry system, measurements are used to express the absorbed irradiation dose of cGrays. All dosimetry measurements are associated with a degree of uncertainty or possible error. The magnitude of the uncertainty depends on the kind of dosimeter used. For most dosimeters, the level is 5% of the measured value. For a central absorbed dose level of 2560 cGy (see theoretical dose map in Fig 2) the value could be as high as 2788 cGy or as low as 2432 cGy. Correspondingly, a measured value of 2400 cGy could be as high as 2520 cGy or as low as 2380 cGy. Because the measured value could in actuality meet the 2500 cGy standard, it is appropriate to accept a value of 2400 cGy as meeting the current standard. The same approach should be used when evaluating the minimum value on a dose map. Albeit arbitrary and cautious, the actual minimum on an irradiation dose map should not be below 1500 cGy.

7. Precautions with free-standing irradiators

It is important periodically to lengthen the time of irradiation to correct for decay of the isotopic source that emits the gamma irradiation. Until recently, this was the only major quality assurance measure that was performed routinely. With the half-life for ^{137}Cs being 30 years, annual lengthening of the timer setting is appropriate.On the other hand, with the half-life of ^{60}Co being only 5,27 years, the time of irradiation should be increased on a quarterly basis. The additional seconds of irradiation that are needed can be calculated using formulae that can be found in any physics text. Alternatively, distributors of irradiators provide a chart that specifies the appropriate setting as a function of calendar time.

7.1 Turntable rotation
For ^{137}Cs irradiators, it is essential that the turntable operates at a constant speed in a circular pattern to ensure that each part of a blood component is exposed equally to the source. Daily verification of turntable rotation is an appropriate quality assurance measure. With some free-standing irradiator models rotation of the turntable can be observed before the door of the compartment in which the canister is positioned is closed. In other models, this can be done only indirectly by ensuring that an indicator light is operating appropriately. With some older models, there have been occasional reports that the turntable failed to rotate because of mechanical problems. Such problems should not be encountered with the newer models because of changes in the turntable mechanisms. In any event, daily verification of turntable rotation is a prudent quality assurance measure.

7.2 Radioactivity leakage
Irradiators are constructed so that the isotopic sources are contained in a chamber heavily lined with a protective lead shield to prevent leakage of radioactivity. Accordingly, gamma irradiators are considered to be very safe instruments. Although there have been no reports of source leakage of radioactivity, periodic measurements are warranted to ensure that this is the case. Attaching a film badge to the outside of the irradiator, using a Geiger counter periodically, and performing a wipe test of the inside of the chamber where the canister is positioned at least semiannually are measures that are being used.

8. Dose mapping with linear accelerators

Linear accelerators that are used therapeutically to provide radiation therapy are carefully monitored to ensure appropriateness of dose to an irradiation field. When blood components are treated with x-rays, the instrument settings are very different than those used to treat oncology patients. Hence, additional periodic quality control measures, primarily to assess the dose delivered to blood components, are needed to ensure that linear accelerators are being operated appropriately when used for blood irradiation. Currently, there are no commercially available systems for assessing the dose delivered throughout the area of an irradiation field in which blood components are placed for treatment with x-rays. An ideal dosimeter for this purpose would be made of a tissue-compatible plastic phantom, containing appropriate dosimeter material and a covering that could be placed at the appropriate distance from the source. An alternative approach might involve the use of a blood bag filled with water (simulating a blood unit) containing TLD chips, as described earlier. In comparative studies using such simulated blood units, it was determined that radiation delivery was more uniform with linear accelerators than with ^{137}Cs free-standing irradiators. This reflects the relative homogeneity of x-ray beams. In the absence of an available system modified for the irradiation of blood bags, the dose delivered throughout an irradiation field should be mapped with the dosimetric measuring system known as an *ionization chamber.* The ionization chamber is used to calibrate linear accelerators for patient use. In addition, on a yearly basis, dose mapping should be performed using a tissue-compatible phantom.

In view of the widely divergent conditions that are used during the operation of linear accelerators, other parameters pertaining to the x-ray beam should be evaluated on at least a quarterly basis to provide assurance that the instrument is being used appropriately for the irradiation of blood components. The goal is to ensure that the instrument is being set in a consistent fashion. When setting a linear accelerator for blood component irradiation, the following should be measured: the distance between the x-ray source and the position where the blood components are to be placed; consistency in the strength of the x-ray beam; and (3) the intensity of the x-ray beam. The distance between the source and position on the table where blood components will be placed (referred to as the *target)* can be evaluated easily with a calibrated measuring device. This is a simple task that can be performed on a routine basis. The consistency of beam output can be evaluated by measuring the beam current. Beam intensity can be evaluated by measuring the ionization current in a monitoring ionization chamber array that can be expressed in terms of the number of photons delivered per square centimeter. These parameters should be assessed routinely as part of quality control programs used by radiation physicists. A code of practice was published in 1994 by the Radiation Therapy Committee of the American Association of Physicists in Medicine for the quality control of radiotherapy accelerators. The described practices are used routinely by radiation physicists. It would he prudent to ensure that an institution using a linear accelerator for blood irradiation follow these quality assurance guidelines and recommendations.

9. Confirming that irradiation occurred

It is important to have positive confirmation that the irradiation process has taken place. This is to identify whether an operator fails to initiate the electronically controlled irradiation process or when the irradiation process is not performed because of instrumentation malfunction. A radiation-sensitive indicator label has been developed specifically for this purpose by International Speciality Products,Wayne, NJ. The label containing a radiationsensitive film strip is placed on the external surface of the blood component. Irradiation causes distinct visually observable changes: The appearance changes from clear red to opaque with obliteration of the word "NOT." When the label is placed on a blood component, there is a visual record that the irradiation process took place. The reliability of this type of indicator was documented recently in a multisite study.

Two versions of the indicator label have been manufactured. The difference is the range of radiation needed to cause a change in the radiationsensitive film. The ratings for these indicators are 1500 cGy or 2500 cGy. The ratings serve as an approximate guideline for the amount of absorbed radiation that will be needed to completely change the window from reddish to opaque with complete obliteration of the word "NOT." Because the indicator labels are designed for and are used to confirm that the irradiation process has occurred, we have concluded that the 1500 cGy label is the most appropriate tool to perform this quality control measure. This is based on the routinely observed pattern of dose distribution to a blood component in a canister of a free-standing irradiator. Despite a targeted central dose of 2500 cGy, there will be spots at which the dose will be less. If the theoretical dose map presented in Figure 2 is used as an example, there will be a spot that will receive only 1800 cGy. If the 2500 cGy-rated label were to be located on the external surface of a component, there may be minimal changes in the appearance of the radiation-sensitive film window.This would result in a judgment that the blood component was not irradiated, when in actuality it was treated satisfactorily.

Dose	
Linear accelerators	Free standing irradiators
2500 cGy to the center of an irradiation field with a minumum of 1500 cGy elsewhere.	2500 cGy to the central midplane of a canister with a minumum of 1500 cGy elsewhere.
Dose mapping	
Linear accelerators	Free Standing irradiators
Yearly dose mapping with an ionization chamber and a water phantom.More frequent evaluation of instrument conditions to ensure consistency of x-rays.	Routinely,once a year Cs-137 or twice a year Co-60 and after major repairs;the irradiation procedure should be tested using a fully filled canister with a dosimetry system to map the distrubition of the absorbed dose.
Correction for radioisotopic decay	
Cs-137; annually	Co-60; every 3 month
Turntable rotation(Free standing Cs-137 irradiators)	
Daily should be checked.	
Storage time (after irradiation)	
Red cells	Platelets
For up to 28 days;total storage time cannot exceed maximum stroga time for unirradiated red cells	No change due to the irradiation.

Table 4. Guidelines for irradiating blood components.

10. References

[1] Anderson KC, WeinsteinHJ: Transfusion-associated graftversus-host disease. New Eng J Med .1994;323:315-321

[2] Roberts GT, Luban NLC: Transfusion-associated graftversus-host disease, in Rossi EC, Simon TL, Moss GC, Goldis A(eds): Principles of Transfusion Medicine. Baltimore MD,Williams and Wilkins, 1996, pp 785-801

[3] Linden JV, Pisciotto PT: Transfusion-associated graftversus-host disease and blood irradiation. Transfus Med Rev .1992;6:116-123

[4] Anderson KC: Clinical indications for blood component irradiation, in Baldwin ML, Jefferies LC (eds): Irradiation of Blood Components, Bethesda, MD, American Association of Blood Banks, 1992, pp 31-49

[5] Brubaker DB: Transfusion-associated graft-versus-host disease, in Anderson KC, Ness PM (eds): Scientific Basis of Transfusion Medicine. Implications for Clinical Practice. Philadelphia, PA, W.B. Saunders Company, 1994, pp 544-573

[6] Williamson LM: UKguidelines for the irradiation of blood components. Transfus Sci .1995;16:135-137

[7] Davey RJ: Transfusion-associated graft-versus-host disease and the irradiation of blood components. Immunological Investigations .1995;24:431-434

[8] Kanter MH: Transfusion-associated graft-versus-host disease disease: Do transfusions from second-degree relatives pose a greater risk than those from first-degree relatives? Transfusion.1992;32:323-327

[9] McMilan KD, Johnson RL: HLA-homozygosity and the risk of related-donor transfusion-associated graft-versus-host disease. Transfus Med Rev.1993; 7:37-41

[10] Petz LD, Calhoun L, Yam P, et al: Transfusion-associated graft-versus-host disease in immunocompetent patients: Report of a fatal case associated with transfusion of blood from a second-degree relative, and a survey of predisposing factors. Transfusion .1993;33:742-750

[11] Williamson LM, Warwick RM: Transfusion-associated graft-versus-host disease and its prevention. Blood Reviews.1995; 9:251-261

[12] Ohto H, Anderson KC: Survey of transfusion-associated graft-versus-host disease in immunocompetent recipients. Transfus Med Rev .1996;10:31-43

[13] Davey RJ: The effect of irradiation on blood components, in Baldwin ML and Jefferies LC (eds): Irradiation of Blood Components, Bethesda, MD, American Association of Blood Banks, 1992, pp 51-62

[14] Fearon TC, Luban NLC: Practical dosimetric aspects of blood and blood product irradiation. Transfusion .1986.26:457-459

[15] Suda BA, Leitman SF, Davey RJ: Characteristics of red cells irradiated and subsequently frozen for long term storage. Transfusion.1993 33:389-392

[16] Miraglia CC, Anderson G, Mintz PD: Effect of freezing on the in vivo recovery of irradiated red cells. Transfusion .1994;34:775-778

[17] Crowley JR Skrabut EM, Valeri CR: Immunocompetent lymphocytes in previously frozen washed red cells. Vox Sang .1974;26:513-517

[18] Akahoshi M, Takanashi M, Masuda M, et al: A case of transfusion-associated graft-versus-host disease not prevented by white cell-reduction filters. Transfusion.1992;32:169-172

[19] Heim MU, Munker R, Saner H, et at: Graft-versus-host Kranldleit(GVH mit letalem ausgang nach der gabe von gefilterten erythrozytenkonzentraten(Ek). Infusionstherapie.1991; 18:8-9

[20] Hayashi H, Nishiuchi T, Tamura H, et al: Transfusion associated graft-versus-host disease caused by leukocyte fltered stored blood. Anesthesiology.1993;79:1419-1421

[21] Anderson KC: Leukodepleted cellular blood components for prevention of transfusion-associated graft-versus-host disease.Transfus Sci .1995;16:265-268

[22] Ramirez AM, Woodfield DG, Scott R, et at: High potassium levels in stored irradiated blood. Transfusion .1997;27:444-445

[23] Rivet C, Baxter A, Rock G: Potassium levels in irradiated blood. Transfusion.1989: 29:185

[24] Swann ID, Williamson LM: Potassium loss from leucodepleted red cells following "v-irradiation. Vox Sang .1996;70:117-118

[25] Strauss RG: Routine washing of irradiated red cells before transfusion seems unwarranted. Transfusion.1990; 30:675-677

[26] Luban NLC, Strauss RG, Hume HA: Commentary on the safety of red cells preserved in extended-storage media for neonatal transfusion. Transfusion.1991; 31:229-235

[27] Benson K, Marks AR, Marshall MJ, et al: Fatal graft versus-host disease associated with transfusions of HLAmatched,HLA-homozygous platelets from unrelated donors. Transfusion. 1994; 34:432-437

[28] Grishaber JE, Birney SM, Strauss RG: Potential for transfusion-associated graft-versus-host disease due to apheresis platelets matched for HLA class`, I antigens. Transfusion.1993; 33:910-914

[29] Wielding JU, Vehmeyer K, Dittman J, et at: Contamination of fresh-frozen plasma with viable white cells and proliferable stem cells. Transfusion .1994;34:185-186

[30] Bernvill SS, Abdulatiff M, Al-Sedairy S, et at: Fresh frozen plasma contains viable progenitor cells-should we irradiate.Vox Sang .1994;67:405

[31] Davey RJ, McCoy NC, Yu M, et al: The effect of pre-storage irradiation on post-transfusion red cell survival. Transfusion .1992;32:525-528

[32] Mintz PD, Anderson G: Effect of gamma irradiation on the in vivo recovery of stored red blood cells. Ann Clin Lab Sci .1993;23:216-220

[33] Moroff G, Holme S, Heaton A, et al: Effect of gamma irradiationon viability of AS-I red cells. Transfusion .1992;32(suppl):70S(abstr)

[34] Friedman KD, McDonough WC, Cimino DF: The effect of pre-storage gamma irradiation on post-transfusion red blood cell recovery. Transfusion.1991; 31:50S(abstr)

[35] Moroff G, Holme S, AuBuchon J, et al: Storage of red cells and platelets following gamma irradiation. Vox Sang.1994; 67:42, 1994 (Abstr, suppl 2)

[36] Moroff G, George VM, Siegl AM, et al: The influence of irradiation on stored platelets. Transfusion.1996;26:453-456

[37] Espersen GT, Ernst E, Christiansen OB, et at: Irradiated blood platelet concentrates stored for five days--evaluation by in vitro tests. Vox Sang .1988;55:218-221

[38] Duguid JKM, Cart R, Jenkins JA, et al: Clinical evaluation of the effects of storage time and irradiation on transfused platelets. Vox Sang.1991; 60:151 - 154

[39] Read EJ, Kodis C, Carter CS, et at: Viability of platelets following storage in the irradiated state. A paired-controlled study. Transfusion.1988;28:446-450

[40] Rock G, Adams GA, Labow RS: The effects of irradiation on platelet function. Transfusion .1988;28:451-455

[41] Sweeney JD, Holme S, Moroff G: Storage of apheresis platelets after gamma irradiation. Transfusion .1994;34:779-783

[42] Seghatchian MJ, Stivala JFA: Effect of 25 Gy gamma irradiation on storage stability of three types of platelet concentrates: a comparative analysis with paired controls and random preparation. Transfus Sci .1995;16:121-129

[43] Bessos H, Atkinson A, Murphy WG, et at: A comparison of in vitro storage markers between gamma-irradiated and non-irradiated apheresis platelet concentrates. Transfus Sci.1995; 16:131-134

[44] Anderson KC, Goodnough LT, Sayers M, et at: Variation in blood component irradiation practice: Implications for prevention of transfusion-associated graft-versus-host disease. Blood .1991;77:2096-2102

[45] Sprent J, Anderson RE, Miller JF: Radiosensitivity of T and B lymphocytes. II Effect of irradiation on response of T cells to alloantigens. Eur J Immuuol .1974;4:204-210,

[46] Valerius NH, Johansen KS, Nielsen OS, et al: Effect of invitro x-irradiation on lymphocyte and granulocyte function.Scand J Hematol .1981;27:9-18

[47] Pelszynski MM, Moroff G, Luban NLC, et at: Effect of irradiation of red blood cell units on T-cell inactivation as assessed by limiting dilution analysis: implications for preventingtransfusion-associated graft-versus-host disease. Blood.1994; 83:1683-1689

[48] Luban NLC, Drothler D, Moroff G, et at: The effect of irradiation on lymphocyte reactivity in platelctpheresis components assessed by limiting dilution analysis. Transfusion.1994; 34:66S(abstr)

[49] Rosen NR, Weidner JG, Bold HD, et al: Prevention of transfusion-associated graft-versus-host disease: selection of an adequate dose of gamma irradiation. Transfusion.1993; 33:125-127

[50] Center for Biologics Evaluation and Research, Food and Drug Administration: Recommendations regarding license amendments and procedures for gamma irradiation of bloodproducts.
http://www.fda.gov/downloads/BiologicsBloodVaccines/GuidanceComplianceR egulatoryInformation/OtherRecommendationsforManufacturers/Memorandumto BloodEstablishments/UCM062815.pdf

[51] Anderson G: Quality assurance of the irradiation process of blood components, in Baldwin JL, Jefferies LC (eds): Irradiation of Blood Components, Bethesda MD, American Association of Blood Banks, 1992, pp 63-75

[52] Masterson ME, Febo R: Pretransfusion blood irradiation Clinical rationale and dosimetric considerations. Med Phys.1992; 19:649-457

[53] Leitman SF: Dose, dosimetry and quality improvements of irradiated blood components. Transfusion.1993; 33:447-449

[54] Perkins JT, Papoulias SA: The effect of loading conditions on dose distribution within a blood irradiator. Transfusion .1994;34:75S(abstr)

[55] Moroff G, Luban NLC, Wolf L, et al: Dosimetry measurements after gamma irradiation with cesium-137 and linear acceleration sources. Transfusion.1993; 33:52S (abstr)

[56] Luban NLC, Fearon T, Leitman SF, et al: Absorption of gamma irradiation in simulated blood components using cesium irradiators. Transfusion.1995; 35:63S(abstr)

[57] Kutcher GJ, Coia L, Gillin M et al.: Comprehensive QA for radiation oncology: report of AAPM radiation therapy committee task group 40. Med Phys.1994; 21:581-618

[58] Nath R, Biggs PJ, Bova FJ, et al: AAPM code of practice for radiotherapy accelerators: report of AAPM radiation therapy task group no 45. Med Phys.1994; 21:1093-1121

[59] Leitman SF, Silberstein L, Fairman RM, et al: Use of a radiation-sensitive film label in the quality control of irradiated blood components. Transfusion.1992;32:4S (abstr)

Part 2

Examples for Different Quality Control Processes

9

Procedures for Evaluation of Slice Thickness in Medical Imaging Systems

Giuseppe Vermiglio, Giuseppe Acri, Barbara Testagrossa,
Federica Causa and Maria Giulia Tripepi
Environmental, Health, Social and Industrial Department –
University of Messina
Italy

1. Introduction

The main goal of a medical imaging system is to produce images to provide more accurate and timely diagnoses (Torfeh et al., 2007). In particular, Computed Tomography (CT), Magnetic Resonance Imaging (MRI), and Ultrasound (US) are resourceful tools in medical practice, and in many cases a life saving resource when rapid decisions are needed in the emergency room (Rehani et al., 2000). The above diagnostic techniques are based on the evaluation of high resolution images from technologically sophisticated equipment. Individual images are obtained by using several different electronic components and considerable amounts of data processing, which affect the quality of images produced and, consequently, make the diagnostic process more complicated (Torfeh et al., 2007).

In this context, to guarantee a consistent image quality over the lifetime of the diagnostic radiology equipment and to ensure safe and accurate operation of the process as a whole, it is necessary to establish and actively maintain regular and adequate Quality Assurance (QA) procedures. This is significant for computer-aided imaging systems, such as CT, MRI and US. The QA procedure should include periodic tests to ensure accurate target and critical structure localization (Mutic et al., 2003). Such tests are referred to as Quality Controls (QCs). They hold a key role within the QA procedure because they enable complete evaluation of system status and image quality (Chen et al., 2004; Vermiglio et al., 2006).

Importantly, QCs permit the identification of image quality degradation before it affects patient scans, and of the source of possible equipment malfunction, pointing to preventive or immediate maintenance requirements. Thus, image QC is crucial to ensure a safe and efficient diagnosis and treatment of diseases (Rampado et al., 2006; Torfeh et al., 2007). For this reason, periodic QCs have been recommended by manufacturers and medical physicists' organizations to test the performance of medical imaging systems. Protocols for QCs and QA in medical imaging systems have been produced by several professional groups (AAPM – NEMA) (Goodsitt et al., 1998). This highlights the extensive role of QA programs, including QC testing, preventive maintenance, etc. (Rampado et al., 2006).

In any clinical imaging study, it is important to have accurate confirmation of several physical characteristics of the medical imaging device. In particular, the slice thickness accuracy represents an important parameter that should be estimated during QC procedures, not only because the signal to noise ratio varies linearly with the slice thickness, but also because clinical image resolution is strongly affected by partial volume effects, thus reducing clinical image quality with increasing slice thickness (Narayan et al., 2005). In addition, to determine the FWHM the AAPM procedure involves the evaluation of a line profile of the slice. So, during QC procedures many and different available test objects are used to assess different physical characteristics of the medical imaging device, including slice thickness, spatial resolution, dark noise, uniformity, etc. AAPM Reports No 1 and No 28 state that the slice thickness can be evaluated from the measure of the full width at half maximum (FWHM) of the response across the slice (Judy et al., 1977; Price et al., 1990). In particular, for a high-accuracy measurement of slice thickness, several test objects inserted in multipurpose phantoms can be used, most of which utilize inclined surfaces (plane, cone or spiral). A typical test object for the slice thickness evaluation is the crossed high signal ramps oriented at a fixed angle (Price et al., 1990).

Whereas in US equipment the slice thickness is typically not measured (Skolnick, 1991), most CT and MRI scanners adopt specific and automated procedures that require the use of dedicated phantoms, coupled with a dedicated imaging software that, however, does not always include line profile tools.

Standard slice thickness accuracy evaluation methods consist of scan explorations of phantoms that contain different specific patterns. These methods are based on manual scans with graphics tools or, alternatively, on automatic scans utilizing encoded masks to determine the Region Of Interest (ROI) for quantization (Torfeh et al., 2007). Therefore, a variety of different phantoms presently exists, but each requiring a specific QC protocol.

Further, even the newest medical imaging software do not allow a direct measurement of the slice thickness accuracy with CT and MRI scanners, but require a complicated procedure to be performed by specialized technicians authorized to enter in the SERVICE menu of medical devices.

To reduce complications and provide a versatile and unique QC procedure to estimate slice thickness accuracy, a novel dedicated phantom and associated procedure is proposed here that is easy to implement and that can be used on both CT and MRI scanners.

Such phantom can be used either with already existing dedicated software or with to this aim dedicated LabView-based tools, to readily measure the slice thickness in real time and/or post-processing operation. The reliability of the innovative technique proposed here has been evaluated with respect to previously validated procedures by conducting statistical analysis, as discussed in detail in the following sections.

Further, this novel technique is suitable also for the evaluation of the elevation resolution in US scanners, and easier to perform than standard techniques. This chapter is structured as follows: a review of the materials and methods commonly used in CT, MRI and US imaging systems, and the novel and versatile methodology proposed here for slice thickness measurements are presented in Section 2. The results obtained from the application of the proposed novel methodology to the three different imaging techniques are discussed in Section 3. The conclusions are drawn in Section 4.

2. Materials and methods

In this section the commonly used methods for determining the slice thickness accuracy in CT, MRI and US scanners are presented. In addition, a novel procedure that uses a dedicated phantom and the following image elaboration by employing LabView based software is also proposed.

2.1 CT scanners

The slice thickness is evaluated by measuring the width (FWHM) of the image of one or more Aluminium ramps at the intersection of the ramp(s) with the scan plane (CEI, 1998), as illustrated schematically in Fig. 1. The sensitivity profile measures the system response to an attenuating impulse as a function of the z-axis position, through the slice plane. The sensitivity profile is a function of pre- and post-patient collimation, and appears as a blurred square wave. The FWHM of this blurred square is defined as the nominal slice width. The slice thickness of a CT scanner is determined by focal spot geometry as well as pre-patient and detector collimation and alignment.

An customary method of monitoring equipment performance is to measure the parameters of interest using test objects. For example, Goodenough et al. proposed an approximate measure of beam width by using a series of small beads positioned across the beam width. However, this phantom is difficult to use for a precise quantitative measure because of the uncertainty on the alignment of the beads with respect to the beam. Indeed, the beam width can be measured directly from a beam profile plot only if care is taken to ensure that the Aluminium piece is oriented at 45 degrees across the width of beam (Judy et al., 1977, as cited in Goodenough et al., 1977).

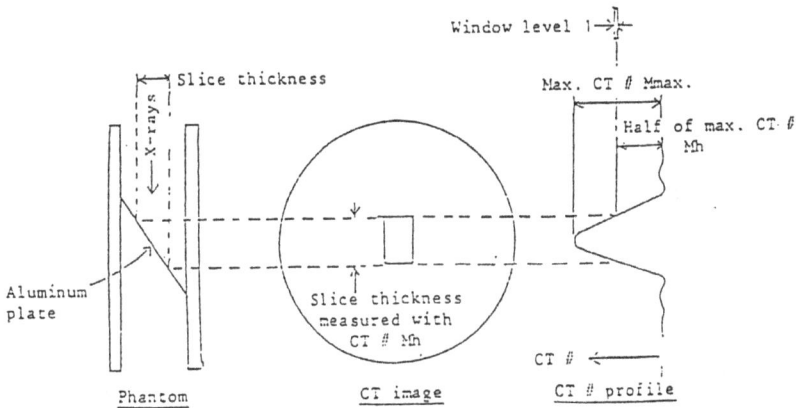

Fig. 1. Principle of slice thickness measurement (Philips, 1997).

A typical performance phantom (Philips, 1997) uses Aluminium plates slanted 26.565 degrees and which are across each other, as shown in Fig. 2. With X-rays irradiated to the Aluminium plates, the axial length of each Aluminium can be measured in CT image. With this method, it is possible to obtain an accurate measurement even if the intersection of the Aluminium plates is not aligned with the X-ray beam, by averaging the two measurements L_a and L_b. The slice thickness L is calculated as follows:

$$L = \left(\frac{L_a + L_b}{2}\right) \cdot \tan 26.565° = \frac{L_a + L_b}{4} \tag{1}$$

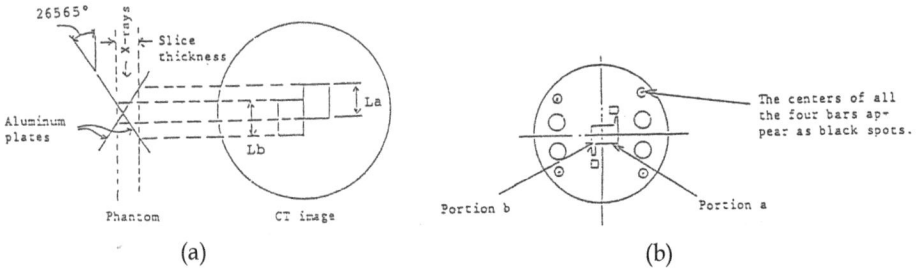

Fig. 2. CT image of the section dedicated to the measurements of slice thickness of a typical performance phantom: (a) layout of the measurement methodology; (b) schematic of resulting image with the centres of the four bars appearing as black spots (Philips, 1997).

Another performance phantom used for CT scanners (Fig. 3) is a poli methyl methacrilate (PMMA) box presenting a pattern of air filled holes drilled 1 mm apart and aligned in the direction of the slice thickness (perpendicular to the scan plane). Each visible hole in the image represents 1 mm of beam thickness (General Electric [GE], 2000).

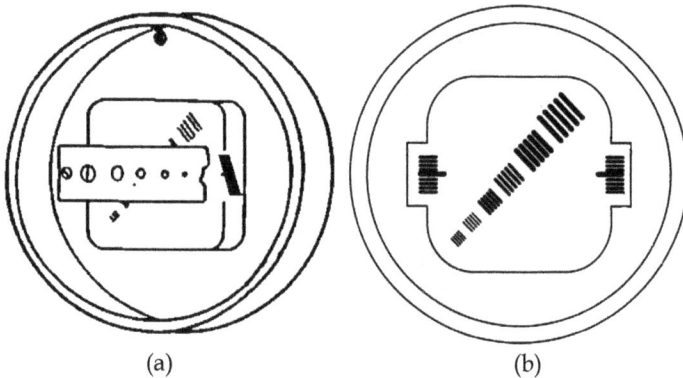

Fig. 3. Phantom with air-filled holes for slice thickness measurement in CT scanners: (a) schematic of the phantom, (b) schematic of the calibration image (General Electric [GE], 2000).

To determine the slice thickness, the image is displayed at the recommended window level and width. The number of visible holes (representing air-filled holes) is counted. Holes that appear black in the image represent a full millimetre slice thickness. Holes that appear grey count as fractions of a millimetre; two equally grey holes count as a single 1 mm slice thickness.

2.2 MRI scanners

The technique of MRI differs from X-ray CT in many ways, but one of the most interesting is perhaps that the slice is not determined primarily by the geometry of the scanning apparatus but rather by electronic factors, namely the spectrum of radio frequency pulse and the nature of the slice selection gradient (Mc Robbie et al., 1986). The slice profile and width of a 2D imaging technique such as MRI is a very important feature of its performance.

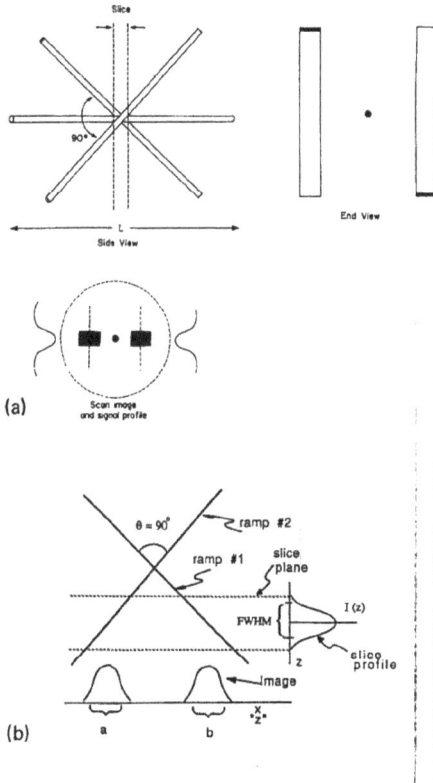

Fig. 4. High signal ramp phantoms: (a) A typical slice-thickness phantom consisting of a set of crossed thin ramps. A ramp crossing angle of 90° yields an angle of 45° between the ramp and the image plane. The phantom length (L) should be greater than twice the maximum slice thickness. An alignment rod placed between the two ramps defines the point where the two ramps cross. When the slice is properly aligned through the intersection of the ramps the images of the ramps and of the rod will be aligned. (b) The slice sensitivity profile is directly proportional to the image intensity profiles if the image plane is perpendicular to the alignment rod (Price et al., 1990).

Since the signal obtained is directly proportional to the thickness of the slice, an inaccurate slice width can lead to a reduced Signal-to-Noise Ratio (SNR) (Lerski, 1992). Partial volume effects can significantly alter sensitivity and specificity. Quantitative measurements such as relaxation time T_1 and T_2 values, are also greatly influenced by slice thickness. Inaccuracies in slice thickness may result in inter-slice interference during multi-slice acquisitions, and invalid SNR measurements (Och et al., 1992). The slice profile, ideally rectangular, may contain side lobes which can produce very confusing effects (Lerski, 1992). In addition, gradient field nonuniformity, radio frequency field nonuniformity, nonuniform static magnetic field, noncoplanar slice selection pulses between excitation and readout, T_R/T_1 ratio (where T_R represents the repetition time), and radio frequency pulse shape and stimulated echoes can also affect the slice thickness accuracy (Price et al., 1990).

A variety of phantoms have been designed to evaluate slice thickness. All are some variation of an inclined surface. These may include wedges, ramps, spirals, or steps. A typically used phantom is the crossed high signal ramps.

High signal ramp (HSR) phantoms generally consist of opposing ramp pairs oriented at a fixed angle θ (Fig. 4). The HSR's should be thin (ideally infinitesimally thin) to quantify the slice profile accurately. In general, the thickness of a (90°) HSR oriented at 45° respect to the image plane should be < 20% of the slice profile FWHM (i.e., for 5-mm slice it is necessary to use a 1-mm ramp) to obtain a measurement with < 20% error.

The FWHM is the width of the slice profile (SP) at one-half of the maximum value. In this case, the SP should be obtained for each ramp. The FWHM then becomes

$$L = FWHM = \frac{(a+b)\cos\theta + \sqrt{(a+b)^2 \cos^2\theta + 4ab\sin^2\theta}}{2\sin\theta} \tag{2}$$

where a and b refer to the FWHM of the intensity profiles measured for ramp 1 and ramp 2, respectively. Note that for θ=90° then Eq. 2 is simplified to:

$$L = FWHM = \sqrt{ab} \tag{3}$$

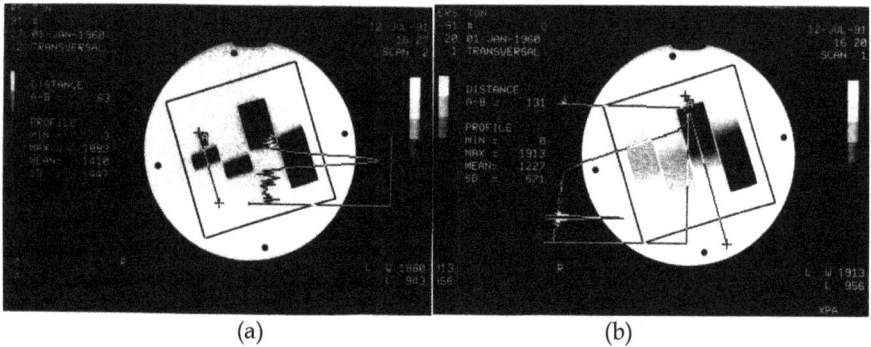

(a)   (b)

Fig. 5. Example of CT scanner image obtained from the assessment of the EUROSPIN phantom (Lerski, 1992).

The EUROSPIN test phantom contains two sets of structures that may be used for slice profile and width measurement. Pairs of angled plates are used to obtain a direct measurement. The additional pairs of wedges are used to calibrate especially thin slices. Typical examples of images obtained for slice width are presented in Fig. 5(a and b). The dark bands on the left hand side of Fig. 5a represent a projection of the slice profile from the angled plates; the shaded region on the right hand side of the Fig. 5b represents the projection of the profile at the wedge.

2.3 US scanners

Ultrasound image resolution depends on beam width in the scan and elevation (section thickness) planes (Skolnick, 1991). On US scanners slice thickness evaluation or elevational resolution is useful to understand some of the problems due to partial volume effect. Section thickness is significantly more complicated to check. This characteristic of the ultrasound beam depends on the focusing effect in the elevation direction, which is perpendicular to the scanning plane (Richard, 1999). In linear, curved linear and phased array sector probes, focus is controlled electronically, but in the elevation plane it is determined mechanically by the curvature of the crystals. The beam in the scan plane can be sharply focused only in a narrow focal range. Thus, beam profiles in the scan plane are not indicative of beam profiles in the elevation plane. As with lateral and axial resolution, elevational resolution can be measured indirectly with anechoic spherical objects or cylindrical plug phantoms. Slice thickness focusing can also be evaluated qualitatively by scanning the anechoic cylindrical objects in an ultrasound QC test phantom with the scan plane along the lengths of the cylinders (e.g., perpendicular to the usual scan direction). Quantitative assessment can be achieved by using an "inclined plane" phantom (Fig. 6) (Goodsitt et al., 1998).

The methodology used with the inclined plane phantom consists in obtaining the elevation beam profile, finding the depth where the image is narrowest. By focusing on that plane, the thickness of the image is measured at the focal plane.

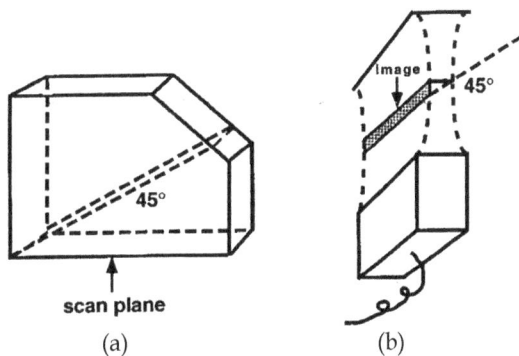

Fig. 6. Phantom used for beam width measurement in the elevation plane: (a) schematic of the phantom; (b) schematic of the procedure to obtain the image at the beam waist (Goodsitt et al., 1998).

This technique enables measurement of the elevation dimension of the beam only at a single depth. To determine the entire profile in the elevation plane, the probe must be moved horizontally along the surface of the phantom to make a series of measurements, with the beam intersecting the inclined plane at different depths (Skolnick, 1991).

Slice thickness focal range, thickness and depth should be recorded on the US unit for each commonly used transducer. Any significant variation from those reference values may indicate a detachment of the focusing lens.

2.4 The novel procedure

A novel and versatile methodology is proposed here to determine the slice thickness accuracy using a novel phantom. The methodology can be applied to any image scanning technique, including CT, MRI and US scanning. The methodology consists of two steps: 1) acquisition of images of the phantom; 2) image elaboration by using the dedicated LabView-based software. To test the proposed procedure and to obtain detailed information about the quality of the obtained results, the acquired and processed images were compared with those obtained by elaborating the same phantom images using commercial software following already validated procedures (Testagrossa et al., 2006).

The novel proposed dedicated phantom consists of a poli-methyl-methacrilate (PMMA) empty box (14.0 cm x 7.5 cm x 7.0 cm) diagonally cut by a septum at 26 degrees (Fig. 7). The PMMA septum is 2.0 mm thick and it divides the box into two sections, thus reproducing both single and double wedges. The two sections can be filled with the same fluid or with fluids of different densities. In particular, to determine the slice thickness accuracy the PMMA box was filled with two different fluids (water and air) for assessment in a CT scanner. To perform the same assessment with an MRI scanner, water was replaced with a $CuSO_4$ $5H_2O+H_2SO_4+1ml/l$ antialga (ARQUAD) liquid solution (T_1=300 ms, T_2=280ms). For US systems the upper wedge was filled with ultrasound gel, as conductive medium.

In addition, a spirit level is used to verify the planarity of the phantom with respect to the beam and the patient couch.

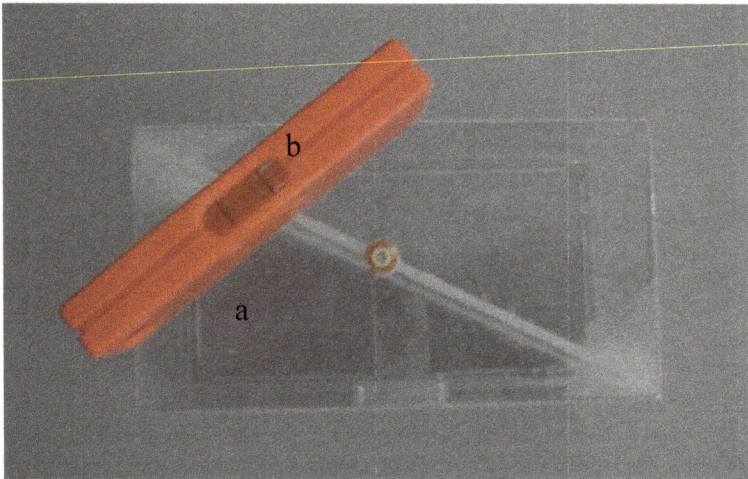

Fig. 7. The novel proposed PMMA phantom for the evaluation of the slice thickness accuracy: (a) proposed for the evaluation of the slice thickness accuracy. (b) the spirit level used to verify the planarity of the phantom with respect to the beam and the patient couch is also shown for completeness.

The test procedure followed for both CT and MRI devices consists of four steps:
1. Placing the slice thickness accuracy phantom in the scanner head holder.
2. Adjusting level position of the phantom if necessary.
3. Moving/positioning the phantom in the gantry aperture.
4. Scanning the phantom with a single slice using the desired slice width available

The phantom images were acquired using standard Head and Body protocols, shown in Table 1 and Table 2 for CT and MRI medical devices, respectively. After, the phantom images were acquired, elaborated and analyzed, they were stored and/or transmitted to a printer. The stored ones were further transferred to a dedicated workstation, whereas the printed ones were acquired by the same workstation using a VIDAR Scanner, for the next elaboration.

Scan Parameters	Head Protocol	Body Protocol
kV	120	120
mA	100	45
Scan Time (s)	3	1
Field of View (mm)	230	360
Reconstruction Matrix	512	512
Filter	None	None

Table 1. Standard Protocol for testing CT medical devices.

Scan Parameters	Head Protocol	Body Protocol
Coil type	Head	Body
Scan mode	SE	SE
Scan technique	MS	MS
Slice orientation	Transversal	Transversal
Numbers of echoes	2	3
Field of View (mm)	250	250
Repetition Time (ms)	1000	1000
Scan matrix	256	256
Water fat shift	1.3	Maximum

Table 2. Standard Protocol for testing MRI medical devices.

The methodology developed here for data elaboration, utilizes a purposely developed LabView-based slice thickness measurement software. LabView is a graphical programming language that uses icons instead of lines of text to create applications. In contrast to text-based programming languages, where instructions determine program execution, LabView uses data flow programming, where the flow of data determines execution (National Instruments [NI], 2003). The software is compatible with both non-standard and standard image formats (BMP, TIFF, JPEG, JPEG2000, PNG, and AIPD) (Vermiglio et al., 2008). To evaluate the slice width the FWHM of the wedge, expressed in pixels, is measured and calibrated with respect to the effective length of the PMMA box, expressed in mm. The result is displayed in real-time at the user interface, known as the software Front Panel (Fig. 8), with the advantage of providing a complete set of data with a user-friendly interface.

By plotting the radiation profile obtained as a system response to an attenuating impulse, as a function of position (z axis), that is through the slice plane, it is possible to estimate the slice thickness accuracy of the acquired image utilising the developed software. This is referred to as the sensitivity profile. Gaussian smoothing is applied to smooth out the sensitivity profile and permit a clearer estimate of the desired FWHM. To evaluate the slice thickness , in real time, the software utilises the following equation:

$$ST = FWHM \cdot \tan(26°) \tag{4}$$

To test the proposed procedure, results are compared with those obtained by elaborating the same phantom images using commercial software, in particular Image-Pro Plus software from Media Cybernetics (Sansotta et al., 2002; Testagrossa et al., 2006).

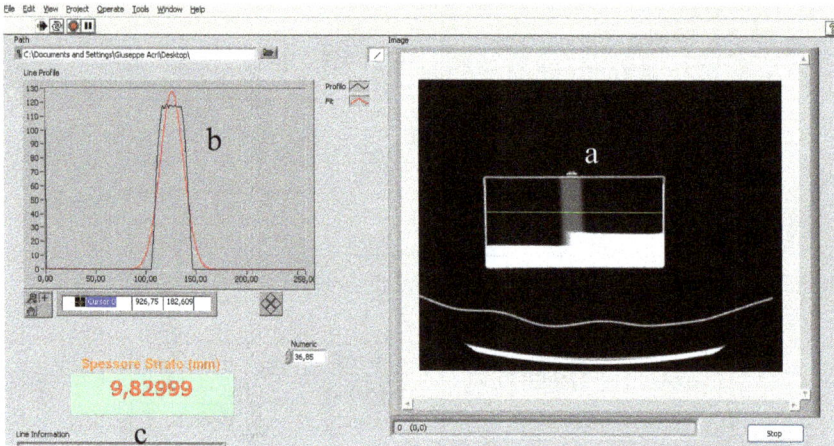

Fig. 8. Front Panel of the dedicated slice thickness LabView software showing (a) X-ray phantom section, (b) the detected line profile and corresponding Gaussian fit, and (c) the resulting slice thickness value. All steps are performed in real time.

The measurements presented here have been conducted on several CT and MRI devices in an extended study from 2006 to 2010. In addition, a statistical analysis was conducted on the resulting datasets to further validate the proposed methodology. The chosen statistical method is the variance analysis, through Fisher's exact test (F-test), to assess if a significant difference exists between datasets obtained following different procedures (C.A. Markowski & E.P. Markowski, 1990). The F-test is useful when the aim of the study is the evaluation of the precision of a measurements technique. In fact, the variance analysis consists in the factorisation of the total variance into a set of partial variances corresponding to different and estimated variations. The statistical analysis was conducted both on CT and MRI datasets.

For CT scanners three different datasets of slice thickness measurements were considered. The first dataset of 16 measurements was done on a reference (RF) value of 10 mm The second dataset of 14 measurements, on a RF value of 5 mm. The third dataset of 10 measurements, on a RF value of 2 mm. The data was obtained using two different procedures.

For MRI scanners slice thickness measurements were done on a RF value of 10 mm. In this case three different procedures were compared and 24 measurements in total were obtained on different MRI systems.

3. Results

The slice thickness measurements results using the novel proposed methodology and the statistical data analysis are presented in this section for CT, MRI and US systems.

3.1 CT medical devices

A two-dimensional image of the wedge of the dedicated phantom of Fig. 7 was acquired using the Standard Head Protocol described in Table 1. The X-Ray image of the phantom is presented in Fig. 9. In this case the line profile tool was available and the related trend was shown in the same figure.

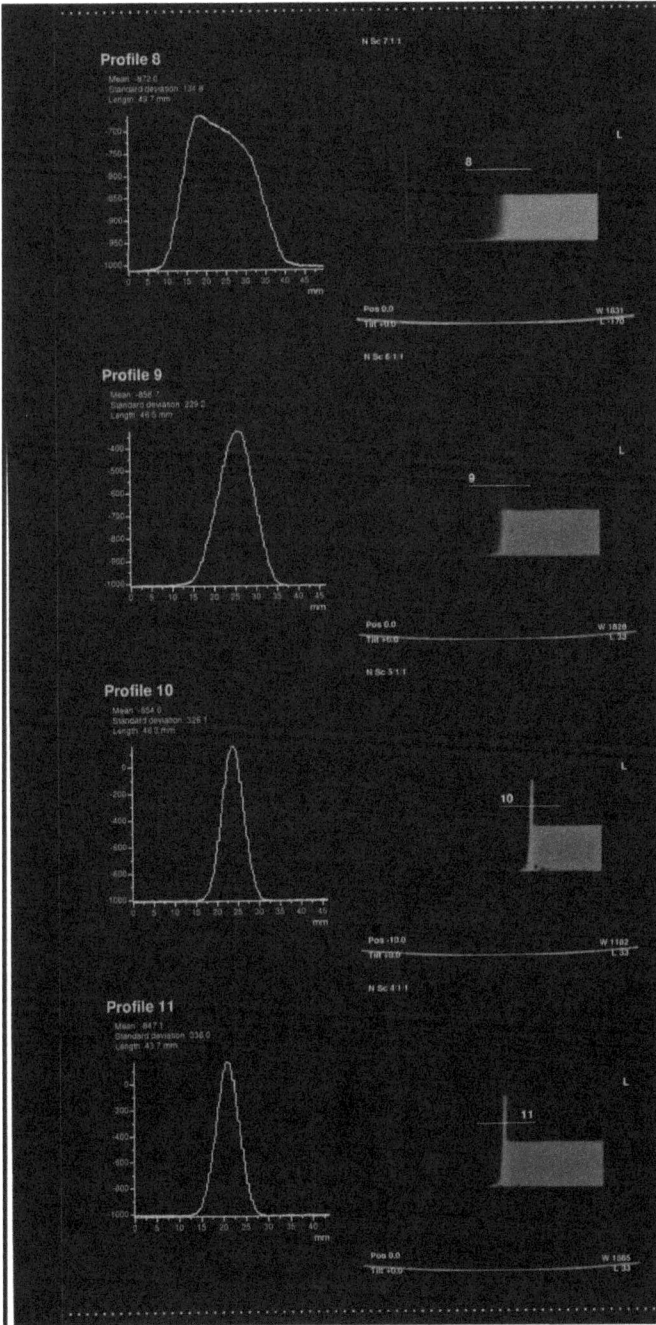

Fig. 9. CT scanner in-built software results displaying the line profiles obtained for different readings of the X-ray image of the dedicated phantom.

The results obtained by measuring the slice thickness accuracy with different CT scanners and by employing the in-house developed LabView (LV) program and the commercial Image Pro Plus (IPP) software are compared with the corresponding 10 mm RF value in Table 3. In the same table mean values and standard deviations are also reported for both procedures.

RF Value (mm)	LV (mm)	LV mean value and standard deviation (mm)	IPP (mm)	IPP mean value and standard deviation (mm)
10	8.61	9.69±0.68	8.07	9.64±1.26
	9.30		10.33	
	9.44		9.44	
	10.80		10.04	
	9.50		8.62	
	10.18		12.00	
	10.26		8.54	
	9.45		10.10	

Table 3. Slice thickness accuracy results obtained with LV and IPP for a 10 mm RF value, using different types of CT scanners.

The data of Table 3 are presented in graphical form in Fig. 10. The slice thickness accuracy results (blue: IPP; red: LV) and deviation from the RF value (blue: IPP; red: LV) obtained using the IPP and LV procedures are presented for the 10 mm RF value.

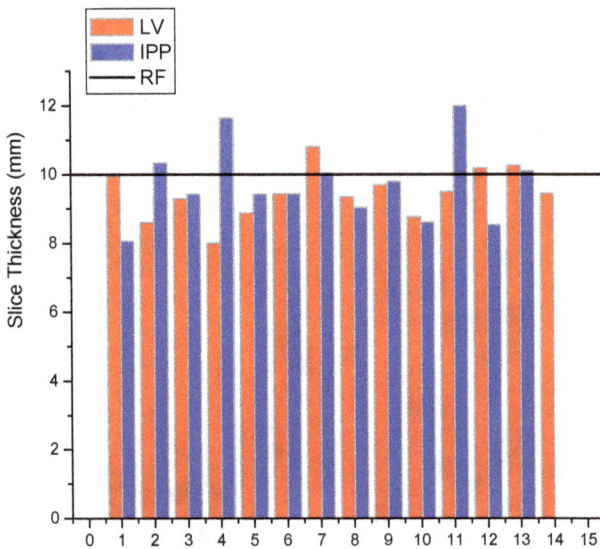

Fig. 10. (a)

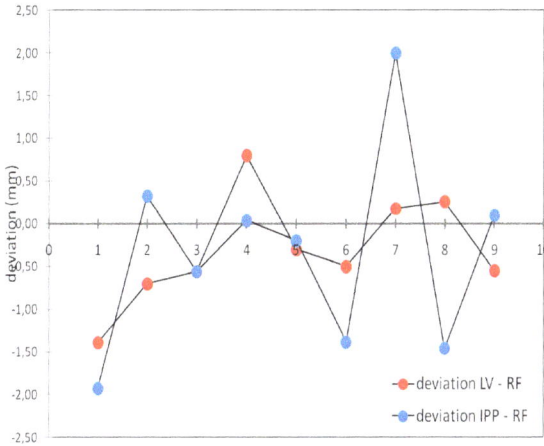

(b)

Fig. 10. CT slice thickness accuracy obtained for a 10mm RF value: (a) data set (blue: IPP; red: LV); (b) deviation from the RF value (blue: IPP; red: LV).

From the results presented in Table 3, it is observed that the mean values calculated for both the IPP and LV procedures, are comparable. However, the standard deviation obtained from the two different procedures is considerably different, with the LV procedure providing a narrower deviation, and hence more accurate results of the performed measurements.

In Table 4, the slice thickness accuracy results obtained with IPP and LV are compared with the corresponding 5 mm RF value. Also in this case, for the sake of completeness, the respective mean values and standard deviations obtained from the IPP and LV datasets are reported.

RF Values (mm)	LV (mm)	LV mean value and standard deviation (mm)	IPP (mm)	IPP mean value and standard deviation (mm)
5	4.50	4.73±0.25	4.03	4.83±0.41
	4.75		4.82	
	4.80		4.83	
	4.36		5.36	
	4.89		4.85	
	4.71		4.80	
	5.13		5.13	

Table 4. Slice thickness accuracy results obtained with LV and IPP for a 5mm RF value, using different types of CT scanners.

Also for the 5 mm RF value case, the mean values obtained from LV and IPP datasets are comparable, with a slightly better estimate for the IPP procedure. However, the standard deviations obtained from the two procedures are significantly different. As in the previous case, the LV procedure provides a narrower deviation, thus enabling a more accurate measurement.

The data of Table 4 are reported in graphical form in Fig 11, where the slice thickness accuracy results obtained with the two procedures (blue: IPP; red: LV) and their deviation from the 5 mm RF value are presented.

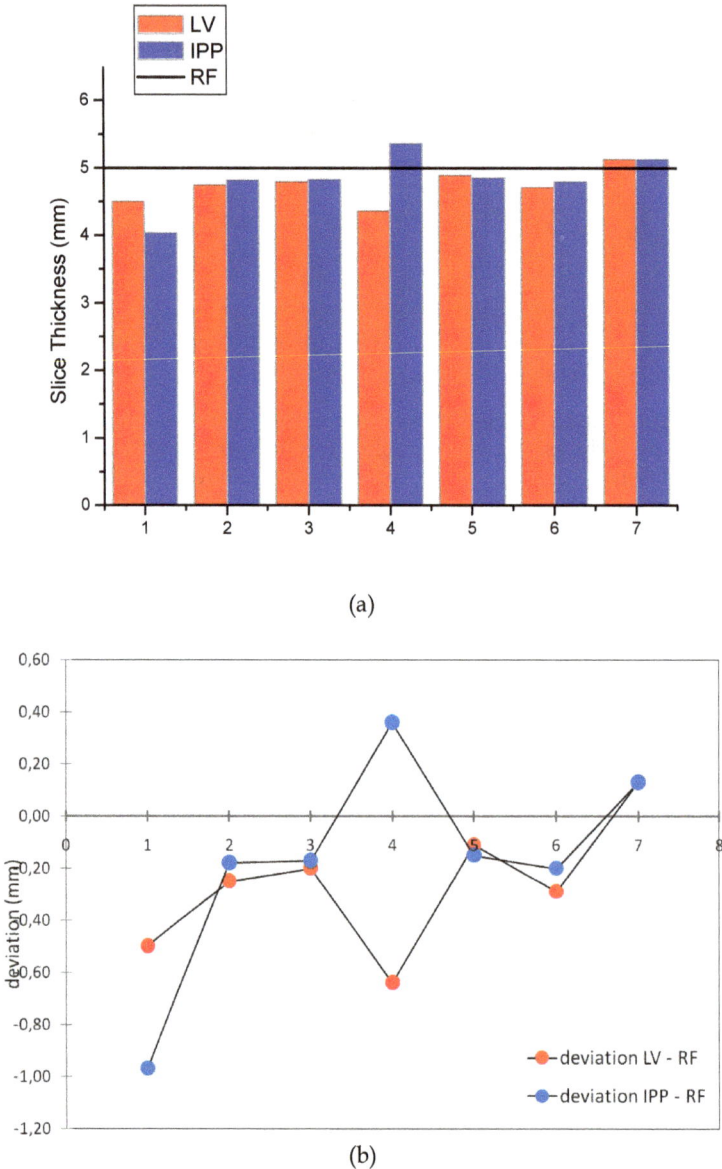

(a)

(b)

Fig. 11. CT slice thickness accuracy obtained for a 5 mm RF value: (a) sample set (blue: IPP; red: LV); (b) deviation from the RF value (blue: IPP; red: LV).

Finally, in Table 5, the results obtained by measuring the slice thickness accuracy by employing the IPP and the LV procedures are compared with the corresponding 2 mm RF value and mean values and standard deviation are also indicated. From the analysis of the data of Table 5, it can be observed that the mean value calculated by the LV dataset is significantly closer to the RF value than that calculated from the IPP dataset. This further supports the validity of the proposed technique.

RF Values (mm)	LV (mm)	LV mean value and standard deviation (mm)	IPP (mm)	IPP mean value and standard deviation (mm)
2	2.66	2.32±0.48	2.79	2.79±0.16
	3.00		2.75	
	2.06		2.83	
	1.96		2.56	
	1.95		3.00	

Table 5. Slice thickness accuracy results obtained with LV and IPP for a 2 mm RF value, using different types of CT scanners.

The data of Table 5 are represented in graphical form in Fig 12, where the slice thickness accuracy results obtained using IPP and LV (blue: IPP; red: LV) and their deviation from the 2 mm RF value (blue: IPP; red: LV) are presented.

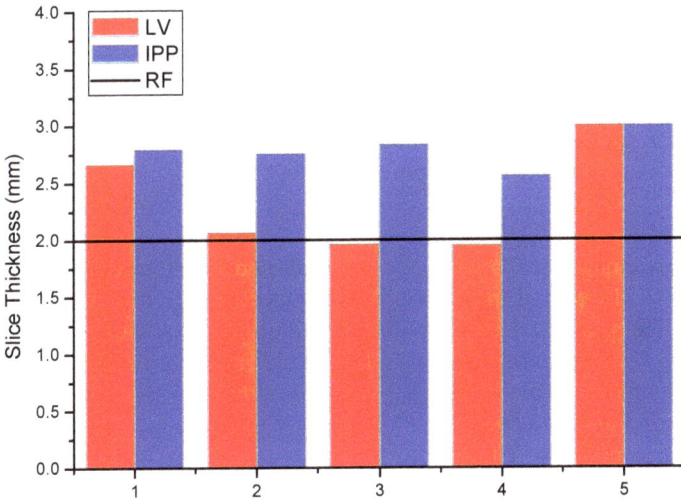

Fig. 12. (a)

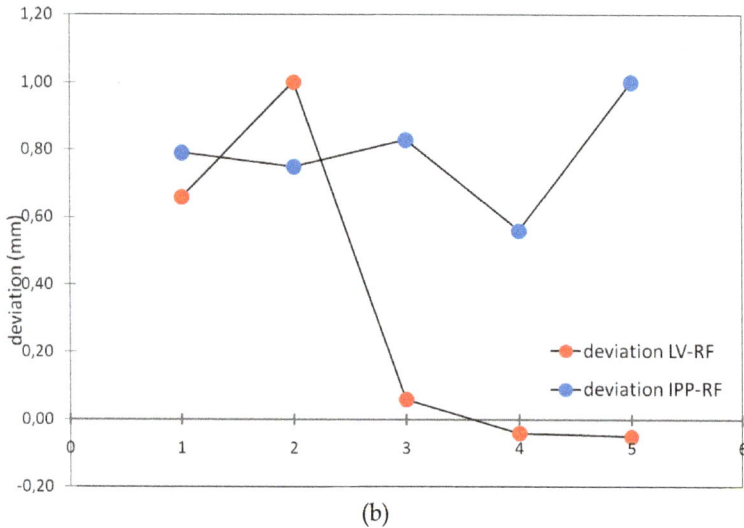

(b)

Fig. 12. CT slice thickness accuracy obtained for a 2 mm RF value: (a) sample set (blue: IPP; red: LV); (b) deviation from the RF value (blue: IPP; red: LV).

Statistical analysis conducted on the datasets shown in Tables 3-5 yielded the F-values reported in Table 6. Such F-values indicate that there is no significant statistical variation between the two different procedures, thus validating the methodology.

RF Values (mm)	F_C	F_T
10	0.011	4.54
5	0.332	4.67
2	4.18	5.12

Table 6. F values calculated for different RF values (F_C). These F-values were compared to the tabulated ones (F_T) for the P=0.05 confidence level.

3.2 MRI medical devices

Two-dimensional images of the wedge of the dedicated phantom were acquired using the Scan Head Protocol reported in Table 2. A typical MRI image of the PMMA box and the corresponding line profile calculated with the in-built MRI software are shown in Fig. 13. The preliminary results obtained using the IPP and LV procedures discussed above applied to MRI are reported in Table 7 for a RF value of 10 mm. In this case the slice thickness accuracy measured using an in-built MRI software (CS) is also included. The corresponding mean value and standard deviation of the three datasets are also reported.

Also in this case, the mean values calculated from the three different datasets of measurements are comparable between them. However, whereas the standard deviations obtained from CS and LV procedures are comparable, those obtained from IPP and LV are significantly different. The LV procedure provides a narrower deviation with respect to that obtained with IPP, which gives evidence for a more accurate measurement.

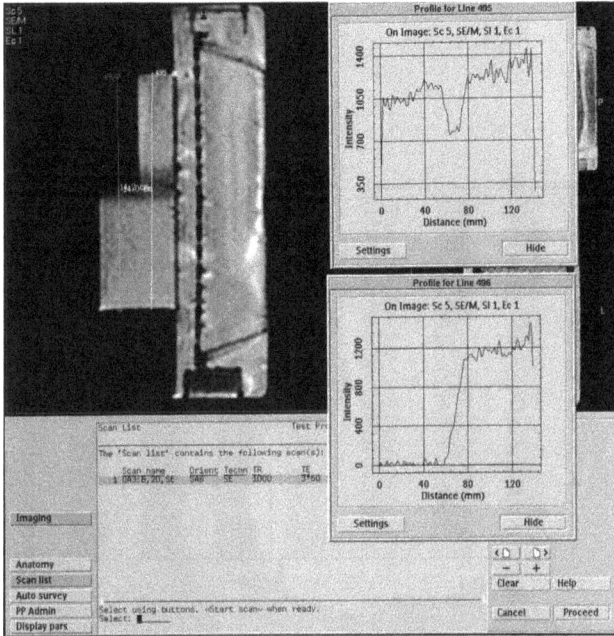

Fig. 13. MRI PMMA box image and corresponding line profile appearing on the MRI display. It is possible to notice the related line profile as obtained directly at the equipment console.

The slice thickness accuracy data of Table 7 are presented in graphical form in Fig. 14(a), where the slice thickness accuracy determined directly at the equipment console and that measured using the IPP and the LV procedures are compared. In Fig. 14(b) deviations from the RF value are presented for the three sets of data.

RF value (mm)	CS (mm)	CS mean value and standard deviation (mm)	LV (mm)	LV mean value and standard deviation (mm)	IPP (mm)	IPP mean value and standard deviation (mm)
10	10.1	10.05±0.17	10.3	10.05±0.15	9.80	9.96±0.31
	9.90		9.92		9.85	
	10.3		10.2		9.80	
	9.80		10.0		10.5	
	10.1		9.94		10.1	
	10.2		10.17		10.3	
	9.90		9.95		9.60	
	10.1		9.90		9.90	

Table 7. Comparison between the slice thickness accuracy obtained from the MRI in-built software (CS), the Image Pro Plus (IPP) procedure and the dedicated LabView (LV) software for a 10.0 mm Reference (RF) value. In the same table mean values and corresponding standard deviations for the three different procedures are also reported.

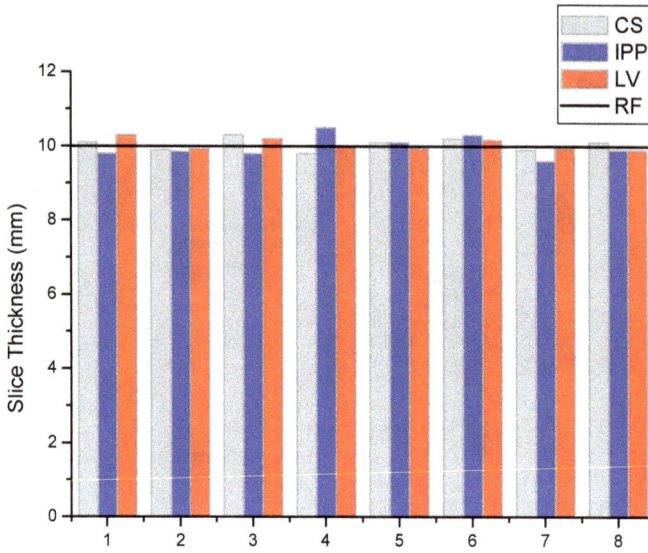

(a)

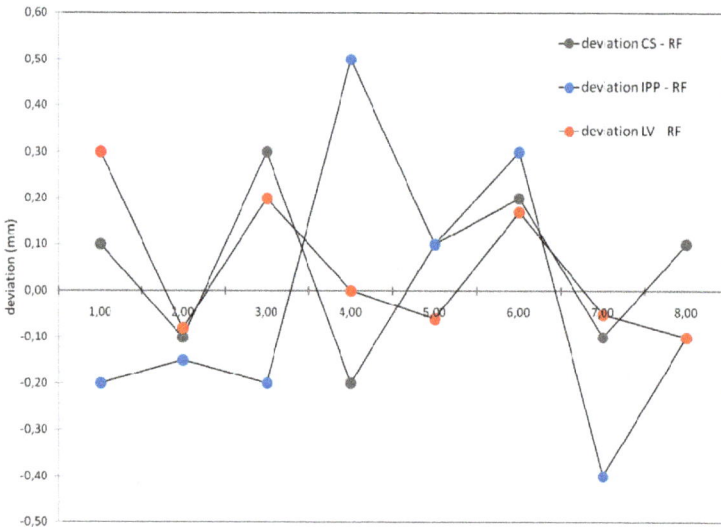

(b)

Fig. 14. MRI slice thickness accuracy obtained for a 10 mm RF value: (a) data obtained directly from the MRI scanner (CS) and measured using the IPP and the LV procedures (grey: CS; blue: IPP; red: LV); (b) deviation from the RF value (grey dots: CS; blue dots: IPP; red dots: LV).

The F-test is then used to verify that the three datasets are comparable (Table 8). As evident from the data of Table 8, the F values calculated for the three sets of data are significantly smaller than that tabulated (F_T) for a P=0.05 confidence level in all cases. Therefore, there is no significant statistical difference between the three different procedures, thus validating the novel LV procedure proposed here also for MRI imaging.

Datasets	F_C	F_T
CS-IPP	0.836	
CS-LV	0.00165	4.54
IPP-LV	0.513	

Table 8. Results of the F-test calculated comparing the CS, IPP and LV datasets (F_C). These F-values were compared to the tabulated ones (F_T) for the P=0.05 confidence level.

3.3 US scanners

The novel PMMA phantom proposed here (Fig. 7) was utilised also to test the elevation resolution in ultrasound systems. Preliminary measurements have been conducted on US probes to evaluate if it is possible to measure the beam width in the elevation plane. To correctly determine the elevation beam profile, it was necessary to slightly modify the phantom. In particular, the septum position was modified and, for the beam width evaluation, the inclined plane was oriented 28 degrees to the top and bottom surface, so the probe intersects the inclined plane at 28 degrees. The upper side of the inclined surface was filled with ultrasound gel, as conductive medium. The echoes reflected from the inclined plane are displayed as a horizontal band. Because the beam intersects the plane at 28 degrees, the beam width in the elevation plane can be calculated as follows:

$$ST = FWHM \cdot \tan(28°) \qquad (5)$$

The above technique enables the measurement of the elevation dimension of the beam at a single depth only. To determine the entire profile in the elevation plane, the probe must be moved horizontally along the surface of the phantom to make a series of measurements, with the beam intersecting the inclined plane at different depths. With the use of the novel phantom described here, the resolution in the elevation plane is completely independent from the lateral resolution in the scanning plane. The new phantom proposed here is easy to use and does not require any additional equipment and the results are immediately displayed on the screen of the dedicated PC.

4. Conclusion

Slice thickness represents an important parameter to be monitored in CT, MRI an US. Partial volume effects can significantly alter sensitivity and specificity. Especially for MRI, quantitative measurements such as T_1 and T_2, are also greatly influenced by the accuracy of the slice thickness. Inaccuracies in the measurement of this parameter may result in inter-slice interference in multi-slice acquisitions, leading to invalid SNR measurements. In addition, for the US scanners, significant differences in image resolution can occur because of variations in the beam width of the elevation plane. So, it is important to know the profiles of the elevation planes of various probes to choose the probe that offers resolution in the elevation plane that is most appropriate for the particular clinical application. Moreover, slice thickness accuracy is an important element also in the QC program for CT scanner, because quantitative CT analysis is heavily dependent on accuracy of slice thickness. Therefore, to provide an adaptable, reliable and robust procedure to evaluate the slice thickness accuracy a novel, dedicated phantom and corresponding LabView-based procedure have been proposed here, up to date applied to both CT and MRI devices. The new PMMA box proposed here to be associated with the dedicated software enables an innovative, accurate, easily applicable and automated determination of this parameter.

The carried on studies have utilised the 2 mm septum because the z axis dimensions for rotating anode x-ray tube foci are typically less than this value, but is our intention reduce the septum width, in order to evaluate the slice thickness also for the helicoidal CT scanners, that can reach slice widths smaller than 2 mm.

The accuracy (standard deviation) obtained with the novel procedure proposed here is significantly higher than that obtained with other procedures (e.g., in-built software, IPP).

This new slice thickness accuracy procedure is proposed as an alternative to the commonly adopted ones, which are typically complicated by the use of ad-hoc software and phantoms distributed by manufacturers and specific to the medical equipment. The proposed method employs a novel universal phantom, coupled with a dedicated LabView-based software that can be used on any CT and MRI scanner in a quick, simple and reproducible manner.

The readiness and applicability of the proposed procedure has been validated by quantitative tests using several different medical devices and procedures. In all cases the results obtained using the novel proposed procedure were statistically compatible with other commonly used procedures, but provided a very immediate determination of the slice thickness for both CT and MRI equipments, thus confirming the flexibility of the described method, its simplicity and reproducibility as an efficient tool for quick inspections.

The same procedure should be suitable also for determining elevation accuracy on US scanner: in fact, preliminary results confirmed that the novel phantom, opportunely modified, when coupled with the LabView dedicated software, allowed measurements of the section thickness

5. References

CEI EN 61223-2-6. (1997). Evaluation and routine testing in medical imaging departments. Part 2-6: Constancy tests – X-ray equipments for computed tomography. CEI, (Ed.), pp. 1-26

Chen, C.C.; Wan, Y.L.; Wai, Y.Y. & Liu, H.L. (2004). Quality assurance of clinical MRI scanners using ACR MRI phantom: preliminary results. *Journal of Digital Imaging*, Vol. 17, No. 4 (December 2004), pp. 279-284, ISSN 0897-1889

General Electric Medical System. (2000). Quality Assurance. In: *CT HiSpeed DX/i Operator manual Rev. 0*. General Electric Company (Ed.). Chapter 6, pp. 1-28

Goodsitt, M.M.; Carson, P.L.; Witt, S.; Hykes, D.L. & Kofler, J.M. (1998). Real-time B-mode ultrasound quality control test procedures. Report of the AAPM ultrasound task group No. 1. *Medical Physics*, Vol. 25, No. 8 (August 1998), pp. 1385-1406, ISSN 0094-2405

Judy, P.F.; Balter, S.; Bassano, D.; McCollough, E.C.; Payne, J.T. & Rothenberg, L. (1977). Phantoms for performance evaluation and quality assurance of CT scanners.

AAPM Report No. 1 (American Association of Physicist in Medicine, Chicago, Illinois, 1977)

Lerski R.A. & Mc Robbie D.W. (1992). EUROSPIN II. Magnetic resonance quality assessment test objects. Instruction for use. Diagnostic Sonar LTD (January 1992), pp. 1-79

Markowski, C.A. & Markowski, E.P. (1990). Conditions for the effectiveness of a preliminary test of variance. *The American Statistician,* Vol. 44, No. 4 (November 1990), pp. 322-326, ISSN 0003-1305

Mc Robbie, D.W.; Lerski, R.A.; Straughan, K.; Quilter, P. & Orr, J.S. (1986). Investigation of slice characteristics in nuclear magnetic resonance imaging. *Physics in Medicine and Biology,* Vol. 31, No. 6 (June 1986), pp. 613-626, ISSN 0031-9155

Mutic, S.; Palta, J.R.; Butker, E.K.; Das, I.J.; Huq, M.S.; Loo, L.N.D.; Salter, B.J.; McCollough, C.H. & Van Dyk, J. (2003). Quality assurance for computed-tomography simulators and the computed tomography simulation process. Report of the AAPM radiation therapy committee task group No. 66. *Medical Physics,* Vol. 30, No. 10 (October 2003), pp. 2762-2792, ISSN 0094-2405

Narayan, P.; Suri, S.; Choudhary, S.R. & Kalra, N. (2005). Evaluation of Slice Thickness and Inter Slice Distance in MR scanning using designed test tool. *Indian Journal of Radiology & Imaging,* Vol. 15, No. 1 (February 2005), pp.103-106, ISSN 0971-3026

National Instruments. (2001). LabView User Manual. National Instruments Corporation (Ed.). Available from
http://www.ni.com/pdf/manuals/320999d.pdf

Och, J.G.; Clarke, G.D.; Sobol, W.T.; Rosen, C.W. & Ki Mun, S. (1992). Acceptance testing of magnetic resonance imaging systems: Report of AAPM Nuclear Magnetic Resonance Task Group No.6. *Medical Physics,* Vol. 19, No. 1 (January-February 1992), pp. 217-229, ISSN 0094-2405

Philips. (1997). Performance Phantom C instruction manual. In: Tomoscan CX/Q technical documents. pp. 1-18

Price, R.R.; Axel, L.; Morgan, T.; Newman, R.; Perman, W.; Schneiders, N.; Selikson, M.; Wood, M. & Thomas S.R. (1990). Quality assurance methods and phantoms for Magnetic Resonance Imaging. Report of the AAPM Nuclear Magnetic Resonance Task Group No. 28. *Medical Physics,* Vol. 17, No. 2 (March-April 1990), pp. 287-295, ISSN 0094-2405

Rampado, O.; Isoardi, P. & Ropolo, R. (2006). Quantitative assessment of computed radiography quality control parameters. *Physics in Medicine and Biology,* Vol. 51, No. 6 (March 2006), pp. 1577-1593, ISSN 0031-9155

Rehani, M.M.; Bongartz, G.; Golding, S.J.; Gordon, L.; Kalender, W.; Albrecht, R.; Wei, K.; Murakami, T. & Shrimpton, P. (2000). Managing patient dose in computed tomography. *Annals of ICRP,* Vol. 30, No. 4 (December 2000), pp. 7-45, ISSN 0146-6453

Richard, B. (1999). Test object for measurement of section thickness at US. *Radiology,* Vol. 211, No. 1 (April 1999), pp. 279-282, ISSN 0033-8419

Sansotta, C.; Testagrossa, B.; de Leonardis, R.; Tripepi, M.G. & Vermiglio, G. (2002). Remote image quality and validation on radiographic films. *Proceedings of the 7th Internet World Congress for Biomedical Sciences, INABIS 2002,* Available from
http://www.informedicajournal.org/a1n2/files/papers_inabis/sansotta1.pdf

Skolnick, M.L. (1991). Estimation of Ultrasound beam width in the elevation (section thickness) plane. *Radiology*, Vol. 180, No. 1 (July 1991), pp. 286-288, ISSN 0033-8419

Torfeh, T.; Beaumont, S.; Guédon, J.P.; Normand, N. & Denis, E. (2007). Software tools dedicated for an automatic analysis of the CT scanner Quality Control's Images, In: *Medical Imaging 2007: Physics of Medical Imaging. Proceedings of SPIE*, Vol. 6510, J. Hsien & M.J. Flynn (Eds.), 65104G, ISBN 978-081-9466-28-0, San Diego, California, USA, March 6, 2007

Testagrossa, B.; Novario, R.; Sansotta, C.; Tripepi, M.G.; Acri, G. & Vermiglio, G. (2006). Fantocci multiuso per i controlli di qualità in diagnostica per immagini. *Proceedings of the XXXIIIth International Radio Protection Association (IRPA) Conference*, IRPA, ISBN 88-88648-05-4, Turin, Italy, september 20-23, 2006

Vermiglio, G.; Testagrossa, B.; Sansotta, C. & Tripepi, M.G. (2006). Radiation protection of patients and quality controls in teleradiology. *Proceedings of the 2nd European Congress on Radiation Protection «Radiation protection: from knowledge to action»*. Paris, France, May 15-19, 2006, Available from
http://www.colloquium.fr/06IRPA/CDROM/docs/P-121.pdf

Vermiglio, G.; Tripepi, M.G.; Testagrossa, B.; Acri, G.; Campanella, F. & Bramanti, P. (2008). LabView employment to determine dB/dt in Magnetic Resonance quality controls. *Proceedings of the 2nd NIDays*, pp. 223-224. Rome, Italy, February 27, 2008

10

Nursing Business Modeling with UML: From Time and Motion Study to Business Modeling

Sachiko Shimizu et al.[1]
Osaka University
Japan

1. Introduction

A nurse is an autonomous, decentralized worker who recognizes goals, his or her environment, the conditions and actions of patients and other staff members, and determines his or her own actions. Put another way, the nurse makes decisions flexibly in the midst of uncertainty. Because of this, nursing work differs from individual nurse to nurse, and understanding this process theoretically is considered to be difficult.

Concerning nursing work analysis, research has been done on task load (time required for tasks). However, there has been scant academic research on work processes in nursing compared with research that has accumulated in other industrial fields, including research on structuralizing work, i.e., defining and visualizing work processes. To improve work processes, it is necessary to understand and clarify work as a chain of theoretically related activities.

Thus in this study, using time and motion study techniques, a method used to measure jobs, we clarify the structure of the work of transporting patients by nurses. We also attempt to visualize it. We use objected-oriented modeling to express the operation visually.

2. From time and motion study to business modeling

Time and motion study is a method that actually measures the movements of a particular person. Its results can be applied not only to measuring the work load of nurses (Van de Werf et al., 2009; Were et al., 2008;Hendrich et al.,2008) and analyzing the workflow(Tang et al., 2007), they can also be used as basic data for task scheduling(Yokouchi et al., 2005) and efficient arrangement of personnel. In addition, the results are being used as indicators to evaluate changes in a hospital brought about by systems deployed (Yen et al., 2009), such as an electronic medical record (EMR) system. Thus many time and motion studies of hospitals have been conducted both within Japan and without.

Specifically, a time and motion study is defined as a study that records the time of occurrences of tasks through continuous observation. A type of measuring technique similar

[1]Rie Tomizawa, Maya Iwasa, Satoko Kasahara, Tamami Suzuki, Fumiko Wako, Ichiroh Kanaya, Kazuo Kawasaki, Atsue Ishii, Kenji Yamada and Yuko Ohno
Osaka University, Japan

to the time study is work sampling, which seeks to comprehend a job by sampling its conditions at predetermined time intervals. Work sampling cannot comprehend a job in its entirety, but it lessens the burden on the measurer. It also makes it possible for the worker himself or herself to record time. In contrast, a time and motion study comprehends the job in its entirety, but the burden on the measurer is great. The differences in results between the two methods have been observed to be large for jobs in which there were few events(Finkler et al., 1993). Currently, the results that come from measuring a job through continuous time and motion observation are said to be the gold standard.

While the breadth of research that utilize measurement results from time and motion studies encompasses all nursing work, individual studies have been limited to examining the amount of work for individual caring assignments, such as cleaning a patient, feeding a patient, and taking care of a patient's toilet needs. There have been especially few studies that evaluate the work amount of a job by focusing on the job and clarifying its work process. While not on concerned with nursing work, the only such study conducted so far in the medical field was visualizing and understanding the amount of work involved in the process of registering cancer patients by Shiki et al. (Shiki et al., 2009). They proposed the method of "time-process study," a method to visualize tasks by adding time information to the process. However, because both the process and amount of work were estimated through interviews, the results can be said to be lacking in objectivity. Thus our study uses the time and motion study method, which actually measures a task. We focus on the job of transporting patients and clarifying its process. We also study the possibility of a method to visualize the work process using the clarified process and time information.

Transporting patients is an operation that is often performed outside hospital wards. It is both physically and mentally demanding of nurses. This job should also be scrutinized because it reduces the number of nursing staff inside the wards, as nurses go outside the wards in order to safely transport patients.

3. Methods

3.1 Study setting

We carried out a time and motion study of nursing work related to transporting patients in four hospital wards of a cardiovascular treatment facility. We tracked our subjects, who were nurses, nursing assistants, and medical clerks, from the time of the start of a task until its end, and recorded the task actions. The record of a task action included the content of the action, the time of its start and end, the person who was the target of the action, and the location of the action. The four wards of the treatment facility consisted of the cardiac failure ward, arrhythmia ward, cardiomyopathy/pulmonary hypertension ward, and cerebral vascular and metabolism ward. The destinations of patient transport included exam rooms for CT, X-ray, MRI, echocardiography, respiratory function testing, cardiac rehabilitation, neurological rehabilitation, cardiac catheterization investigation, and dialysis.

3.2 Business modeling with UML

From the time and motion study records we obtained, we created a use case diagram and activity diagram. Use case diagrams and activity diagrams are types of diagrams created using Unified Modeling Language (UML). UML is the de facto standard objected-oriented modeling language, and was developed for software development. In recent years,

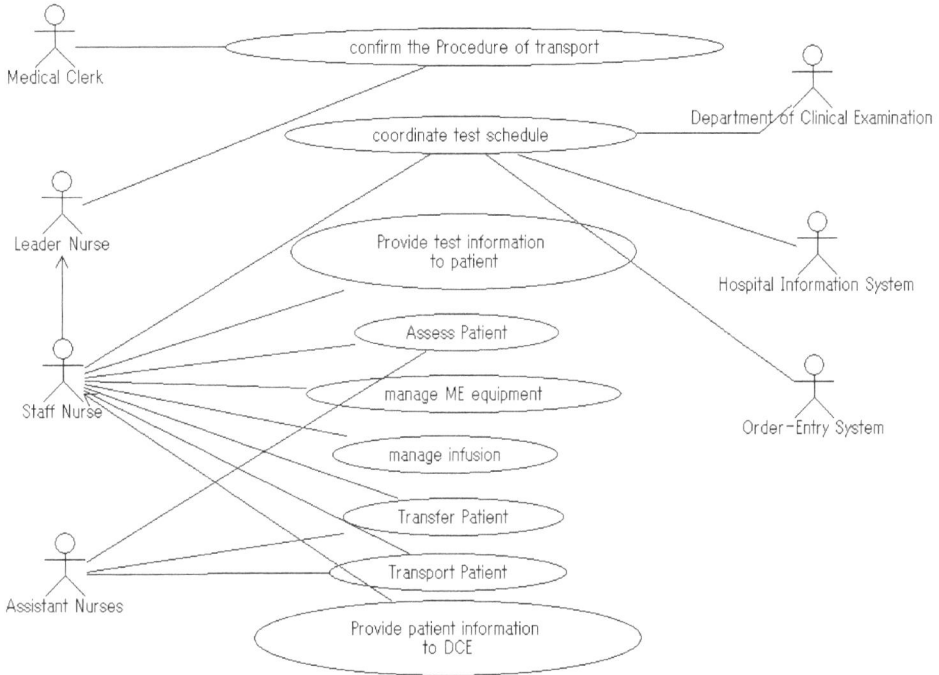

Fig. 1. Functional diagram-patient transports system.

however, its use for business modeling has been suggested (Eriksson and Penker, 2000). The reason is that the structure of a job can considered oriented-oriented in nature. The content of a job can be treated as exchanges of messages between objects, such as materials and users. Thus UML as a descriptive method can allow one to intuitively understand the job. In this study, we elucidated the functional aspect of the operation of transporting patients. We also used an activity diagram to visualize the work process of transporting patients. Finally, we discussed the work load and its time efficiency by adding time information to the activity diagram. This study was approved by the ethics committee of the hospital we studied.

4. Results

From the time and motion study, we observed and recorded 213 jobs of transferring patients. Overall, the number patient transfer assignments recorded was 3,775. Of these records, 387 records were not jobs related to transporting patients, so they were removed from our analysis.

A use case diagram extracted from the results of the time and motion study is shown in Figure 1. There were seven types of actors involving in transporting patients: nurses, head nurses, medical clerks, nursing assistants, the central medical examination department, the order entry system, and the hospital information system. The nurses were divided into two groups: head nurses, who had the responsibility of being in charge of nursing duties, and staff nurses, who received patients and provided care for them. The head nurse and the medical clerk received communication about the transport of a patient, and confirmed the predetermined method of transport care. In addition, the head nurse made adjustments such as changing the transport personnel and finding appropriate personnel. Of the tasks related to transport care, the nurse and nursing assistant handled tasks that had direct bearing on the patient. In the hospital of this study, patients undergoing oxygen therapy, patients being monitored by EKG, and patients undergoing transfusion were the responsibility of nurses, not nursing assistants.

Task	TOT	Frequency	Median	Range
T01 Coordinate time for examination	0:33:27	28	58	(5-273)
T02 Confirm schedule of examination	0:05:24	10	29	(4-100)
T03 Accept call for examination	0:31:30	45	34	(1-324)
T04 Look for patient record	0:04:32	11	18	(2-64)
T05 Check bed rest level	0:09:11	10	36	(6-186)
T06 Identify care-giver	0:00:58	3	21	(4-32)
T07 Prepare map	0:08:27	20	23	(3-70)
T08 Prepare patient consultation card	0:14:37	31	18	(1-108)
T09 Prepare patient record	0:28:41	42	31	(5-187)
T10 Find care-giver	0:01:59	3	42	(16-60)
T11 Find patient	0:07:33	11	17	(4-116)
T12 Wait for care-giver	0:00:21	1	21	(21-21)
T13 Relay examination information to patient	0:29:55	43	34	(1-144)
T14 Hand necessary materials to patient	0:00:21	3	6	(2-13)
T15 Change care-giver assignment	0:00:37	1	37	(36-36)
T16 Relay exam information to nurse	0:26:48	38	21	(1-384)
T17 Prepare film	0:00:44	2	22	(15-29)
T18 Prepare materials to be brought	0:04:02	3	38	(6-198)
T19 Prepare transport care equipment	0:22:38	46	20	(1-139)
T20 Carry transport care equipment	0:21:27	40	26	(1-88)
T21 Assess situation	0:24:48	17	26	(2-382)
T22 Confirm patient name	0:02:45	10	16	(6-30)
T23 Prepare to move ME devices	0:13:50	19	31	(7-237)
T24 Prepare to move medical supplies	0:16:43	23	42	(2-117)
T25 Assist in excretion	0:05:16	5	52	(10-152)
T26 Assist in changing of clothes	0:12:35	19	25	(10-127)
T27 Prepare for transfer	0:10:22	13	29	(5-199)
T28 Carry patient	1:46:59	83	43	(3-707)
T29 Transport patient	9:15:49	109	292	(1-866)
T30 Go through reception procedures	0:08:56	34	9	(1-90)
T31 Hand-over patient	0:01:55	8	13	(2-34)
T32 Hand-over necessary supplies	0:10:31	30	15	(1-89)
T33 Relay information	0:33:09	31	63	(3-156)
T34 Prepare for examination	0:27:16	26	32	(1-370)
T35 Assist in examination	0:42:01	41	28	(6-255)
T36 Standby at destination	1:57:19	35	92	(1-1612)
T37 Receive patient	0:06:37	7	20	(6-208)

Task	TOT	Frequency	Median	Range
T38 Reattach ME devices	0:41:25	18	82	(6-766)
T39 Reattach medical supplies	0:21:23	14	69	(2-396)
T40 Secure consultation card	0:04:35	23	9	(1-44)
T41 Secure patient record	0:23:02	30	19	(1-560)
T42 Clear away film	0:00:28	4	5	(3-16)
T43 Clear away transport care equipment	0:25:52	40	34	(2-115)
T44 Clear away map	0:01:54	5	11	(1-78)
T45 Finish clean up	0:13:24	15	33	(1-159)
T46 Record the transfer	0:11:10	11	32	(3-247)
M Move	4:36:03	119	95	(2-1068)

TOT: time on task.

Table 1. Identified tasks and their descriptive statistics.

The dynamic aspect of transporting patients is shown as an activity diagram (see Figure 2). The head nurse, who is in charge of communication in the hospital ward, and the medical clerk receive a call for a patient from the central medical examination department. They confirm the bed rest level of the patient from his or her chart. If the patient can walk outside the ward by himself or herself (self-reliant), the person in charge of communication prepares the chart, the patient's exam ticket, and the map to the exam room. He or she searches for the patient, relays the call for examination to the patient, and hands over necessary items. If the bed rest level is escort (transport in a wheelchair) or litter care (transport on a stretcher), the person in charge of communication searches for the transport personnel and hands over the exam call. The transport personnel prepare the patient's chart, the exam ticket, and the instrument for transport care such as a wheelchair or stretcher, and move to the patient's location. They relay the exam call to the patient, and assess the patient's conditions to determine if transport is possible. If the transport personnel determine that the patient can be transported, he/she/they prepare oxygen or transfusion devices for transport, and perform excrement care and assist the patient in changing clothes. Next, the transport personnel move the patient from the bed to the transport instrument, and transport the patient to the exam room. After the patient arrives in the examination room, the transport personnel notify the exam receptionist of the patient's arrival, hand over the patient, and hand over items brought along, such as the patient chart and the exam ticket. If the exam takes only a short time, e.g. in the case of an x-ray exam, the transport personnel wait in the exam room, assist with preparing the patient for examination, and assists in the examination. If the exam takes a longer period of time, the transport personnel return to the hospital ward and perform other tasks. When communication comes from the examination room, the transport personnel receive the message and move to the exam room. After the exam has completed, the transport personnel receive the call from the patient, transfer the patient to the transport instrument, transport him or her back to the ward, and again move him or her to the hospital bed. The transport personnel prepare medical electronic equipment and medical devices attached to the patient so that subsistence in bed is possible. After assessing the patient's conditions, the transport personnel puts away the items brought along, such as the exam ticket and the patient chart, and record the transport. As shown in the activity diagram, we clarified that the process of transporting a patient was composed of 47 tasks.

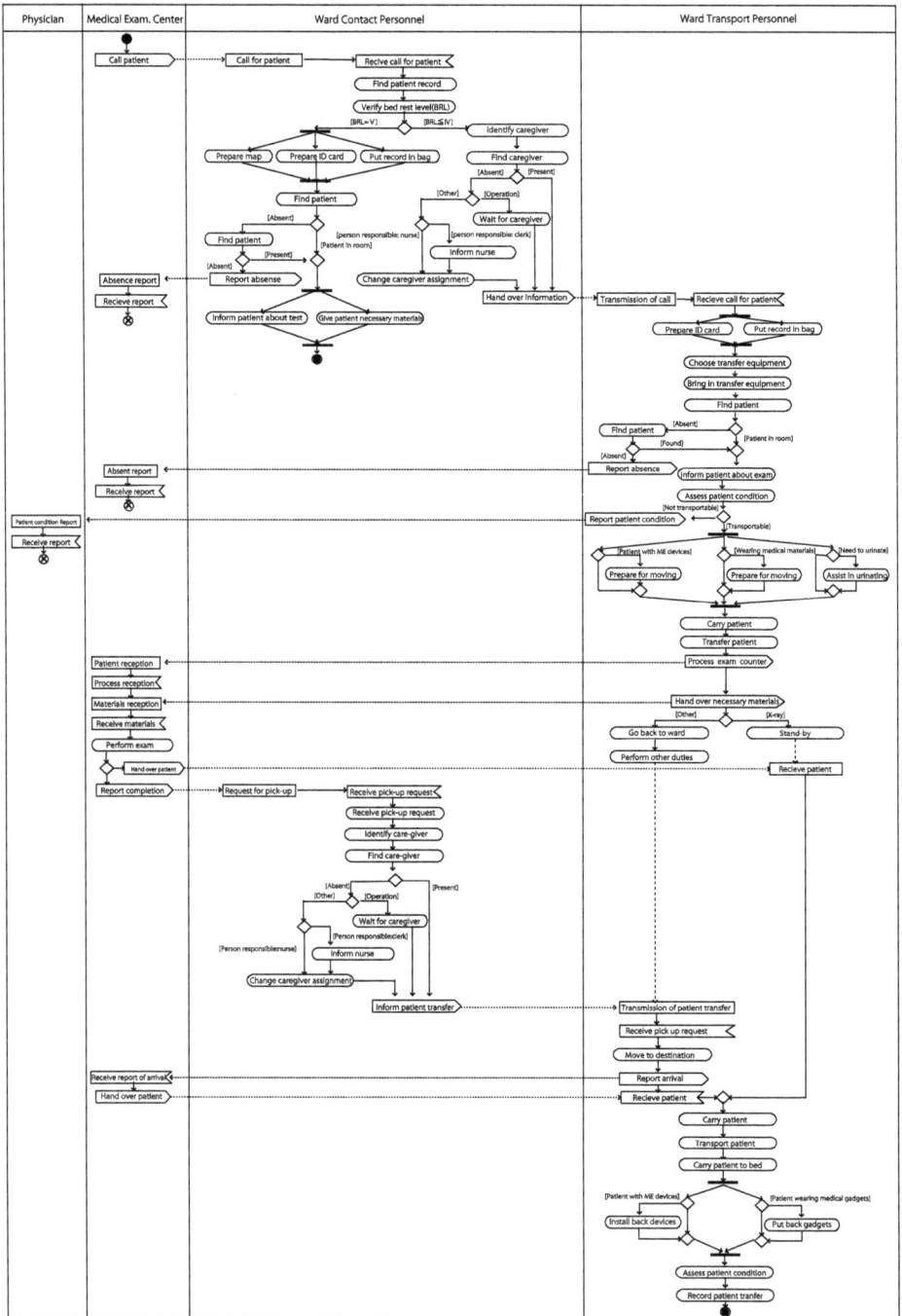

Fig. 2. Dynamic diagram-patient transports.

Table 1 shows the total time on task during a day in the four wards for each of the 47 tasks shown in the activity diagram. Also shown are the number of occurrences of each task, the median value, and the range. The task that took up the most total time was "T29 Transporting patient" (9:15:49). It took about 5 minutes on average for the nurse(s) to transport a patient. Of the 213 patient transport jobs observed, 109 actually involved transporting the patient. Patient transport jobs that did not involve transport were only those to support self-reliant patients and to adjust the scheduled time of exams. After T29, the task that took the most time was "T36 Standing by at the destination" (1:57:19), followed by "T28 Transferring the patient" (1:46:59). On the other hand, there were few occurrences of tasks related to searching for or changing transport personnel, such as "T06 Identifying care provider," "T12 Waiting for care provider," and "T15 Changing care provider." Comparing the coefficient of variance, we found that the coefficient of variance for "T41 Putting patient chart away," "T16 Conveying exam information to nurse," "T36 Standing by at destination," and "T21 Assessing conditions" was high. On the other hand, the coefficient of variance of "T29 Transporting patient" and "T43 Putting instruments for transport care" was relatively low.

The time on task for each type of task is shown in Table 2. Direct tasks are those that deal directly with the patient. Indirect tasks are tasks carried out without direct contact with the patient, including preparatory tasks for direct tasks and cleaning tasks. Direct tasks, which involve transporting the patient, made up about 60 percent of all tasks, and indirect tasks made up 14 percent of all tasks.

Task category	No. of task	Time on Task	(%)
Indirect care	21	3:56:23	(14.1)
Direct care	21	16:08:27	(58.0)
Communication	2	0:59:57	(3.5)
Waiting	1	1:57:19	(7.0)
Record	1	0:11:10	(0.6)
Move	1	4:36:03	(16.5)
Total	47	27:49:19	(100.0)

Table 2. Time on task by each task category.

5. Discussion

First, we clarified the location and roles of persons in charge of tasks by making use of time and motion study data to visualize the object-oriented work process. From a functional point of view, the main persons in charge of the job of transporting patients were nurses. However, we understood that medical clerks participated in coordinating communication and that nursing assistants participated in transporting patients who did not need custody or attachment of medical electronic or transfusion devices. We understood that while medical clerks received communication about exams and confirmed the method of transport care on the patient chart, they did not have privilege to change the transport personnel or delegate the task, so they turned the task over to lead nurses. Furthermore, in the case of self-reliant patients, the person in charge of communication in a ward had the responsibility of transmitting the exam information to the patient regardless of whether he or she was a medical clerk or nurse. Furthermore, in the case of patients who needed wheelchair or stretcher transport, the person in charge of communication had the responsibility of sending information about the exam call to the transport personnel after receiving the communication about the exam. Our study showed that if the person in charge of communication was a medical clerk, he or she turned the task over the head nurse, because he or she did not have the privilege to change the care provider. The task that took the most time in this process was "Conveying exam information to the patient," followed by "Preparing patient chart" and "Preparing exam ticket." Use of the exam ticket was limited to outpatient exams of hospitalized patients and during the medical exam, so the repositories of the tickets were fixed. In contrast, because patient charts were used for a variety of purposes by physicians, nurses, medical clerks, and many other hospital employees, search for the charts took place, and the time required to prepare the charts grew longer. After information was conveyed to the transport personnel by the person in charge of communication, the transport personnel handled all responsibilities, including the final task of recording the transport.

Second, we understood the divergence between the work process specified in the hospital procedures manual and the actual work process. The manual used in the hospital of our study did not specify tasks such as "Searching for the patient," "Searching for the transport personnel," "Changing the transport personnel," "Preparing the exam (in the exam room)," and "Assisting in the exam." This reason is that the work procedures manual contains standard procedures. Irregular events and redundant tasks that should be kept in mind were not included. Also, the procedures manual was written to describe work procedures for individual nurses, so the location and role of workers described above were not clarified.

Third, from the work process diagram based on actual work records collected by this study and by adding time information to the process, we understood the efficiency with which tasks were carried out. By understanding the time used for each task and the variability of time, we clarified the time element that makes up the care of transporting patients. In the future, we seek to understand in detail how time on task changes depending on constraints.

Fourth, our study suggests that the data can be used for risk analysis. Our study extracted 47 tasks that made up the transport of patients, and listed their sequential order from time study records. Through our study, we clarified the input and output of each task, as well as the frequency of irregular events. Irregular events such as "Searching for the patient" and "Searching for the nurse" can be considered risks recorded by this study that prevent the work goal from being achieved. Although not carried out in this study, each task can be scrutinized to clarify factors that hinder each of their output. Doing this can draw out the risks associated with the work of transporting the patient, and produce discussions about concentrating risks and avoiding risks.

6. Future outlook

In this study, the structure of the work of transporting patients was visualized. The study suggests that the work of transporting patients has great differences in the objects, the process, and time efficiency depending on the conditions of the patients, type of exam, and occurrence of the work. Also, because many work occurrences were irregular and required quick responses, we learned that nurses must make adjustments with other tasks while at the same time accomplishing the task of transporting patients.

This study showed the usefulness of time and motion study for clarifying not only work load but also work structure and work processes. In the future, we seek to confirm the applicability of this study by conducting similar studies based on other jobs and records of jobs in several other facilities.

7. Acknowledgment

We wish to express our sincere gratitude to all the nurses, nursing assistants, and medical clerks who cooperated with us in this study. We also received assistance from research seminar students of the Osaka University Graduate School of Medicine, Division of Health Sciences. They conducted the time and motion study measurements. Once again, our deep gratitude to all involved.

8. References

Eriksson , H.E., Penker, M. (2000). *Business modeling with UML*. Wiley, New Jersey, 9780471295518.

Finkler, S.A., Knickman, J.R., Hendrickson, G. et al.(1993). A Comparison of work-sampling and time and motion techniques for studies in health services research. *Health Service Research,* Vol.28, No.5, pp.577-597, 0017-9124.

Hendrich, A., Chow, M.P., Skierczynski, B. A. et al.(2008). A 36-Hospital Time and Motion Study: How Do Medical-Surgical Nurses Spend Their Time? *The Permanente Journal*, Vol.12, No.3, pp.25-34, 1552-5767.

Shiki, N., Ohno, Y., Fujii, A. et al.(2009).Time process study with UML a new method for process analysis. *Methods of informatics in Medicine*, Vol.48, No.6, pp.582-588, 0026-1270.

Tang, Z., Weavind, L., Mazabob, J. et al.(2007).Workflow in intensive care unit remote monitoring: a time and motion study. *Critical Care Medicine*, Vol.35, No.9, pp.2057-2063, 0090-3493.

Van de Werf, E., Lievens, Y., Verstraete, J. et al.(2009).Time and motion study of radiotherapy deliverly: economic burden of increased quality assurance and IMRT. *Radiotherapy and Oncology* ,Vol.93, pp.137-140, 0167-8140 .

Were, M.C., Sutherland, J.M., Bwana, M. et al.(2008).Patterns of care in two HIV continuity Clinics in Uganda, Africa:a time motion study. *AIDS care*, Vol.20, No.6, pp.677-682, 0954-0121.

Yen, K., Shane, E.L., Pawar, S.S. et al.(2009).Time motion study in a pediatric emergency Department before and after computer physician order entry. *Annals of emergency medicine*, Vol.53, No.4, pp.462-468, 0196-0644.

Yokouchi, M., Ohno, Y., Kasahara, S. et al.(2005). Development of Medical Task
    Classification for Job Scheduling. *Medical and Biological engineering*, Vol.43, No.4,
    pp.762-768, 1347-443X.

11

R&D: Foundation Stone of Quality

Petr Košin, Jan Šavel and Adam Brož
Budweiser Budvar, N.C.
Czech Republic

1. Introduction

There are many definitions of quality. The oldest based on simple fulfilling of desired technical parameters developed into nowadays agreement, that the pivot of interest should be customer and his needs. Modern quality systems define quality as a degree of fulfilling of customers´ demands, or the degree of customers´ satisfaction by goods or services that he had paid for (Juran, 2000).

1.1 The role of quality in success on the market

Quality management systems have been developing since the beginning of industrial goods production. Producers of goods with higher quality had advantages for fight with competitors and high quality products were the main way how to satisfy customers. Although quality products are still necessary for success on the market, quality itself does not make nowadays business. Since the end of eighties new kind of companies appeared. These companies were orientated not only on quality goods production, but mainly on branding.

For these companies quality product was just one fourth of their marketing mix, whose management under certain brand was to satisfy customers. The resting parts of marketing mix are e.g. the price, place of purchase and accompanying services, or promotion quality. The original sense of brand was to label product with the place of origin and name of producer, which together had served as a guarantee of quality.

Due to sophisticated marketing methods can brand nowadays bring to customer whole battery of emotional values, which can sort customers into varying social groups. The satisfaction of customers is not brought only by the product quality, but also by emotions connected with social enlistment of brand (Klein, 2005; Olins 2009).

Brand management added to product new value, which allowed brand keepers to sell for much higher prices than would be the prices of their products without brand. Because not all customers could afford to pay extra price for emotional values and engaging to higher social classes, the byproduct of branding was the polarization of marked and origin of new distribution channels selling products without strong brands. These products are sold for much lower prices and usually have good enough quality to satisfy demands of its users. These so called cheap brands e.g. made up 40 % of German beer marked in 2005 (Verstl, 2005).

Brand management is also the reason why contact of quality department with customers is nowadays mediated by marketing department and quality improvement is often managed by marketing manager.

1.2 Quality improvement

Quality manager can never be satisfied with quality. It is because quality is not static; it is developing together with the development of customers´ needs. The process of developing quality is called quality improvement and it is integral part of modern quality management. As any other management systems, the management of quality improvement is based on the flow of information. There are several model systems for information flow in quality improvement management; of the most famous is Deming´s PDCA cycle (fig. 1) or Juran´s quality spiral (fig. 2).

These model systems have in common four basic steps, which also represent four levels where quality is managed in practice. The first level of quality management called "plan" or "product development" step is usually secured by the R&D department or by external consulting expert. This step includes designing of the product with all technical parameters and proposing of production processes with all operation steps and control points.

The second step ("Do" or "Production & process control), usually secured by production department in close cooperation with quality control department, includes production of products by production processes, which are carefully operated by feedback regulation in originally proposed control steps.

The third step where quality is managed is the "Check" or "Final inspection". This step is usually secured by the Quality control department and sampling or evaluation of the results are usually planed and processed with the help of statistical tools, like control charts or histograms.

The last step of information flow at the quality improvement management ("Act" or "Market research") is secured by quality assurance or marketing department. At this step customers´ satisfaction with product is measured. Method can be common market research, like statistic study with questionnaires, or with direct interviews.

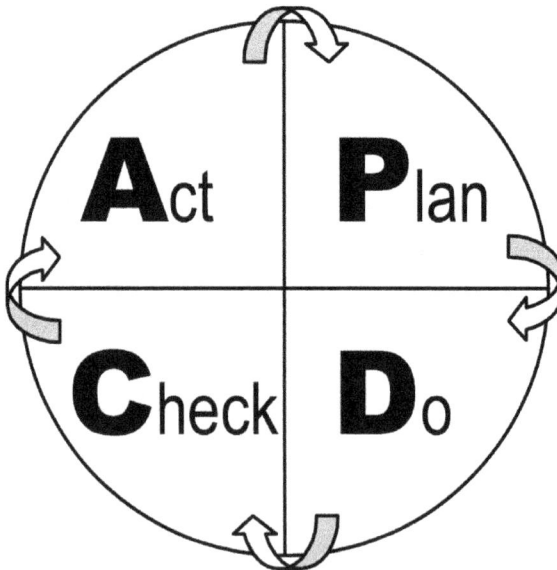

Fig. 1. PDCA cycle.

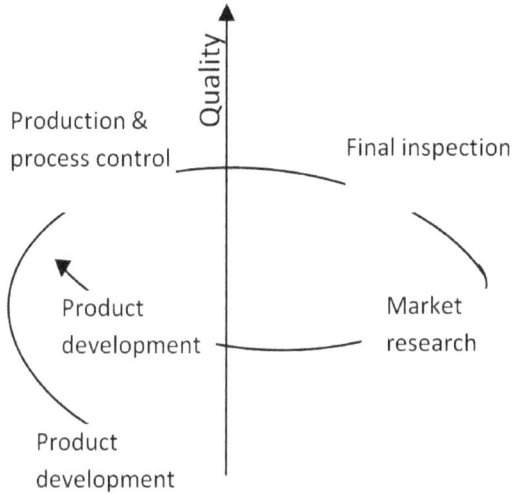

Fig. 2. Juran´s quality spiral.

1.3 Quality characteristics

Quality characteristics can be defined technically as inherent property of product that serves for identification, description and differentiation of product form other products and has a quantity and unit. Customer orientated definition of quality characteristic would be a property of product, that satisfy customer.

Quality characteristics can be divided into two groups: real characteristics and measurable attributes. Real characteristics directly correspond to the customer orientated definition of quality and are the reason why customers buy selected product. These characteristics should be evaluated in the fourth step of PDCA or Juran´s quality spiral. Their disadvantage is problematic measurement and evaluation and that is why real characteristics are usually translated into measurable attributes in the first step of quality improvement process.

Measurable attribute suit more the technical definition of quality and although not corresponding directly to customers' satisfaction they can be quite easily measured and evaluated during production and by feedback effect serve in the second and third step of quality management.

2. Case study: beer foam stability

The case study of this chapter for illustration of quality improvement in practice will be beer foam stability. It is a measurable attribute, which closely describes real quality characteristic called foam appearance. Foam appearance is one of the most important quality parameters of beer, because it is a visual parameter and visual parameters are ease to evaluate by almost all customers, who are much surer by what they see than what they taste.

Foam stability is not the only attribute describing foam appearance; the others are e.g. foam density, creaminess, color, or ability to cling on beer glass. Although these technical parameters have their meaning for foam appearance, stability is the most important parameter because when foam is not stable, it disappears and there is nothing to judge.

Assignment for this study is was to improve beer foam stability without changing any other beer quality and production parameters, which most often include raw materials. Complexity of this quality parameter much differs from common example situations like screw production.

2.1 Procedure of foam quality improvement

The general procedure of quality improvement has several steps. Whole process starts with bibliographic research, because many of the basic questions have been solved before by someone else.

The second step should be the verification of bibliographic result under the specific conditions of the company. With good luck this can be the end of quality improvement.

If bibliographic results do not help, the third step should be the start of own primary research. There cannot be given general instruction on this step, but there is one way that occasionally helps and was useful also in the case study of this chapter. It is to develop your own analytical method, of which results closely corresponds to customers sensation of the quality parameter and simultaneously can be used all over whole production line to evaluate how the quality attribute develops during production.

The fourth step can then be to use the new analytical method to find weak points of the production line and find a way how to control these processes to improve the quality problem. Results of the method can then be used for feed-back regulation of selected processes parameters in second step of the quality improvement information flow described by PDCA or Juran´s quality spiral.

The last step of the research would be identical with the third or fourth step of quality management. With the help of sophisticated statistical tools should be precisely evaluated the extent of quality improvement and the economical balance of quality profits and costs.

2.2 Bibliographic search

The research usually starts with bibliographic search. In many cases the same problem has already been discussed either in academic or applied research.

Academic sources

Academic research offers several solutions for foam quality improvement, mostly based on reductionist analytical approach. The idea is that foam stability can be increased by addition of foam stabilizing material to beer. There have been described several foam stabilizing substances, but less methods how to increase the content of these and not change beer taste or composition. Of the most discussed are bitter acids and proteins (Evans, 2002).

All of hop bitter acids can increase foam stability, but the most effective are chemically reduced derivates. Their production stars with extraction of α-bitter acids from hop by organic solvent or supercritical CO_2. The second step is isomerization of α-bitter acids in alkali and high temperature conditions and the third step chemical reduction of iso-α-bitter acids into di-, tetra- or hexa-hydro-iso-α-bitter acids. Most often discussed are tetrahydro-iso-α-bitter acids, produced under the brand Tetrahop.

Tetrahop is used in downstream processes as additive in milligrams per liters of beer. There are several problem of this way of improving foam quality. Major is that although foam stability is increased, foam structure at the end of foam collapse has unnatural appearance resembling polystyrene foam. Next problem is harsh character of Tetrahop's taste, which is far away from fine taste of natural hops. Probably the least important problem is that the chemical preparation of Tetrahop collides with Reinheitsgebot, German beer purity law saying that beer can be only made from water, barley malt and hops. Improving foam quality by simple increase of natural hop components would have negative effect in change of bitterness intensity, one of the most sensed sensory attribute of beer.

The role of proteins in foam stability has been the most studied part of foam quality in academic research. There have been described several proteins that influence foam quality, mainly hydrophobic proteins like protein Z, or lipid transfer proteins (LTP). Protein Z represents proteins with high molecular weight (relative molecular weight 35 000 – 50 000) and LTP have relative molecular weight 5 000 – 15 000. Proteins, which together with bitter acids and ions build up the framework of foam bubble walls, come to beer from malt.

A lot of studies on which malt contains more of these foam promoting proteins were driven by the idea, that change of malt specifications could be a way for a brewer how to fix problems with foam. The problem of this approach is that changing malt specifications can substantially change some of the other important parameters of beer, e.g. color, fermentability and final degree of attenuation, or the essential character of beer taste, which is hidden in the unfermented remainder of malt in the beer body.

Although foam stability has been in focus of academic research for quite a long time, there have not been found a practical recipe how to improve foam quality and not change any of the other beer parameters, including beer raw materials and composition.

Applied sources

There are far less papers written from applied research compared to academic research. The reason is not only the evaluation of academic research quality by the quantity of published papers, but also historic transfer of applied research from goods producers to service and suppliers companies, who more carefully guard their knowhow and do not publish much of technical papers.

Although it is quite hard to come across this kind of publication, they are of great use because they usually look for practical solutions. Contrary to academic research, which usually looks for answers on questions "how does it work", applied research usually solves questions concerning what can one do to economically solve a problem.

For our case study of foam stability can be found sporadic publications recommending some practical solutions like optimization of the malt grinding, correct choice of lauthering tun, sufficient separation of sediment after wort boiling, or consistent rinsing of bottles at the end of washing (Haukeli, 1993).

3. Improvement of foam stability

The assignment of the research was to improve foam without changing any other quality characteristics, especially beer appearance and taste, which is secured by constant specifications of raw materials. That is why trials with alternative malt specification, hop dosage or use of any additive to beer as discussed in the academic research was excluded from the design of this study.

3.1 Foam stability measurement
Foam collapse can be divided into three stages from the macroscopic point of view. The first is the drainage of beer out of wet foam, where significant upward movement of beer-foam interface can be observed. The second stage is the collapse of dry foam and is accompanied by significant decrease of foam surface. The third stage is the break-up of the last foam layer, which results in the appearance of a "bald patch" on the surface of the beer.

This division corresponds to measurement strategy focused on the second stage of foam decay. The collapse of the foam accordance to first order kinetic equation is usually expressed as the time dependency of beer volume remaining in dry foam after initial beer drainage.

Kinetic equation can also connect the first two phases of foam collapse as expressed in formula (1),

$$c = c_{\infty} - a_0 \cdot \frac{k_2}{k_2 - k_1} \cdot e^{-k_1 \tau} + (a_0 \cdot \frac{k_1}{k_2 - k_1} - b_0) \cdot e^{-k_2 \tau} \tag{1}$$

where c is the beer volume or its height under the foam, c_{∞} is the total volume of beer after complete foam decay, a is volume of beer bound in dry foam, b is beer freely present in the foam and τ is time. The constants k_1, k_2 describe the foam collapse in the first and second stage of decay, index 0 indicates the beginning of foam degradation (Savel, 1986).

How customers evaluate foam quality was uncovered by qualitative research at which 30 random customers were asked about their satisfaction with beer foam on the beer they were drinking in pubs or bars in the Czech Republic. In contrast to similar investigations, interviewees were not asked a long series of questions about foam quality or served any adjusted beer samples. The intention was to discretely interview the drinkers in their normal pub or bar drinking situation and gauge their opinion of the foam on the beer they were being served. The only question asked by the interviewers during drinking was the unforced question "is everything OK or not with the foam?" At this point, according to the interviewee's opinion, if there was something wrong with the foam, we visually evaluated the stage of foam collapse, in particular noting the presence of a "bald patch" in the foam, on the surface of the beer.

According to this qualitative assessment of customer perception of foam quality, customers in did not pay much attention to the foam until a problem with the foam is perceived. The beer was seen as problem free so long as there was a sufficient amount of foam to cover the beer surface in the glass. Customers start to be concerned about the foam quality in their glass only once they perceive that there was something wrong with the foam in their glass. This was at the end of foam collapse, when bald patches start to appear on the beer surface. At this point, approximately a quarter of the customers started to pay attention to the quality of beer foam in their glasses. The other three quarters of customers did not have any problems with the foam quality, even at this point. As the break-up of the last foam layer proceeded to produce a substantial bald patch on the beer surface, more customers started to be concerned with foam quality. Once the beer surface was almost completely bald, almost all customers commented that the foam quality was not satisfactory. Thus it can be concluded that for beer drinkers, the early appearance of this bald patch indicates a poor quality beer.

Close to this sensation of foam quality is a method for foam stability measurement called pouring test, which measures foam stability through whole collapse curve and includes the last collapse stages where bald patch appears. It is based on pouring atempered (8 °C) beer from the bottle to a standard tasting glass and time from pouring to the first bald patch larger than 5 mm appearance is recorded. Although this test is principally very close to the customer sensation of foam quality, it has a disadvantage of low reproducibility.

One of the most spread methods among brewing laboratories is a method called NIBEM. This method is based on recording of the speed of downward movement of foam surface in the second stage of foam collapse. This method has much higher reproducibility than the pouring test, but is quite far away from the customer perceived foam stability, as can be seen from the low correlation with pouring test (fig. 3).

Disadvantage of both of these methods, NIBEM and pouring test, is that it cannot be used to measure foam stability of samples that do not contain sufficient amount of CO_2 to create foam. NIBEM is slightly less sensitive to CO_2 content of sample than pouring test, because foam is created by flushing of the sample through the jet.

3.2 Matrix foaming potential

The new method for foam stability measurement, which was optimized and tested for foam stability improvement, is called the matrix foaming potential (MFP) and is measured by Foam stability tester type FA by 1-CUBE, Havlickuv Brod, Czech Republic (fig. 4). Foam is created by introducing of a gas into liquid sample and mixing with stirrer. By the combination of gas type, gas flow rate and revolution speed of the mixer there can be prepared foam of various structures, eg. by introducing 0,25 mL/min of air and mixing at 1200 RPM creates very fine foam resembling foam created on draught beer, or introducing 0,5 ml/min of air and mixing at 900 RPM creates medium coarse foam resembling foam on beer poured from the bottle (fig. 5).

Foam stability is evaluated as a time from the end foam generation to the decrease of foam surface over a set distance, which is a distance of electrodes that are in the place of measurement. The MFP value expressed in seconds covers the height of created foam under standard conditions, which corresponds to foaming ability of the sample, time to foam drainage in the first stage of foam collapse and the whole second stage of foam collapse.

As can be seen from the measurement principal, MFP can be used all over the whole beer production line, because even samples without CO_2 can be evaluated. Samples that contain CO_2 have to be degassed prior to the measurement. The MFP measurement has lower reproducibility due to various reasons, e.g. the temperature sensitivity (fig. 6). Regardless the reproducibility this method is much close to real foam quality as sensed by consumers, as can be seen from satisfactory correlation with pouring test (fig. 7).

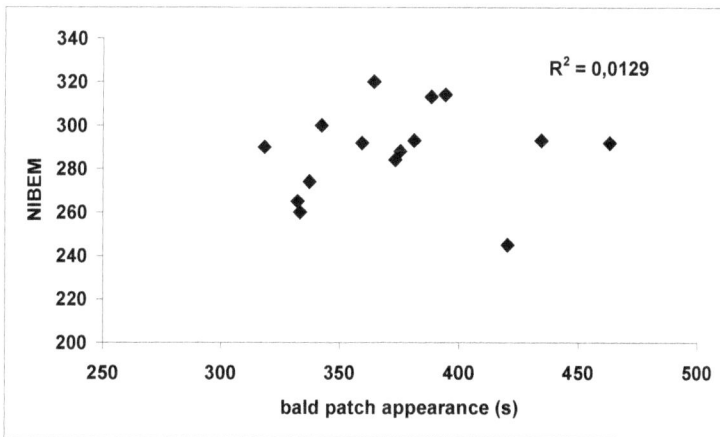

Fig. 3. Scatter plot and regression analysis of NIBEM with customer perceived stability measured by the pouring test.

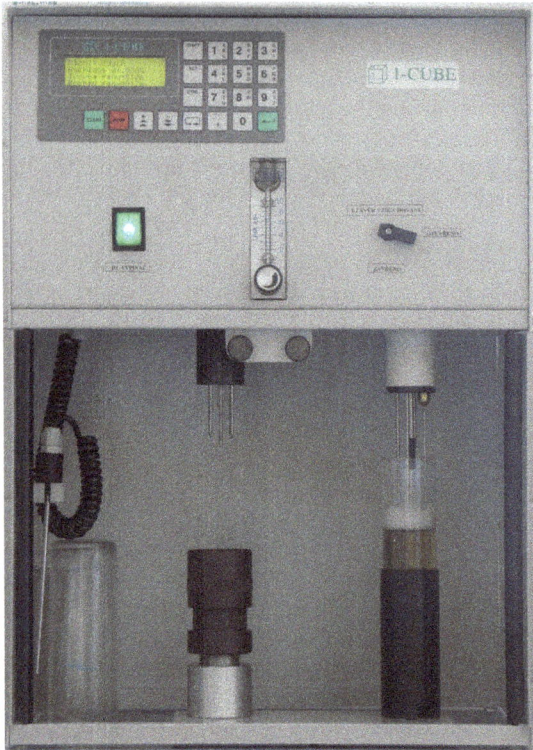

Fig. 4. Foam stability tester type FA by 1-CUBE.

Fig. 5. Fine (right) and medium coarse foam.

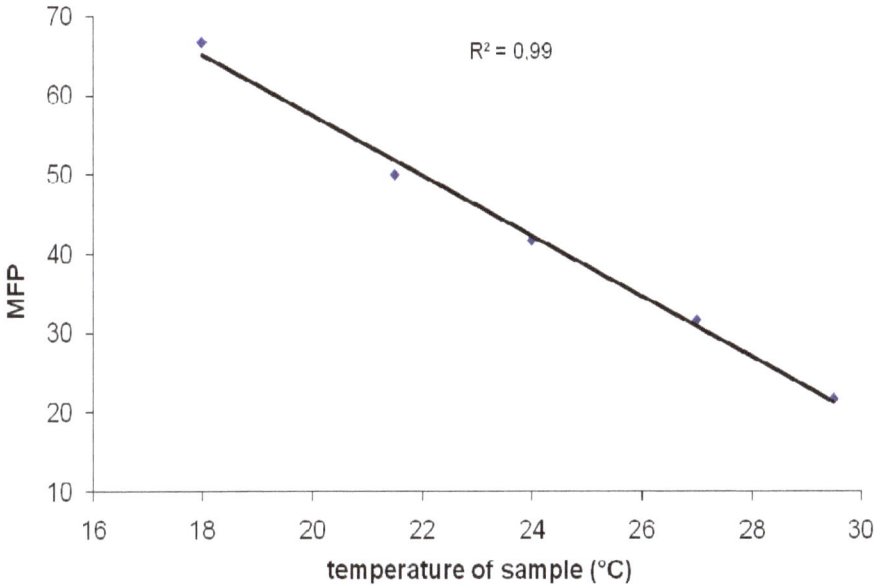

Fig. 6. Temperature dependance of Matrix foaming potential.

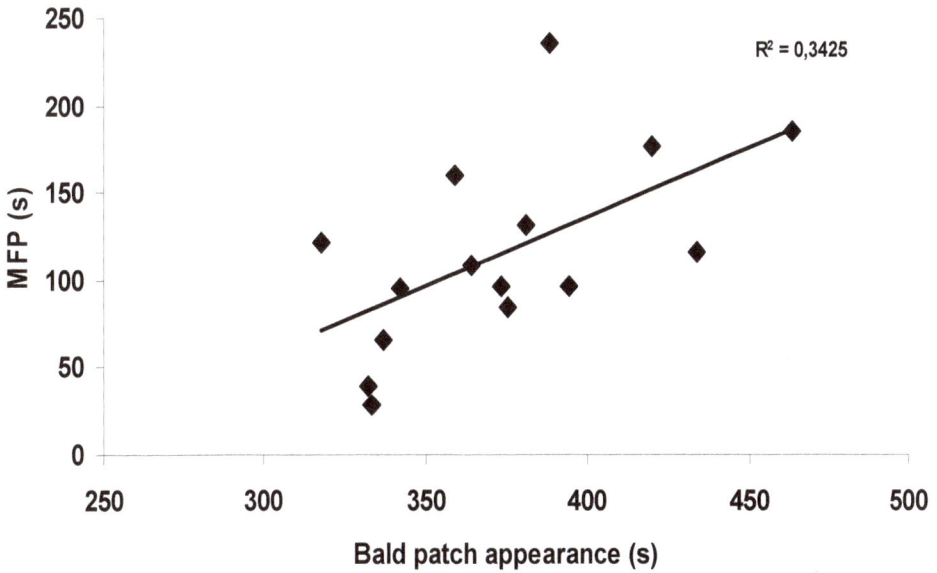

Fig. 7. Correlation of pouring test with Matrix Foaming Potential (MFP).

3.3 Foam positive and negative substances

As discussed above, brewers or researchers looking to improve foam quality typically take a reductionist analytical approach. Accordingly, the quality of beer foam generated is tried to be evaluated by measuring the content of foam positive components in beer, and then attempting to modify the brewing process to increase the content of these foam positive compounds to improve foam quality. Most often targeted with such an approach are foam positive components including protein Z, LTP1 and other proteins, and iso-α-acids or their reduced forms. Much more infrequently, the role of foam negative components such as lipids is considered in the technical literature.

It was observed that beer, even with the lowest content of proteins, could be foamed to 100 % of volume of relatively stable foam by simple foaming technique (Fig 8). A simple approach was to correlate the content of foam positive proteins assessed by the Bradford Coommassie blue binding assay (CBB) with foam stability measured by both NIBEM value and by pouring the beer to a glass from the bottle and measuring the time to bald patch appearance (Fig 9). This experiment was conducted with 15 brands of commercial lagers and showed no association between the level of foam positive proteins in beer or its foam stability with NIBEM (Fig 9A), although there was some association with bald patch formation (Fig 9B) although the slope was relatively low. On the basis of these results it was questioned whether the content of foam positive compounds was as important for the beer foam quality of beer as found in previous studies.

This suggests that the content of foam positive proteins/components may not be limiting with respect to foam quality, a similar conclusion that can perhaps be drawn from the beer dilution experiments of Roberts over 30 years ago (Roberts, 1978).

Fig. 8. whole volume of low protein beer converted into foam.

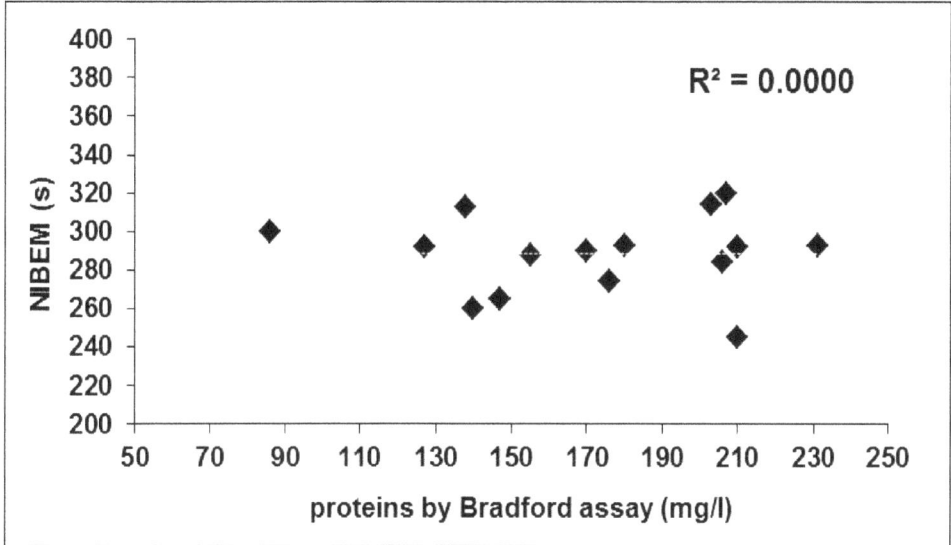

(A)

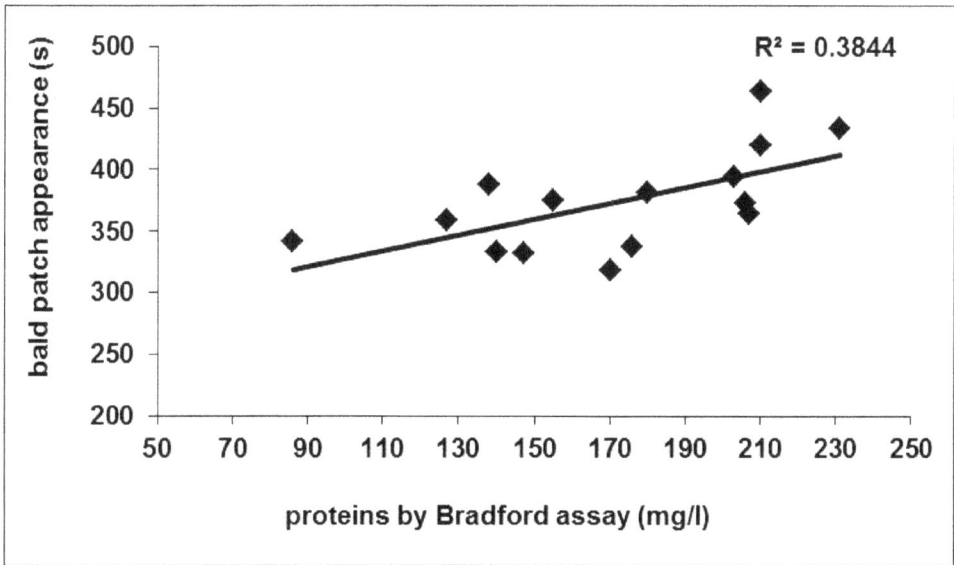

(B)

Fig. 9. Scatter plot and regression analysis of protein content of beer with NIBEM value (A), and time to bald patch appearance (B).

Previous investigations using a "foam tower" have shown that hydrophobic and foam positive components such as LTP1 and iso-α-acids are preferentially concentrated in the foam. To study if the content of beer foam positive proteins/components were limiting, a serial re-foaming experiment as depicted in figure 10 was designed. Degassed beer, created foam by stirrer as in MFP measurement with foaming time kept constant so as to measure the quantity of foam produced. The foam and beer phases were separated by pouring beer from under the foam, refilled the beer phase to original volume by "fresh" degassed beer to keep standard foaming conditions and again created foam with the mixer. The refilled amount was less than 10 % of the total volume. This cycle of foaming and separation was repeated 15 times.

The basic premise of the experiment was that if the content of foam positive proteins/components content was limiting in beer foam, foam capacity (amount of generated foam) and foam stability would decrease with sample order number in the experiment as these foam active components were concentrated in the foam and depleted from the beer. Thus foam positive proteins/components would migrate and concentrate in the foam phase in the earlier foaming and separation steps and so that there would not be a sufficient amount in beer phase in the later steps to generate sufficient amounts of stable foam.

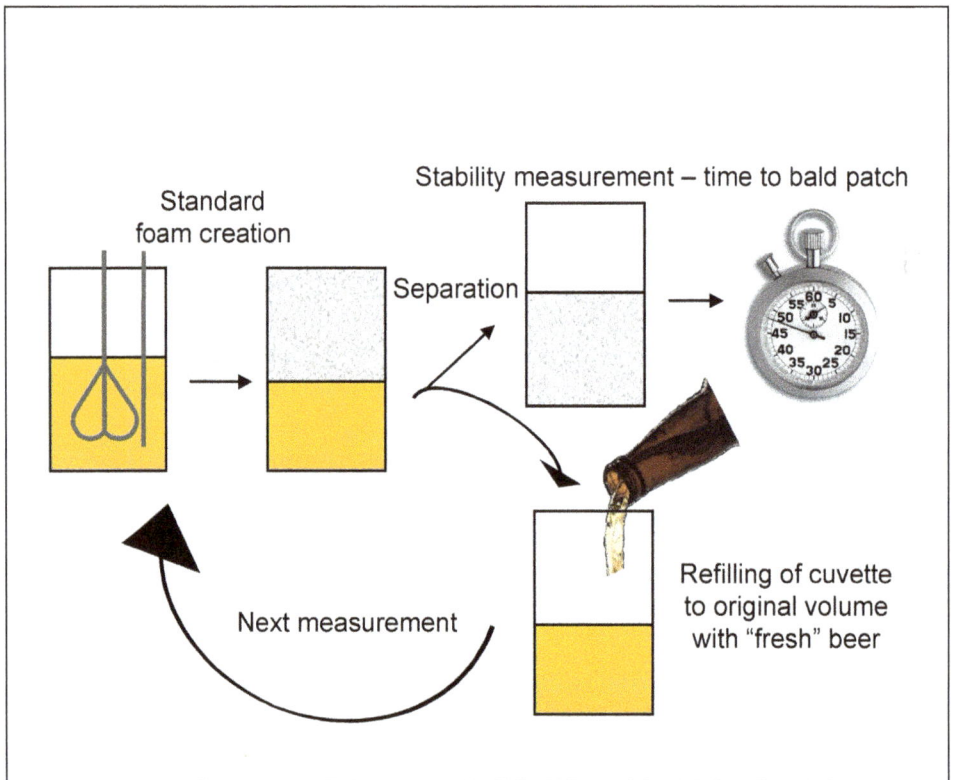

Fig. 10. Schematic of the design of experiment for the serial re-foaming of beer. Fifteen iterations of re-foaming of the beer were undertaken.

Figure 11 clearly shows that with serial re-foaming, foam capacity throughout the experiment was unchanged and foam stability, in terms of time till bald patch formation, was substantially increased, being five times higher at the end of experiment than at the beginning. It follows that foam stability was not just determined by the level of foam positive compounds, but it was more the result of compromise or balance between foam positive and negative components. Moreover, as the foam stability increased with re-foaming, it was apparent that both negative and positive foam components were presumably concentrated in the foam, thus separated from the beer to be re-foamed in the next cycle. This unexpected and contrary result could be explained by the following hypotheses. Firstly, Bamforth proposed that "hydrolyzed hordein appears to selectively enter beer foams at the expense of the more foam-stabilising albuminous polypeptides" such as protein Z (Bamforth, 2004). As such, as the level of hydrolyzed hordein is depleted relative to the albuminous polypeptides, foam stability would be seen to improve. However, Lusk et al. found in their foam tower experiments, as the content of LTP1 was depleted, the foam became less "creamy" and contained coarse bubbles, that were not observed in this experiment (Lusk, 1999). Secondly, as LTP1 is concentrated in beer foam and is thought to play a lipid-binding role in beer, both the LTP1 and the foam stabilizing lipids would be removed with the separated foam. An improvement in foam stability would occur if the level of lipids were limiting in the beer relative to LTP1, other lipid binding components and foam positive proteins/components.

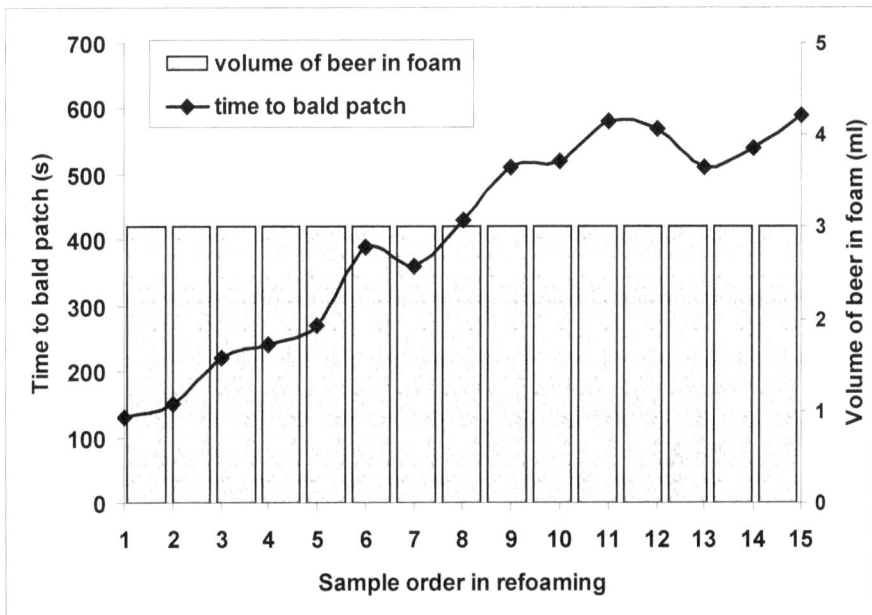

Fig. 11. Results of foaming capacity and foam stability during a serial refoaming experiment with 15 iterations. The volume of beer foam formed was exactly the same for each refoaming iteration, when using a constant time of foaming.

3.4 Practical approach to foam stability
The insights gained from these experiments recommend several practical approaches to foam quality improvement in commercial production. These are based on the premise that

by measuring the MFP during and within each production stage, critical points can be identified in the process that reduce foam stability and indicates process parameters that can be modified to improve foam stability, particularly the limiting of the inclusion of foam negative components. One example was to apply MFP measurement during the course of lautering and sparging process. Figure 12 shows, that extended sparging was one of the steps that reduces foam stability. It has long been known that although extended sparging recovers more extract, it also results in the extraction of increasing amounts of undesirable substances such as polyphenols, husk bitter substances, etc , and foam negative materials. Similarly, the MFP analysis was applied during the course of main fermentation (Fig 13). During fermentation, foam stability was decreased to almost a third.

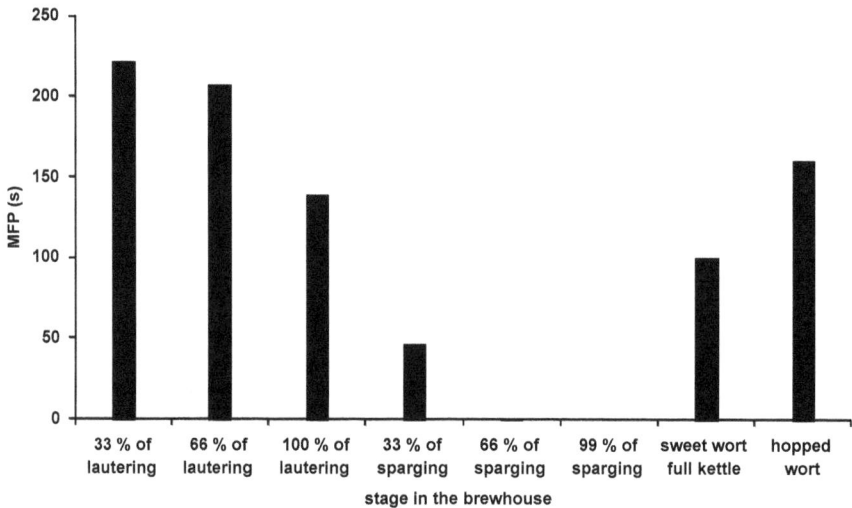

Fig. 12. Matrix foaming potential of wort samples taken during the course of lautering and sparging.

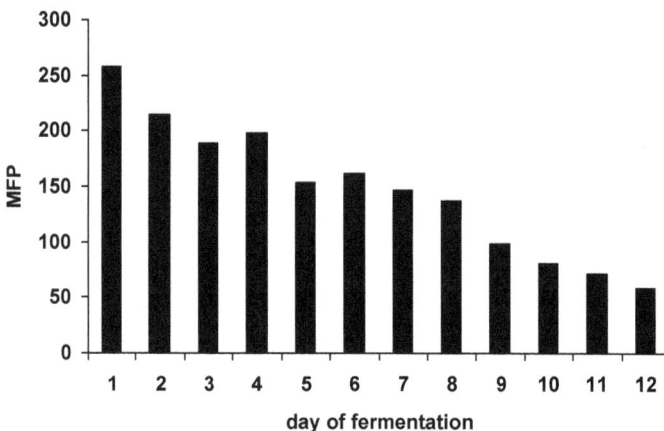

Fig. 13. Matrix foaming potential of "beer" samples during the course of main fermentation.

By this approach there were found several control points for foam stability, of which some fulfilled the demand for not changing of any other beer quality parameter, including raw materials. The effectiveness of these new control points has to be validated by statistical methods, which also serve for control of constancy of other quality parameters after setting new control points.

The most suitable statistic tool is regulation chart, which illustrates the change of foam stability in figure 14 by NIBEM value. Foam stability increased from values around lower specification limit into optimal central zone for NIBEM, MFP increased approximately three times.

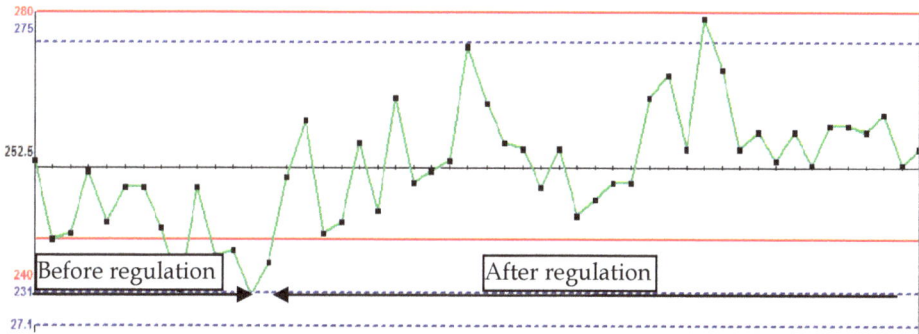

Fig. 14. Regulation chart of foam stability measured by NIBEM during process optimization.

4. Conclusion

Integral part of modern quality management is improving quality. For the task of improving beer foam quality, more successful strategy appeared to be employing own R&D, compared to external consulting expert, even with deep knowledge of bibliographic results on given topic.

The quality improvement procedure included optimization of new method with results close to customer sensation of quality, which could be used even for evaluation of intermediate product all over the production line. By this method weak points were discovered and the success of new regulation was evaluated by control charts.

5. References

Bamforth, C.W. (2004). A critical control point analysis for flavour stability of beer. *Master Brewers Association of the American Technical Quarterly*, Vol.41, (2004), pp. 97-103

Evans, E. & Sheehan, M. C. (2002). Don´t be fobbed off: The substance of beer foam - a review. *Journal of the American Society of Brewing Chemists*, Vol.60, No.2, (2002), pp. 47-57

Haukeli, A. D.; Wulff, T. O. & Lie, S. (1993). Practical experiments to improve foam stability, *Proceedings of 1993 24th European Brewery Convention Congress*, p. 40, Oslo, Norway, 1993

Juran, J.M. & Godfrey, A.B. (2000). *Juran´s quality handbook, fifth edition*, McGraw Hill, ISBN 0-07-116539-8, Singapore, Singapore

Klein, N. (2005). *Bez loga*, Dokořán, ISBN 80-7363-010-9, Prague, Czech Republic

Lusk, L.T.; Ting, P.; Goldstein, H.; Ryder, D. & Navarro, A. (1999). Foam tower fractionation of beer proteins and bittering acids, In: *European Brewery Convention Monograph*, 166 – 187, ISBN 3-418-00774-0, Amsterdam, Netherlands

Olins, W. (2009) *O Značkách*, Argo, ISBN 978-80-257-0158-4, Prague, Czech Republic

Roberts, R.T.; Keeney, P.J. & Wainwright, T. (1978). The effects of lipids and related materials on beer foam. *Journal of the Institute of Brewing*, Vol.84, (1978), pp. 9-12

Verstl, I. (2005). The servant of two masters or how to keep your customers satisfied : the European brewing industry, the Zeitgeist and Tradition, *Proceedings of 2005 30th European Brewery Convention Congress*, p. 171, Prague, Czech Republic, May 14-19, 2005

Šavel, J. (1986). Dva modely rozpadu pivní pěny, *Kvasný Průmysl*, Vol.32, No.4, (1986), pp. 76-78

12

Herbal Drug Regulation Illustrated with Niprifan® Antifungal Phytomedicine

Sunday J. Ameh[1], Obiageri O. Obodozie[1], Mujitaba S. Abubakar[2],
Magaji Garba[3] and Karnius S. Gamaniel[1]

[1]Department of Medicinal Chemistry and Quality Control,
National Institute for Pharmaceutical
Research and Development (NIPRD), Garki, Abuja,
[2]Department of Pharmacognosy and Drug Development, Ahmadu Bello University, Zaria,
[3]Department of Pharmaceutical and Medicinal Chemistry, Ahmadu Bello University,
Zaria,
Nigeria

1. Introduction

Quality system is defined as arrangements, procedures, processes and resources; and the systematic actions necessary to ensure that a manufactured product will meet given specifications. On the other hand, quality control is defined as measures taken, including sampling and testing, to ensure that raw materials, intermediates, packaging materials and finished goods conform to given specifications. Quality specification refers to a written procedure and requirements that a raw material, intermediate or finished good must meet for approval. On the other hand, standard operating procedure (SOP) refers a written procedure, giving step-by-step directions on how a particular operation is to be carried out. Quality manual means, a document that describes the various elements of the system used in assuring the quality of results or products generated by a laboratory or factory. The term quality assurance refers to the totality of all the arrangements made with the objective of ensuring that products are of the quality required for their intended use. Good manufacturing practice (GMP), on the other hand, is that aspect of quality control that deals directly with manufacturing and testing of raw materials, intermediates and finished goods to ensure a product of consistent quality. Essentially, GMP involves two types of control - analytical and inspection, and both require: i) clear instructions for every manufacturing process; ii) a means of controlling and recording every manufacturing process; iii) a means of ensuring that the complete history of a batch can be traced; iv) a mechanism for recalling any batch of product from circulation; v) a system for attending to complaints on quality of product or service; and vii) a programme for training operators to carry out and to document procedures. The foregoing definitions and description of GMP conform to those of WHO (2000). It is also clear from the foregoing that GMPs are not prescriptive instructions on how a manufacturer can produce, but are rather a series of principles that must be observed for quality products, services or results to emerge. Invariably, GMPs are approved and enforced by an appropriate National Agency, but the onus of preparing and executing GMPs rests with the manufacturer. In Nigeria (population ~ 150 million), GMPs are enforced by NAFDAC – established by decree in 1992/93.

The tests carried out for this study were according to official procedures - mostly BP (2004) and WHO (1998). The results are discussed within the context of requirements for herbal drug regulation as per WHO, EMEA and NAFDAC. It is noted that herbal drug regulation in Nigeria (Table 1) as compared that in Europe (Table 2) is paradoxically hampered not by the rigor and "stringency" of rules, but by the fact that the rules are only merely cumbersome, being neither adequate nor enforceable (Table 3), unlike those of EMEA.

S/No.	Regulatory aspect	Requirement
1	Legal status of applicant, who may be: Manufacturer. Marketer. Distributor.	The applicant must be certified by the Corporate Affairs Commission as a registered business in Nigeria. A marketer or distributor must show evidence of Power of Attorney issued by the manufacturer.
2	Analytical status of the product for registration.	The product must : Have a certificate of analysis. Be accompanied by a dossier with information on: Ingredients. Method of analysis. Stability. Dosage. Safety, among others.
3	Pre-registration inspection of premises.	Manufacturing, storage and distribution premises must be GXP compliant. Marketers must provide convincing evidence of GDP and GSP.
4	Post marketing surveillance plan/ report	Applicant may be required to provide a plan for reporting on: The use of the product. Any adverse reactions.

The above information were drawn from NAFDAC's leaflets and website: www.nafdacnigeria.org

Table 1. NAFDAC requirements for registering herbal medicines.

Essentially, the NAFDAC rules seem perhaps affected, or rather made cumbersome without being truly rigorous or "stringent" as actually claimed in the Agency's website. It is further noted that, like in China, where the head of the drug regulatory agency was sentenced to death for corruption (Gross and Minot, 2007), drastic actions, including the wholesale reorganization of NAFDAC management, had to take place in 2000 to straighten things out. The high frequency of confiscation and public destruction of counterfeit products by NAFDAC strongly testifies to the inadequacy of the rules and policies guiding drug regulation in Nigeria. Unfortunately, this worrisome state of affairs is equally true of many countries, as stated in the case of china. There is thus, the need for drug regulatory agencies in these countries to brace up. The aim of this article therefore, is to further an earlier advocacy (Ameh et al., 2010a) that includes alerting and encouraging Drug Regulatory Agencies, Health Ministries, and Parliamentary Health Committees, especially those in developing countries, to enact laws and evolve policies that will better regulate the

S/No.	Regulatory aspect	Requirement
1	**Product information**: Summary of product characteristics	Name of the product. Strength. Dosage form. Quantity of active ingredient (Example: 25 mg *P. guineense*). List of excipients (Example: *P. guineense, E. caryophyllata* etc., etc). Shelf life. Posology and method of administration. Indications. Contraindications/ special warnings. Precautions for use. These data are used as the basis for inserts, packaging, or advertisement. Inserts must pass "readability testing."
2	**Quality control data**: Refer to GMP requirements for production.	Production must be in a GMP compliant facility Drug must be produced with validated/ reproducible formula method. There must be a finished product specification. The product must be manufactured at least on pilot scale and three batches used for stability studies. Stability studies should be carried out on the product packaged in the container proposed for marketing. A summary of the stability studies undertaken must be provided. From stability data shelf life/ storage precautions should be proposed. A quality dossier must be provided for raw and finished materials The product must be produced from herbs that have been cultivated and harvested in accordance with GACP. 10. The starting material must be evaluated for risk of any environmental contamination.
3	**Safety data requirements**: Refers to safety pharmacology, including animal and human studies	Published animal or human studies. Review of any potential interactions with other drugs, side effects, and any proposed contraindications/ precautions in the product information. Recognized monographs on the material or product with information on safety. Any information concerning special groups such as children, the elderly or pregnant women. It is interesting to note that in the US, where herbal medicines are regulated as dietary supplements, manufacturers are not required to prove safety or efficacy, but the FDA can withdraw a product from sale if it proves harmful.
4	**Traditional use evidence**: Refers to history and prevalence.	There is no requirement to prove efficacy. Instead data must provide reference that the product has been in use as medicine for 30 years or more. Of 30 or more years, the last 15 must be in Europe. The data must be presented in a special format, called: *Common Technical Document Format*.

The Table was drawn based on data gathered from references including (DSHEA 1994; Goldman, 2001; De Smet, 2005; Ann Godsell Regulatory, 2008).

Table 2. EMEA requirements for registering herbal medicines.

S/No.	Extra requirement	Remark
1	Five (5) copies of the product dossier.	Probably unreasonable
2	Three (3) packs of the products samples.	Probably reasonable
3	Notarized original copy of the duly executed Power of Attorney from the product manufacturer.	Clearly unreasonable for all categories of applicants
4	Certificate of Manufacture issued by the competent health or regulatory authority in country of origin and authenticated by the Nigerian Mission in that country. Where there is no Nigerian mission, The British High Commission or an ECOWAS country Mission will authenticate.	Probably unreasonable for all categories of applicants
5	If contract-manufactured, Contract Manufacturing Agreement, properly executed and notarized by a Notary Public in the country of manufacture.	Clearly unreasonable for all categories of applicants
6	Current World Health Organization Good Manufacturing Practice Certificate for the manufacturer, authenticated by the Nigerian Mission.	Clearly unreasonable for all categories of applicants
7	Certificate of Pharmaceutical Products (COOP) duly issued and authenticated.	Clearly unreasonable for all categories of applicants
8	Current Superintendent Pharmacists license to practice issued by the Pharmacists Council of Nigeria (PCN).	Only probably reasonable
9	Premises Registration License from PCN	Only probably reasonable
10	Certificate of Registration of brand name with trademark registry in the Ministry of Commerce here in Nigeria; Letter of invitation from manufacturer to inspect factory abroad, stating full name and location of plant.	Probably unreasonable for all categories of applicants
11	The applicable fee payable only if documents are confirmed to be satisfactory.	Likely to be abused if the amount is high. The fee should be a token amount paid by all applicants
12	Nutraceuticals, medical devices and other regulated drug products have similar requirements, with minor variations. Specific details can be obtained from NAFDAC.	A sketch of the minor variations should be provided in print no matter how brief. Any information provided by NAFDAC should be printable for sake of transparency.

The information on NAFDAC were drawn from leaflets and NAFDAC's website (2010):
www.nafdacnigeria.org/ The remarks are informed by current affairs and public perception
of NAFDAC's role and activities including the wholesale reorganization of its Management
in 2000.

Table 3. NAFDAC's extra requirements for registering herbal medicines.

production, distribution and use of herbal drugs. This is in view of the ever increasing use of
herbs notably after the Alma-Ata Declaration (Ameh et al, 2010b) which paved the way for
the stupendous growth of herbal drug use worldwide, particularly in North America where
that growth had been stymied by the Flexner Report of 1910. That Report, which coincided
with Paul Ehrlich's introduction of Salvarsan and the term "chemotherapy" in 1909, had
favoured chemical medicine over herbal (Pelletier, 2006). Furthermore, apart from the said

growing use in the West, it is held that some 80 % of the populace in many developing countries still relies predominantly on herbs and other alternative remedies (WHO, 2008). Indeed, in some parts of Africa, for example, Ethiopia, a dependence of up to 90% has been claimed (BBC, 2006).

2. Experimental

The study applied official procedures – mainly WHO (1998) and BP (2004) to: evaluate the quality parameters of the raw materials and their extracts; and the changes in these parameters during dark, dry storage in capped glass bottles under tropical room temperature and humidity (RTH) as obtain in a typical Nigerian Traditional Apothecary (NTA). The parameters evaluated were: appearance, loss on drying, ash values, extractability, solubility, pH, TLC features, light absorption and foaming index. Basic morphological studies were carried out as per WHO (1998). Appropriate phytochemical tests were also conducted by official methods as described elsewhere (Ameh et al., 2010c; 2010d).

2.1 Treatment and sampling of material

The aerial parts of *Mitracarpus scaber* obtained during the months of October and November from the botanical garden of the National Institute for Pharmaceutical Research and Development (NIPRD) were air-dried in a well-ventilated shade, designed for drying medicinal plant materials. The materials were subsequently comminuted to coarse powder with a grinding machine. The procedure for sampling was as per WHO (1998) as had been described in detail earlier (Ameh et al., 2010c). Three (3) original samples from each batch or container were combined into a pooled sample and subsequently used to prepare the average sample. The average sample was prepared by "quartering" the pooled sample as follows: each pooled sample was mixed thoroughly, and constituted into a square-shaped heap. The heap was then divided diagonally into 4 equal parts. Any 2 diagonally opposite parts were taken and mixed carefully. This step was repeated 2 to 4 times to obtain the required quantity of sample. Any material remaining was returned to the batch. The final samples were obtained from an average sample by quartering, as described above. This means that an average sample gave rise to 4 final samples. Each final sample was divided into 2 portions. One portion was retained as reference material, while the other was tested in duplicate or triplicate. The samples for stability study were stored at room temperature and humidity (RTH) in capped glass bottles and placed in a shelf protected from light.

2.2 Macroscopic examination and phytochemical tests on the fresh and air-dried materials

The procedures adopted were as per WHO (1998). Shape and size were determined with the aid of a ruler and a pair of calipers. Diffuse day light was used on the untreated sample to determine its colour. The texture and surface/ fracture characteristics of the untreated sample were examined, where necessary, with x10 magnification hand lens to reveal the characteristics of cut surfaces. The material was felt by touch, to determine if it was soft or hard. Or was bent and ruptured, to obtain information on brittleness and appearance of fracture planes – whether it was fibrous, smooth, rough or granular. Odour was determined by placing a pinch in a 50-ml beaker, and then slowly and repeatedly the air above the material was inhaled. If no distinct odour was perceived, the material was crushed between the thumb and index finger, and inhaled as above. The strength of the odour was determined as: odourless, weak, distinct, or strong. The sensation of the odour was determined as: aromatic, fruity, rancid, etc. etc. When possible, the odour was compared with that of a defined substance, such as menthol, sulphur dioxide, eugenol, etc. etc. Taste: In tasting the material, as recommended by our experience with the material, the following procedure was applied: a pinch of the material was mixed with water and savored, or chewed without swallowing, to determine the strength and the sensation of the taste. The strength is recorded as: tasteless, weak, distinct, or strong; and the sensation, as: sweet, sour, saline, or bitter. Phytochemical tests for tannins, saponins, terpenoids, anthraquinones and alkaloids were carried out on samples by procedures as described in detail elsewhere (Ameh et al., 2010c, d).

2.3 Loss on drying
This was carried out using a minimum of 0.5 – 1.0 g of material. Drying was effected in a Lindberg/Blue M gravity-convention oven maintained at 105-110 °C, for 3 h, after which the sample was allowed to cool to room temperature in a desiccator, and subsequently weighed. The time interval from the oven to point of weighing was usually about 30 minutes. The results are expressed as a range or as mean ± standard deviation.

2.4 Evaluation of extractive matter
About 4 g of accurately weighed coarsely powdered, air-dried sample was transferred into a glass-stoppered, 250-ml reflux conical flask, followed by the addition of 100 ml of solvent. The flask was weighed along with its contents, and recorded as W1. The flask was well shaken, and allowed to stand for 1 h. Subsequently a reflux condenser was attached to the flask, and gently brought to boiling and maintained thereat boiled 20 – 60 minutes depending upon the solvent. The mixture was subsequently cooled and weighed again. The weight was recorded as W2, and then readjusted to W1 with the solvent. The flask was shaken well once again and its contents rapidly filtered through a dry filter paper. By means a pipette, 25 ml of the filtrate was transferred to a previously dried and tarred glass dish and then gently evaporated to dryness on a hot plate. Subsequently, the dish was dried at 105 °C for 1-6 hours, cooled in a desiccator for 30 min, and weighed. The extractable matter was calculated as %w/w of the air-dried sample.

2.5 Determination of solubility of material in a given solvent – Methods I and II
The solubility of a material was determined at room temperature ~ 25°C and expressed in terms of "parts', representing the number of milliliter of solvent, in which 1 g of the material is soluble. Vials of appropriate sizes: ~4-ml, ~ 12-ml and ~20-ml capacities were used. The mixtures were thoroughly shaken for at least 30 min before inspection for un-dissolved solute. In methods I, each vial received 100 mg of sample and the volume of solvent indicated. In method II, a vial received 100 mg and increasing volumes of solvent. The methods give the same results.

2.6 Light absorption and thin layer chromatography (TLC)
UV-VIS Spectrophotometer (Jenway or Shimadzu) and quartz 1-cm cells were used for the study. Solutions of herb and extract were made by thoroughly mixing 1 part of the solute

and with 100 parts of solvent methanol: water 80:20, v/v filtering, and diluting the filtrates by 150x with the same solvent. Absorbencies were measured at λ227 nm, using the solvent as the blank. Florescent, precoated plates were used for normal phase TLC, utilizing silica K6, and hexane: ethylacetate as mobile phase. Solutions of analytes were prepared and applied as follows: To 1 mg of the analyte, 2 drops of ethanol were added and mixed well (~1 %w/v solution). The plates used were 5 cm wide x 20 cm long. With a ruler and a pencil, a distance of 5 mm was measured from the bottom of the plate, and a line of origin was lightly drawn across the plate, without disturbing the adsorbent. The analyte was applied to the origin as a 1 μl droplet. The spot was allowed to dry. Subsequently, the plate was developed in a developing tank saturated with the vapour of the solvent system to be used as mobile phase. The level of the solvent in the tank was adjusted to a level 2 to 3 mm below the line of origin on the plate. The plate was considered developed when the solvent front reached a predetermined line, not less than 5 mm below the top of the plate. The air-dried plate was visualized using a viewing cabinet (Cammag) and a UV-lamp (Cammag – equipped to emit light at 254 or 366 nm). The resulting chromatogram was photographed and subsequently drawn to scale.

2.7 Determinations of pH of preparations – herb and the dry extract
Determination of pH values was with a Jenway pH Meter. Standard pH solutions: 4, 7 and 10; and freshly distilled water were used for the study.

2.8 Determination of foaming indices of preparations – herb and the dry extract
Decoctions of plant materials foam due to the presence of saponins. This ability is measured as foaming index, and is an important quality control parameter. The requirements for the test include: conical flasks (500-ml); volumetric flasks (100-ml); test tubes (16cm x 16mm); ruler; and stop-clock. The procedure was as follows: Exactly 1.0 g of powdered material was accurately transferred into a 500-ml conical flask containing 100 ml of boiling water, and maintained at moderate boiling for 30 minutes. The mixture was then cooled and filtered into a 100-ml volumetric flask. The volume was made up to 100 ml with water. Successive portions of 1 ml, 2 ml, 3 ml etc up to 10 ml of the filtrate was poured into ten stoppered tubes having the following dimensions: height, 16 cm; diameter, 16 mm. Subsequently, the volume of each tube was adjusted to 10 ml with water, stoppered and shaken in lengthwise motion for 15 seconds, at 2 shakes per second. The tubes were allowed to stand for 15 minutes, and the height of the foam in each tube was measured. The results were assessed as follows:

Foaming index is ≤ 100, if the height of foam in all the tubes is less than 1 cm.

If a height of 1 cm is obtained in any tube, the volume [V] of the decoction in that tube, is used to determine the foaming index, as = 1000/V.

But if the tube above is the first or the second in the series, prepare an intermediate dilution to obtain a more precise result.

If the height of the foam is > 1 cm in every tube, the foaming index is over 1000.

To obtain a more precise result, repeat the determination using a new series of dilutions of the decoction. Note the tube in which the height of foam is 1 cm, and the volume [V] of the decoction therein, and calculate the foaming index, as = 1000/V. Results are expressed as expressed as a quantity [Q] per ml or as [Q]ml-1.

3. Results

3.1 Results of botanical examination / phytochemical tests on the herb and extract

The key botanical and phytochemical characteristics of *Mitracarpus scaber* Zucc (Family: Rubiaceae) with Voucher specimen number: NIPRD/H/4208, preserved in the Institute's Herbarium are indicated in Table 4. The plant grows erect, up to 55 cm high, usually branched; the leaves are lanceolate, 2-4 cm long, with the upper surface bearing minute hairs. The plant manifests dense clusters of inflorescence, 6-14 mm across, with minute white flowers. The fruits are dehiscent capsules, about 0.5-1 mm long. Both the fresh plant and air-dried weed are practically odourless but possess a slight warm taste. Tannins, saponins and anthraquinones were detected in the weed. The extract also contained tannins and anthraquinones but not saponins. Tests for alkaloids were negative for the weed and extract.

Characteristic	Live Sample	Air-dried Sample
General appearance	M. scaber is an annual, with erect stems, up to 55 cm high, and often branched. The leaves are lanceolate, 3-5 cm long, with the upper surface scabrous. The inflorescence consists of clusters of small white flowers. The fruits are dehiscent capsules, up to 1 mm long. The plant is of the Family, Rubiaceae, reproduces by seeds, and is found in the tropics.	The air-dried sample consists of brownish green twigs and other parts that can readily be ground in a mortar or comminuting machine. The air-drying process takes about a week during the months of October to December, at NIPRD Herbarium, Abuja. The extracts obtained with various solvents yield a black, odorless and sticky mass.
Odor	Odourless	Odourless
Taste	Very slightly warm	Slightly warm
Phytochemicals	Tannins, saponins and anthraquinones were detected. The tests for alkaloids were negative.	Tannins and anthraquinones were detected. The tests for alkaloids and saponins were negative.

The samples described above were obtained from the NIPRD Botanical Gardens at Idu Industrial Area, Idu, Abuja, Federal Capital Territory, Nigeria.

Table 4. Some key characteristics of *Mitracarpus scaber* and its aerial parts.

3.2 Results of physicochemical tests on the herb and extract of *Mitracarpus scaber*

Typical results of loss on drying (LOD as %w/w was 10.29 ± 1.81 for the herb; and 15.86 ± 0.72 for the extract) and total ash (TA as %w/w was 12.44 ± 2.95 for the herb; and 0.40 ± 0.09 for the extract) are shown in Table 5. The Table also shows that a 5%w/v mixture of the herb had pH of 5.7 ± 0.3, while that of the extract was 6.9 ± 0.3. The herb in water foamed slightly (that is: Foaming Index [FI] ≤ 100), but the extract did not foam at all (that is: FI = 0). Table 5 further shows that the herb in methanol/water (80/20: v/v) had an A1%1cm of 325.8 ± 15.6, while that of the extract was 349.5 ± 14.1. The Table also shows the extractabilities of the herb in various solvents. The extractability results expressed as (%w/w) were as follows: acetone, 6.89 ± 0.89; water, 28.37 ± 1.77; ethanol, 11.72 ± 0.81; ethylacetate, 14.02 ± 1.89; hexane, 4.11 ± 0.47; and methanol, 15.11 ± 1.07. The extractability of the air-dried weed was highest in water and least in hexane. Among the organic solvents, extractability was lowest in hexane, and highest in methanol, followed by ethylacetate, ethanol and acetone.

Parameter (Mean ± SD)	Air-dried Herb	Ethylacetate Extract
Loss on drying (LOD: % w/w)	10.29 ± 1.81 (n=12)	15.86 ± 0.72 (n=12)
Total ash (TA: % w/w)	12.44 ± 2.95 (n=12)	0.40 ± 0.09 (n=11)
pH of 5 % w/v in water	5.7 ± 0.3 (n=5)	6.9 ± 0.3 (n=5)
Foaming Index (FI: as ml^{-1})	Slight foam. FI ≤ 100 (n=5)	No foam. FI = 0 (n=5)
A 1%1cm at λ227 nm (MeOH/ H$_2$O: 80/20 v/v)	325.8 ± 15.6 (n=5)	349.5 ± 14.1 (n=5)
Extractability (% w/w) in: **Solubility** (ml/g) in:	Extractive value (n=8-12) -	- Solubility (n=2-6)
Water	28.37 ± 1.77	>10^3
Methanol	15.11 ± 1.07	25.0 ± 0.0
Ethylacetate	14.02 ± 1.89	25.0 ± 5.0
Ethanol	11.72 ± 0.81	25.0 ± 0.0
Acetone	6.89 ± 0.89	15.0 ± 0.0
Hexane	4.11 ± 0.47	55.0 ± 5.0

The LOD results were validated by concurrent determination of the LOD of copper sulphate, which result (mean ± SD) was 36.12 ± 0.19 %. The results prove that the extract was quite hygroscopic. The low TA results for the extract but not the herb probably suggests a high presence of high bio-minerals. The high water extractability result agrees with the high TA of the herb, and the fact that hexane, the least polar solvent, produced the lowest extractability result. Among the organic solvents the solubility of the extract was least in hexane (55 ml/g), but higher in ethanol, ethylacetate and methanol (15-25 ml/g). The extract was practically insoluble in water (>10^{3}ml/g). The colour of the solution obtained from the herb using different organic solvents was clear and greenish-brown in each case, but that obtained with water was yellowish brown, and slightly cloudy, with no tinge of green.

Table 5. Various physicochemical parameters of herb and extract of *Mitracarpus scaber*.

3.3 Results of thin layer chromatographic (TLC) studies on the herb and extracts of *M. scaber*

Figure 1 is a normal phase TLC of the herb and extract developed with hexane-ethylacetate. The Figure indicates the following: The herb in acetone (A) or ethanol (C) yielded 5 identical principal spots, while the herb in water (B) yielded only 2 principal spots – Rf1 and Rf4. The herb in ethylacetate (D) or hexane (E) yielded 3 spots, while the herb (F) in methanol yielded 4. The dry hexane extract (G) re-dissolved in hexane yielded 7 spots, while the ethylacetate extract (H) re-dissolved in ethylacetate yielded 10. Notably, Rf4 was present in all the chromatograms, while Rf2 and Rf3 were present only in the H chromatogram. On the other hand Rf1 was present only in the B, G and H chromatograms.

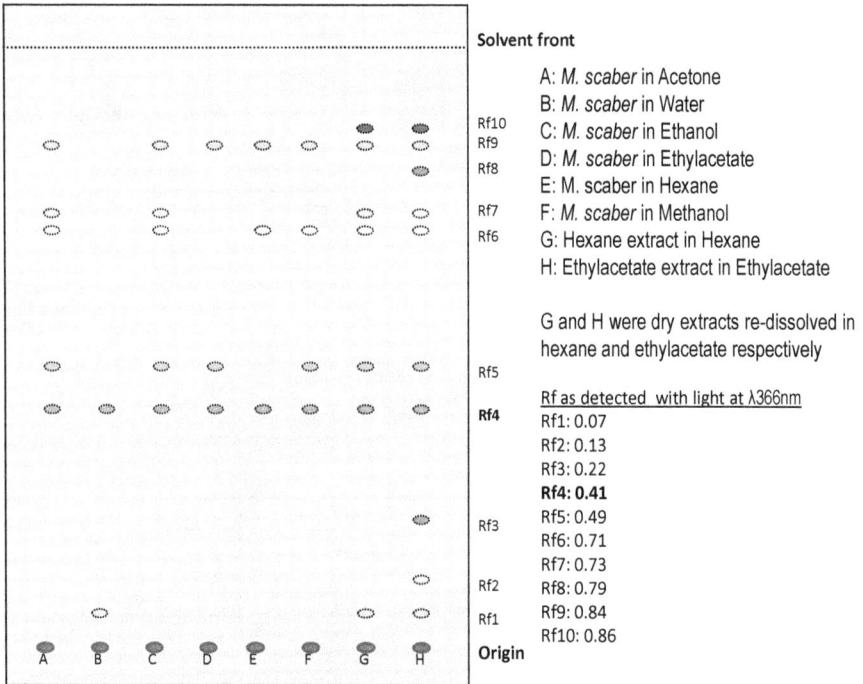

Solvent front

A: *M. scaber* in Acetone
B: *M. scaber* in Water
C: *M. scaber* in Ethanol
D: *M. scaber* in Ethylacetate
E: M. scaber in Hexane
F: *M. scaber* in Methanol
G: Hexane extract in Hexane
H: Ethylacetate extract in Ethylacetate

G and H were dry extracts re-dissolved in hexane and ethylacetate respectively

Rf as detected with light at λ366nm
Rf1: 0.07
Rf2: 0.13
Rf3: 0.22
Rf4: 0.41
Rf5: 0.49
Rf6: 0.71
Rf7: 0.73
Rf8: 0.79
Rf9: 0.84
Rf10: 0.86

The above diagram is of a normal phase TLC (K5 Silica, using hexane: ethylacetate at 60:40 v/v as mobile phase) of samples of samples of herb and extracts in various solvents as follows. Types of samples/ solvents: A: Herb in Acetone, B: Herb in Water, C: Herb in Ethanol, D: Herb in Ethylacetate, E: Herb in Hexane, F: Herb in Methanol, G: Hexane extract in Hexane, H: Ethylacetate extract in Ethylacetate. The samples in G and H were dry extracts re-dissolved in hexane and ethylacetate respectively. Rf as detected at λ366nm: Rf1: 0.07, Rf2: 0.13, Rf3: 0.22, Rf4: 0.41, Rf5: 0.49, Rf6: 0.71, Rf7: 0.73, Rf8: 0.79, Rf9: 0.84, Rf10: 0.86. Descriptions/interpretations: The herb in acetone (A) or ethanol (C) yielded 5 identical principal spots, while the herb in water (B) yielded only 2 principal spots – Rf1 and Rf4. The herb in ethylacetate (D) or hexane (E) yielded 3 spots, while the herb (F) in methanol yielded 4. The dry hexane extract (G) re-dissolved in hexane yielded 7 spots, while the ethylacetate extract (H) re-dissolved in ethylacetate yielded 10. Notably, Rf4 was present in all the chromatograms, while Rf2 and Rf3 were present only in the H chromatogram. On the other hand Rf1 was present only in the B, G and H chromatograms.

Fig. 1. Diagrammatized normal phase TLC of *M. scaber* extracts showing up to ten principal spots.

Months of storage in capped glass bottles at RTH	Appearance		Extractability of herb in different solvents			Solubility of extract in different solvents			Loss on drying (mean ± SD) %w/w	
	Herb	Extract	Ethyl-acetate	n-C₆H₁₄	H₂O	Ethyl-acetate	n-C₆H₁₄	H₂O	Herb	Extract
0	Wrinkled, brownish green/ grey leaves and twigs	Sticky mass becoming slightly gritty upon exposure	14.0 ± 1.8	4.2 ± 0.5	36.4 ±2.2	25.0 ± 5.0	55.0 ± 5.0	>10³	10.51 ± 0.74 [a] (7)	14.07 ± 1.73 [b] (7)
3	Practically unchanged from above	Practically unchanged from above	15.3± 2.2	4.3 ± 0.5	33.9 ± 2.4	25.0 ± 5.0	45.0 ± 5.0	>10³	9.11 ± 0.67 [a] (6)	12.84 ± 1.18 [b] (6)
9	Practically unchanged from above	Practically unchanged from above	14.4 ± 1.9	4.1 ± 0.5	30.2 ± 2.4	25.0 ± 0.0	55.0 ± 0.0	>10³	10.44 ± 0.51 [a] (6)	13.14 ± 1.11 [b] (7)
21	Practically unchanged from above	Practically unchanged from above	14.7 ±1.9	4.1 ± 0.5	30.7± 2.2 [a]	25.0 ± 5.0	50.0 ± 5.0	>10³	10.87 ± 0.73 [a] (7)	14.84 ± 0.99 [b] (5)
39	Practically unchanged from above	Practically unchanged from above	15.0 ± 2.0	4.3 ± 0.5	27.8 ±2.0 [a]	25.0 ± 0.0	55.0 ± 0.0	>10³	8.96 ± 0.59 [a] (5)	11.44 ± 0.86 [b] (6)

The results show that storage of both the herb and the extract in capped glass bottles at room temperature and humidity for up to 39 months, produced no consistent or statistically significant change in moisture content. Both ([a]) and ([b]) indicate that the figures bearing them do not differ statistically from one another (P > 0.05).

Table 6. Effect of storage on herb and ethylacetate extract of *M. scaber* as evaluated by appearance, extractability, solubility and loss on drying.

3.4 Effect of storage on herb and ethylacetate extract of *M. scaber* as evaluated by appearance, extractability, solubility and loss on drying

Table 6 shows that the general appearance of the herb as wrinkled, brownish green/ grey leaves and twigs remained essentially unchanged up to the 39th month of storage. However, the extractability of the herb in water fell slightly but significantly as from after the 3rd month of storage. Table 6 also shows that neither the solubility profile nor the appearance of the extract and the solutions made from them in different solvents changed with storage. The same Table 6 shows that storage of the herb and the extract in capped glass bottles at room temperature and humidity (RTH) for up to 39 months produced no consistent or statistically significant changes in moisture content.

3.5 Effect of storage on light absorption and TLC features of the herb and extract

Table 7 presents the effect of storage on light absorption and TLC features of the herb and extract. It shows the following: that the difference in absorbance between 0th month and the 21st/ 39th months was insignificant for the herb (P > 0.05). By contrast, the corresponding difference for the extract was significant (P < 0.05). Table 7 also shows that the number of TLC spots observed for the herb and extract at every stage of storage was unchanged up to the 39th month.

Months of storage in capped glass bottles at RTH	Herb							Extract		
	Abs. at λ227 nm	Types of solvent/ number of TLC spots						Abs. at λ227 nm	TLC spots	
		A	B	C	D	E	F		G	H
0	2.172 ± 0.104 (5)	5	2	5	3	3	4	2.330 ± 0.094 (5)	7	10
3	2.221 ± 0.114 (6)	5	2	5	3	3	4	2.174 ± 0.107 (5)	7	10
9	2.144 ± 0.098 (5)	5	2	5	3	3	4	2.104 ± 0.070 (5)	7	10
21	2.322 ± 0.117^a(5)	5	2	5	3	3	4	2.039 ± 0.104^b(5)	7	10
39	2.233 ± 0.114^a (5)	5	2	5	3	3	4	2.084± 0.111^b(6)	7	10

For the herb, the difference in absorbance between 0th month and the 21st/ 39th months, denoted by (a), was insignificant (P > 0.05). By contrast, for the extract, the difference in absorbance between the 0th or 3rd month and the 21st or 39th month, denoted by (b), was significant (P < 0.05). Notably, the number of TLC spots observed for both the herb and extract remained unchanged up to the 39th month.

Table 7. Effect of storage on light absorption and TLC characteristics of herb and extract of *M. scaber*.

3.6 Effect of storage on pH and foaming indices of herb and ethylacetate extract of *M. scaber*

Table 8 shows that the pH of a 5 %w/v mixture of the herb or extract in water did not change significantly with storage for up to 39 months. However, although the 5 %w/v mixture of the fresh plant material in water did foam slightly; this property diminished rapidly, and was totally lost after the 3rd month of storage. By contrast, the ethylacetate extract never foamed at any stage of storage.

Months of storage in capped glass bottles at RTH	pH of mixture (5% w/v)		Foaming index	
	Herb in water	Ethyl acetate extract in water	Herb in water	Ethylacetate extract in water
Within 1st day of harvest or preparation	5.7 ± 0.3 (5)	6.9 ± 0.3 (5)	Foam: Slight ≤ 100 (5)	Foam: Nil (5)
0	5.9 ± 0.2 (6)	6.1 ± 0.3 (5)	Foam: Slight < 100 (5)	Foam: Nil (6)
3	5.6 ± 0.2 (5)	6.2 ± 0.4 (6)	Foam: Slight < 100 (5)	Foam: Nil (5)
9	5.4 ± 0.2 (5)	6.1 ± 0.3 (5)	Foam: Nil (6)	Foam: Nil (5)
21	5.9 ± 0.3 (7)	6.7 ± 0.3 (5)	Foam: Nil (5)	Foam: Nil (5)
39	5.9 ± 0.3 (5)	6.5 ± 0.3 (5)	Foam: Nil (5)	Foam: Nil (5)

Both (a) and (b) indicate that the pH of the 5 %w/v aqueous mixtures at every stage fell within the mean values of 5.8 ± 0.2a and 6.4 ± 0.4b - they indicate that any deviations from these mean values were insignificant (P > 0.05). The freshly harvested samples foamed measurably, but the ability was totally lost after the 3rd month of dry storage. In all cases however, the dry ethylacetate extract was virtually insoluble in water, and did not foam at all.

Table 8. Effect of storage on pH and foaming indices of the fresh plant material, the dry herb and the ethylacetate extract of *M. scaber*.

4. Discussions

The aim of this study was to apply official methods of WHO (1998) and BP (2004) to study the key quality attributes of the air-dried weed and the ethylacetate extract of *Mitracarpus scaber*, for the purpose of quality control, GMP production and registration of Niprifan by NAFDAC. WHO had defined "Herbal Substance" as "Material derived from the plant(s) by extraction, mechanical manipulation, or some other process" (WHO, 2005). Thus, either the ethylacetate extract, or even the comminuted, air-dried vegetable matter, may rightly be termed the "Herbal Substance" of Niprifan. Since the advent of the Alma-ata Declaration in 1978, many developing countries opted to adopt the WHO model in developing their National Traditional Medicine, especially phytotherapy (Ameh et al., 2010b). NIPRD's adherence to the WHO model had resulted in the sickle cell drug – Niprisan, developed from Yoruba Traditional Medicine (Wambebe et al., 2001). It is generally held that in most countries, especially in Africa, the populations depend greatly on herbal remedies, up to 90 % in some instances like Ethiopia (BBC, 2006). Such high dependence calls for a system or mechanism for harnessing and optimizing all or most of such plant resources. That means that every effort must be made to obtain maximum benefits from them. One way to do this is to standardize the raw materials used in producing the remedies, by studies such as this one. Such studies will at least help to minimize waste, and even lead ultimately to conservation of endangered plants. Indeed, efforts at conservation are more likely to succeed when the value of what is to be conserved is proven.

Our immediate interest however, is in the need to entrench the use of these resources by taking appropriate actions, which, in this case is - an application to NAFDAC to consider the registration of Niprifan, based on folkloric use evidence, pertinent literature, and the experimental data provided in this study. These three lines of evidence can be summarized as follows. At the peak of British colonialism in Africa considerable effort was made to harness the continent's wealth in herbal traditions. Thus at as far back as the 1930s a team of British scientists had combed the entire West Africa to research traditional herbal remedies. Thus, for *Mitracarpus scaber*, Hutchinson and Dalziel (1948) reported a number of findings that have subsequently been confirmed by work in NIPRD and elsewhere (Benjamin et al., 1986; Irobi and Daramola, 1994; Cimanga et al., 2004; Abere et al., 2007a, 2007b). These include the following: that *M. scaber* was widely distributed and used topically in all of West Africa for various skin infections; and orally for various internal conditions. Among the traditional indications mentioned, and which have since been confirmed by NIPRD's Ibrahim Muaazzam (ethnobotanist and consultant on TM) are: leprosy, lice, ringworm, eczema and craw-craw. Currently, the plant is used orally for sore throat, for which purpose it is wholly macerated in water.

Among the vernacular names of *M. scaber* are: Hausa (*goga masu*); Fulani (*gadudal*); Yoruba (*irawo-ile*); and Ibo (*obu obwa*). Professor Ogundiani (2005) in his inaugural address at the University in Ile-Ife commented on Niprifan,

stressing the antimicrobial potency of *M. scaber*. Ogundiani, as stated in the lecture, had been unaware of the NIPRD's work on Niprifan, until shortly before the inaugural, since that work, led by Professors Wambebe, Okogun and Nasipuri, had been unpublished. Therefore, in this paper we elect to present not only these historical antecedents, but also to furnish the results of our evaluation of the key quality variables of the herb and extract of *M. scaber*, with a view to advancing the registration of Niprifan (for skin infections) by NAFDAC. The results here presented probably suffice for quality control and GMP production, particularly if more emphasis is placed on technical requirements than on bureaucracy. It must be remarked at this juncture that NAFDAC only belatedly recognized the sickle cell drug, Niprisan, after the US-FDA and EMEA had granted it orphan status (Pandey, 2003). One may wonder - What a paradox! Why should the US and Europe that need herbal drugs far less than Nigeria be keener in their regulation? Therefore, from the foregoing, it seems that the key to this Nigerian enigma lies not in the technical but in the non-technical differences between NAFDAC and EMEA as depicted in Tables 1-3. The said differences which hinge on NAFDAC's extra requirements (Table 3) suggest that NAFDAC needs to re-strategize for efficient discharge of its Mandate. For example, despite the widespread use of herbal medicines in Nigeria and the Federal Policy on TM (2007), NAFDAC is not known to have "fully registered" a single herbal medicine since its creation in 1992/3, whereas it should. This is the puzzle this article had hoped to address.

5. References

Abere , T. A.; Onyekweli, A. O. & Ukoh, G. C. (2007a). *In vitro* Antimicrobial Activity of the Extract of Mitracarpus scaber Leaves Formulated as Syrup. *Tropical Journal of Pharmaceutical Research* Vol. 6, No. 1, pp. 679-682.

Abere, T. A.; Onwukaeme, D. N. & Eboka, C. J. (2007b). Pharmacognostic evaluation of the
leaves of *Mitracarpus scaber* Zucc (Rubiaceae). *Tropical Journal of Pharmaceutical
Research* Vol. 6, No. 4, pp. 849-853.

Ameh, S. J.; Obodozie, O. O.; Inyang, U. S.; Abubakar, M. S. & Garba, M. (2010a). Current
phytotherapy – an inter-regional perspective on policy, research and development
of herbal medicine. *Journal of Medicinal Plants Research* Vol. 4(15), pp 1508-1516, 4
August, 2010.

Ameh, S. J.; Obodozie, O. O.; Inyang, U. S.; Abubakar, M. S. & Garba, M. (2010b). Current
phytotherapy - A perspective on the science and regulation of herbal medicine.
Journal of Medicinal Plants Research; Vol. 4(2): 072-081.

Ameh, S. J.; Tarfa, F. D.; Abdulkareem, T. M.; Ibe, M. C.; Onanuga, C. & Obodozie, O. O.
(2010c). Physicochemical Analysis of the Aqueous Extracts of Six Nigerian
Medicinal Plants. *Tropical Journal of Pharmaceutical Research*, 9 (2): 119-125.

Ameh, S. J.; Obodozie, O. O.; Inyang, U. S.; Abubakar, M. S. & Garba, M. (2010d). Quality
Control Tests on Andrographis paniculata Nees (Family: Acanthaceae) – an Indian
'Wonder' Plant Grown in Nigeria. *Tropical Journal of Pharmaceutical Research*, Vol. 9,
No 4, pp. 387-394

Ann Godsell Regulatory (2008). Pharmaceutical Good Manufacturing Practice for Herbal
Drug Substances 2008 [cited 2010 April 8]. Available online at:
http://www.pharmaceutical-int.com/article/category/treatment-herbal-
medicines

BBC News (2006). Can herbal medicine combat Aids? Wednesday, 15 March, 13:10 GMT.
http://newsvote.bbc.co.uk/mpapps/pagetools/print/news.bbc.co.uk/2/hi/Afric
a/4793106.stm .

Benjamin, T. V.; Anucha, T. C. & Hugbo, P. G. (1986). An approach to the study of medicinal
plants with antimicrobial activity with reference to *Mitracarpus scaber*. *In*: Sofowora,
A. (Ed.) *The State of medicinal Plants Research in Nigeria*, pp. 243-245, Ibadan
University Press, ISBN 978-30285-0-2, Ibadan, Nigeria.

Bisignano, G.; Sanogo, R.; Marino, A.; Angelo, V.; Germano, M.; De Pasquale, R. & Pizza, C.
(2000). Antimicrobial activity of Mitracarpus scaber extract and isolated
constituents. *Letters in Applied Microbiology*, Vol. 30, pp. 105-108. doi:10.1046/j.1472-
765x.2000.00692.x.

Cimanga, R. K.; Kambu, K.; Tona, L.; Bruyne, T.; Sandra, A.; Totte, J.; Pieters, L. & Vlietinck,
A. J. (2004). Antibacterial and antifungal activities of some extracts and fractions of
Mitracarpus scaber Zucc. (Rubiaceae). *Journal of Natural Remedies*, Vol. 4, No. 1, pp.
17-25

De Smet, P. N. (2005). Herbal medicine in Europe – relaxing regulatory standards. *New
England Journal of Medicine*, Vol. 352, No. 12, pp. 1176-78.

DSHEA (1994). Dietary Supplements Health Education Act of 1994 [cited 2010 April 8].
Available at: http://fda/Food/DietarySupplements/ucm109764.htm

Gbaguidi, F.; Accrombessi, G.; Moudachirou, M. & Quetin-Leclercq, J. (2005). HPLC
quantification of two isomeric triterpenic acids isolated from Mitracarpus scaber
and antimicrobial activity on Dermatophilus congolensis. *Journal of Pharmaceutical
& Biomedical Analysis*, Vol. 39, No. 5, pp. 990-995.

Germanò, M. P.; Sanogo, R; Costa, C; Fulco, R; D'Angelo, V.; Viscomi, E. G. & de Pasquale, R. (1999). Hepatoprotective Properties in the Rat of *Mitracarpus scaber* (Rubiaceae). *Journal of Pharmacy & Pharmacology*, June 1999. Vol. 51, No. 6, pp. 729-734.

Goldman, P. (2001). Herbal medicines today and the roots of modern pharmacology. *Annual Internal Medicine*, Vol. 135, No. 8, Pt 1, pp. 594–600.

Gross, A. & Minot, J. (2007). Chinese Manufacturing: Scandals and Opportunities. Published in MX, November/ December, Pacific Bridge Medicals.
http://www.pacificbridgemedicals.com/

Houghton, P. J.; Ibewuike, J. C.; Mortimer, F.; Okeke, I. N. & Ogundaini, A. O. (2002). Antibacterial quinones from Mitracarpus scaber. *Journal of Pharmacy & Pharmacology*, 52 (Suppl.) 261.
www.kcl.ac.uk/content/1/c6/01/66/69/PUB82008.pdf

Hutchinson, J. & Dalziel, J. M. (1948). *The flora of West Tropical Africa*. Crown Agents for the Colonies. 4 Millbank, Westminster, London, SW1., London. Vol 11. 1948. p. 222.

Irobi, O. N. & Daramola, S. O. (1994). Bactericidal properties of crude extracts of Mitracarpus villosus. *Journal of Ethnopharmacology*, Vol. 42, No. 1, pp. 39-43.

NAFDAC's website (2010). Website of the National Agency for Food and Drug Administration and Control. *www.nafdacnigeria.org/*

Ogundaini, A. O. (2005). From Greens into Medicine: Taking a Lead From Nature. Inaugural Lecture Series 176. Ile-Ife. Tuesday 26, April. Obafemi Awolowo University Press Limited

Okunade, A. L.; Clark, A. M.; Hufford, C. D. & Oguntimein, B. O. (1999). Azaanthraquinone: an antimicrobial alkaloid from Mitracarpus scaber [letter] *Planta Medica,*Vol. 65, No. 5, pp. 447-8.

Pandey, R. C. (2003). Xechem's Sickle Cell Drug, NIPRISAN – HEMOXIN Granted Orphan Drug Status by the FDA NEW BRUNSWICK, N.J.-- (BUSINESS WIRE)--Sept. 2, 2003--Xechem International, Inc. (OTC BB: XKEM).

Pelletier, K. R. (2009). Guide to Herbal Medicine (cited 2009 June 2). Available from: http://www.ca.encarta.msn.com/sidebar_701509401/Guide_to_Herbal_Medicine. html

Wambebe, C; Khamofu, H.; Momoh, J.; Ekpeyong, M.; Audu, B.; Njoku, B. S.; Nasipuri, N. R.; Kunle, O. O. et al. (2001). Double-blind, placebo-controlled, randomized cross-over clinical trial of NIPRISAN in patients with sickle cell disorder. Phytomedicine, 8(4):252-61.

WHO (1998). Quality control methods for medicinal plant materials. World Health Organization, WHO, Geneva, pp. 1-115.

WHO (2000). General guidelines for methodologies on research and evaluation of traditional medicine (Document WHO/EDM/ TRM/2000.1). World Health Organization, WHO, Geneva. pp. 1-184.

WHO (2005) APPENDIX 2 Glossary of key terms. In: Information needed to support Clinical Trials of herbal products. TDR/GEN/Guidance/05.1Operational Guidance: Special Programme for Research and Training in Tropical Diseases, WHO, Geneva, pp. 1-16.

13

Practical Quality Control: the Experiences of a Public Health Laboratory

Francesc Centrich[1], Teresa Subirana[1],
Mercè Granados[2] and Ramon Companyó[2]
[1]Laboratori de l'Agència de Salut Pública de Barcelona;
[2]Universitat de Barcelona
Spain

1. Introduction

In the 1930's W.A. Shewhart pioneered the application of statistical principles to the quality control (QC) of production processes, eventually publishing the landmark book "Economic Control of Quality of Manufactured Products" (Shewhart, 1931). In this book, he states that a phenomenon is *under control* if its future variation can be predicted (within limits) based on previous experience. This is precisely the idea behind the control charts used in measurement processes—specifically, for chemical analysis. The International Organization for Standardization (ISO), in its standard ISO 9000 (ISO, 2005a), defines *quality control* as "the part of quality management focused on fulfilling quality requirements". According to the standard, quality management also includes quality planning, quality assurance and quality improvement. The above definition is rather vague, because quality management systems based on the ISO 9000 family of standards can be applied to any kind of organization regardless of its field of activity, its size or whether it is from the public or private sectors. Testing laboratories typically distinguish between *internal* and *external* QC. In this context, the International Union of Pure and Applied Chemistry (IUPAC, 1998) gives a definition of internal QC that is well-suited to an analytical laboratory: "the set of procedures undertaken by laboratory staff for the continuous monitoring of operation and the results of measurements in order to decide whether results are reliable enough to be released". Although the aforementioned document does not formally define *external QC*, it does mention that *external control* may be done by submitting blind samples to the measuring laboratory. This activity can be organized in the form of a collaborative test.

The aim of these QC activities is to verify that the quality parameters of an analytical method ascertained in the method validation are maintained during its operational lifetime. Thus, method validation or revalidation tasks are periodic activities that end with a validation report, whereas QC activities are recurrent activities implemented in routine work. Apart from the use of fully validated methods, QC assumes the use of properly maintained, verified and calibrated equipment, reagents and consumables with the proper specifications; standards with well-established traceability; and qualified technicians working in suitable environmental conditions. However, fulfilling all these requirements is not enough to ensure the delivery of appropriate quality results over time: a laboratory's capacity to produce technically correct results must be continuously monitored. Indeed, according to Thompson *et al.* (Thompson & Lowthian, 1993), QC is the only quality management measure that provides a high level of protection against the release of inaccurate data. The authors demonstrate a significant relationship between the efficacy of a laboratory's QC and its subsequent performance in proficiency tests. They also consider that the implementation of QC activities and the participation in proficiency tests are two sides of the same coin: a laboratory's commitment to quality.

Once a laboratory has implemented a method in its routine work, is performing adequate QC, has taken any appropriate corrective and/or preventive actions, and its staff has acquired sufficient expertise, it may consider including this method in its scope of accreditation. Figure 1 shows these activities in the context of the operational lifetime of an analytical method.

This chapter was written to explain, in practical terms, the QC activities and management at an analytical laboratory—namely, the Chemical Analysis Service at the Laboratory of the Public Health Agency of Barcelona (*Laboratori de l'Agència de Salut Pública de Barcelona*; hereafter, *LPHAB*).

2. History and present context of the LPHAB

The LPHAB has its origin in the Municipal Laboratory of Barcelona, a microbiology laboratory created in 1886 to provide support to the sanitary authorities in their efforts to prevent rabies; since its inception, the Municipal Laboratory of Barcelona was a reference laboratory in Spain. In 1907, owing to its ever-increasing activities, it was given a new structure that led to creation of a section dedicated to chemical analysis of foods, with the then innovative objective of studying health problems attributable to the presence of hazardous chemicals in foods.

From the 1950's onwards, the section on chemical analysis of foods underwent major development. This stemmed from advances in knowledge on food chemistry and was catalyzed by various international food crises caused by chemical pollutants such as mercury and methanol. A case of widespread food poisoning in Spain in 1981, traced to denatured rapeseed oil, triggered the modernization of many Spanish public health laboratories, including the Municipal Laboratory of Barcelona. The Laboratory's equipment was soon updated, and its organization and management were overhauled. These changes enabled the Municipal Laboratory of Barcelona to face new analytical challenges. In addition to assessing the nutritional properties of food, it also focused on detection and determination of additives, residues and contaminants in food. The Municipal Laboratory of Barcelona began serving customers outside of the municipal administration; the challenge of providing these customers with the data they sought at specific analysis costs and response times proved highly stimulating. By the year 2000, it had analyzed 20,000 samples. In 2003 the Municipal Laboratory of Barcelona merged with the Public Health Laboratory of the Autonomous Government of Catalonia (*Generalitat de Catalunya*) in Barcelona to form the LPHAB. This union led to significant investments in instrumentation and to the recruitment of new staff; consequently, the newly formed LPHAB became one of the strongest laboratories in Spain for food analysis.

The LPHAB currently comprises four departments: two technical departments (the Chemical Analysis Service [CAS] and the Microbiological Analysis Service) and two management & support departments (the Quality Assurance Unit [QAU] and the Logistics & Services Unit). It presently employs 65 people, 31 of which work in the CAS (11 senior technicians and 20 mid-level technicians and support staff). The CAS encompasses four areas: two dealing with applications (food analysis and environmental analysis) and two dealing with analytical techniques (spectroscopic analysis and chromatographic analysis).

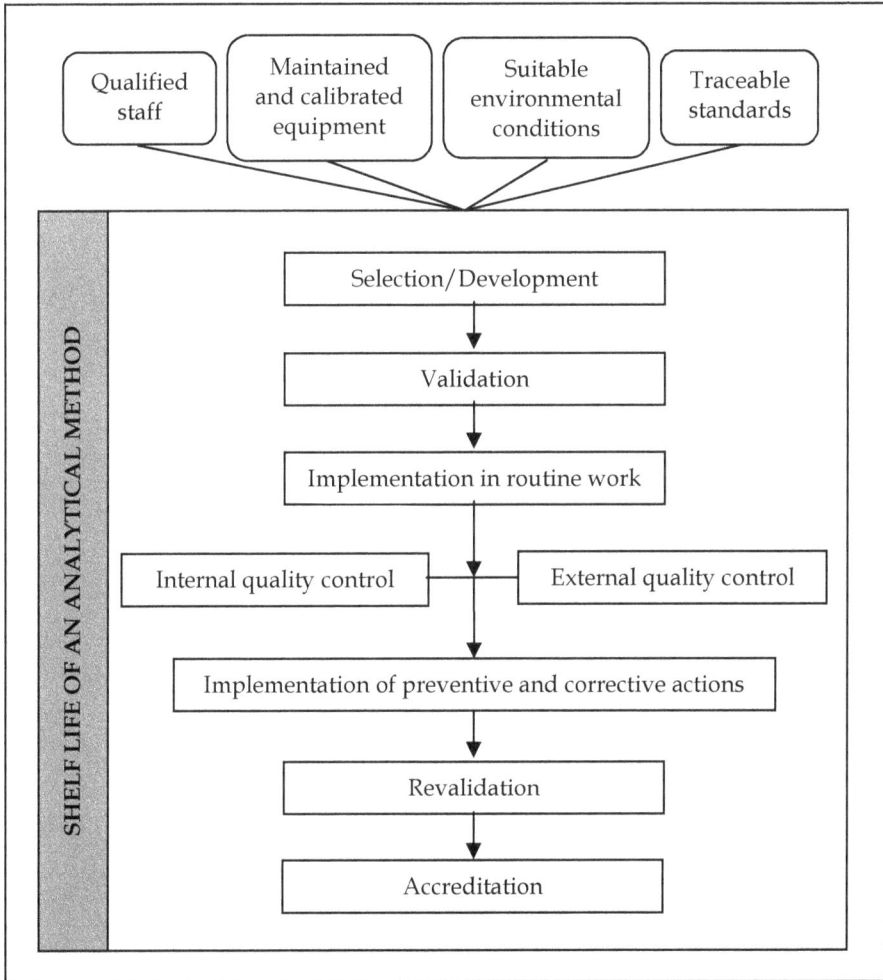

Fig. 1. Activities that determine the reliability of test results.

The LPHAB features a broad array of state-of-the-art equipment: roughly 500 instruments, including those for sample treatment, chromatography and spectroscopy. These include various gas and liquid chromatographs coupled to tandem mass spectrometry plus two inductively coupled plasma spectrometers, one equipped with photometric detection, and the other, with mass spectrometry detection. The LPHAB also uses a laboratory information management system (LIMS).

To date, the CAS has implemented about 110 analytical methodologies included in the scope of accreditation according to the requirements of the ISO 17025 standard (ISO, 2005b). In 2010, the CAS portfolio included approximately 1,800 different determinations, 1,400 of which correspond to its scope of accreditation. Moreover, the flexible scope includes some 55 analytical methods, grouped according to instrumental techniques, and numerous analytes.

In 2010 the LPHAB tested 32,225 samples, for which it performed some 550,000 determinations. Roughly half of these samples were food samples, and the other half, environmental samples (chiefly, potable water and filters for atmospheric control). The LPHAB's main customers are the Public Health Agency of Barcelona (which owns it), and the inspection bodies of the Catalonian and Spanish governments.

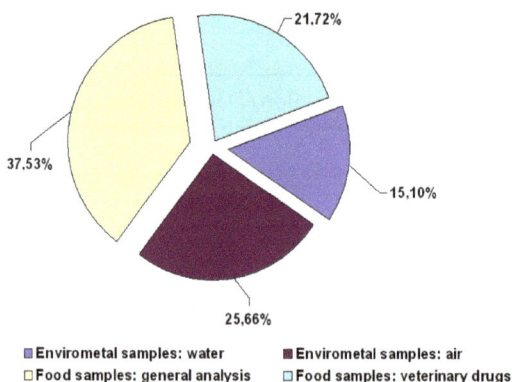

Fig. 2. Breakdown of analyses performed at the LPHAB in 2010, by sample type.

In 2010, the LPHAB's budget, excluding staff costs, was €1.2 million. This includes consumables, gases, reagents, culture media, equipment maintenance, participation in proficiency testing, and small investments. Its revenue contracts and invoices totaled €7 million.

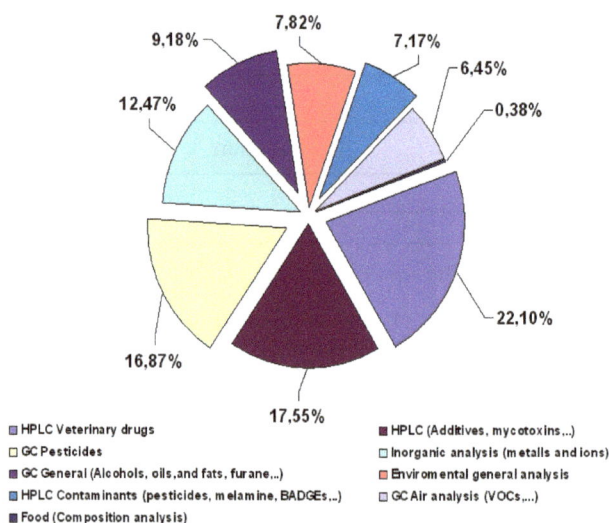

Fig. 3. Breakdown of analyses performed at the LPHAB in 2010, by analytical technique

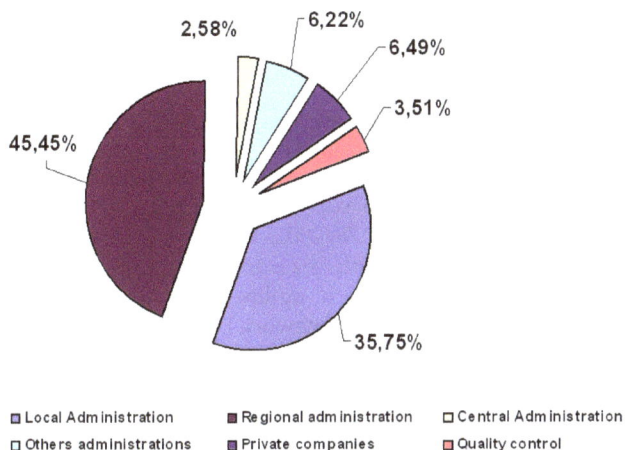

Fig. 4. Breakdown of the LPHAB's customers

The LPHAB performs research on developing and improving analytical methodology, both on its own and in collaboration with various universities. Its staff members often participate as experts in training courses organized by universities or government bodies, and some of its senior technicians are regularly asked by the Spanish Accreditation Body to participate as technical experts in laboratory accreditation audits for the food sector. Lastly, the LPHAB regularly hosts university or vocational students for training stays and internships.

3. QC within the framework of the ISO/IEC 17025 standard

Since it was issued in 2005, the ISO/IEC 17025 standard (ISO, 2005b) has been the international reference for accreditation of the technical competence of testing and calibration laboratories. The requirements of ISO/IEC 17025 (ISO, 2005b) concerning QC are concisely set out in Section 5.9 of the Standard, entitled "Assuring the Quality of Test and Calibration Results". Briefly, the Standard states that QC activities are mandatory and dictates that their results must be recorded. It also mentions the most frequent internal QC and external QC activities, without excluding other possible activities:
"The laboratory shall have QC procedures for monitoring the validity of tests and calibrations undertaken. The resulting data shall be recorded in such a way that trends are detectable and, where practicable, statistical techniques shall be applied to the reviewing of the results. This monitoring shall be planned and reviewed and may include, but not be limited to, the following:
a) Regular use of certified reference materials and/or internal QC using secondary reference materials;
b) Participation in proficiency test or proficiency-testing programs;
c) Replicate tests or calibrations using the same or different methods;
d) Retesting or recalibration of retained items;
e) Correlation of results for different characteristics of an item."
The standard goes on to state that the results of the monitoring activities performed must be analyzed and that appropriate measures should be taken:

"QC data shall be analyzed and, where they are found to be outside pre-defined criteria, planned action shall be taken to correct the problem and to prevent incorrect results from being reported."

4. Legislative requirements

Food safety and environmental protection are top priorities in the EU, which has implemented widespread legislation to support its policies in these fields. Noteworthy examples include Regulation (EC) No 178/2002, which establishes a legal framework for food; Directive 2000/60/EC, which establishes a framework for actions in the EU's water policy; and Directive 2008/50/EC, which outlines measures on ambient air quality. Currently, there is also a proposal for a framework Directive to create common principles for soil protection across the EU.

The EU has high standards for food safety and environmental protection. For instance, Regulation (EC) No 1881/2006 defines maximum levels for certain contaminants (*e.g.* mycotoxins, dioxins, heavy metals, and nitrate) in foodstuffs; Regulations (EU) No 37/2010 and (EC) No 830/2008 stipulate maximum residue levels of pharmacologically active substances or pesticides, respectively, in foodstuffs; and Directive 98/83/EC defines values for several microbiological and chemical parameters for water intended for human consumption. Regarding the environment, Directive 2008/50/EC defines objectives for ambient air quality and establishes limits on the concentration levels of air pollutants; Water Framework Directive 2000/60/EC presents a list of 33 priority pollutants based on their substantial risk; and Directive 2008/105/EC establishes environmental quality standards for these 33 pollutants.

Laboratories in charge of official controls provide essential support for these policies, by proficiently monitoring environmental and food samples. These laboratories should be equipped with instrumentation that enables correct determination of maximum levels as stipulated by EU law. According to Regulation (EC) No 882/2004, the laboratories designated for official controls in feed and food samples must operate and be assessed and accredited in accordance with ISO/IEC 17025 (ISO, 2005b). Likewise, Directive 2009/90/EC establishes that laboratories that perform chemical monitoring under Water Framework Directive 2000/60/EC must apply quality management system practices in accordance with the ISO/IEC 17025 standard or an equivalent standard accepted at the international level. Moreover, the laboratories must demonstrate their competence in analyzing relevant physicochemical parameters or compounds by participating in proficiency testing programs and by analysis of available reference materials representatives of the monitored samples. In Spain, Royal Decree 140/2003 stipulates that laboratories designated for official controls of water intended for human consumption that analyze more than 5,000 samples per year must be accredited in accordance with ISO/IEC 17025 (ISO, 2005b), and that other laboratories, if they are not accredited as such, must be at least certified according to ISO 9001 (ISO, 2005a).

There has been a shift from using official analytical methods to a more open approach that allows the laboratories involved in official controls to use validated analytical methods that have been proven to meet established performance criteria. Thus, different scenarios are presently possible: in very few cases, such as Commission Regulation (EEC) 2676/90, on the analysis of lead in wine, the method is defined; more frequently, as in Directive 2008/50/EC on air quality or in Directive 98/83/EC on water intended for human consumption, although methods are explicitly specified, laboratories are allowed to use alternative methods, providing they can demonstrate that the results are at least as reliable as those produced by the specified methods. Another approach is that of Decision 2002/657/EC, concerning analytical methods for the analysis of residues and contaminants in food products, which establishes the performance criteria for methods. Directives 2009/90/EC and 98/83/EC establish analogous analytical method criteria for monitoring water status, sediment and biota, as do Regulation (EC) 333/2007 (on sampling and analytical methods for the control of some contaminants in foodstuffs), to SANCO/10684/2009 (on method validation and quality control procedures for pesticide residues analysis in food and feed), or to Regulation (EC) 401/2006 (on methods of sampling and analysis for the control of mycotoxins in foodstuffs). Representative examples of performance criteria for methods used to analyze patulin in foodstuffs are shown in Table 1.

This flexible approach to method performance criteria allows laboratories to quickly incorporate advances in analytical techniques and to apply new methods to address new problems when required. The crucial issues here are that the required performance criteria are met and that the method has been properly validated.

Level (μg/kg)	RSD_r % (a)	RSD_R % (b)	Recovery %
< 20	≤ 30	≤ 40	50 to 120
20 to 50	≤ 20	≤ 30	70 to 105
< 50	≤ 15	≤ 25	75 to 105

Table 1. Performance criteria for methods of analysis of patulin in foodstuffs, from Regulation 401/2006. (a: Relative standard deviation, calculated from results generated under repeatability conditions, b: Relative standard deviation, calculated from results generated under reproducibility conditions.)

Since its publication, Decision 2002/657/EC has been a key document for analytical laboratories involved in food analysis and has proven utile for laboratories in other fields, such as environmental analysis. It introduced a change of mindset, replacing reference methods with the criteria approach, and launched new definitions, such as *minimum required performance limit (MRPL)*, *decision limit (CCα)* and *detection capability (CCβ)*. Decision 2002/657/EC determines common criteria for the interpretation of test results, establishes the performance criteria requirements for screening and confirmatory methods, and presents the directives to validate the analytical methods. However, it is a complex document, and guidelines for its implementation have been published (SANCO/2004/2726-rev-4-December-2008). The most relevant aspects of Decision 2002/657/EC are further described below.

Minimum required performance limit is defined as the minimum content of an analyte in a sample that has to be detected and confirmed. It is intended to harmonize the analytical performance of methods for banned substances. The minimum required performance level for a method of a banned substance should be lower than the MRPL; however, very few MRPL values have been established to date.

The decision limit is the limit at and above which one can conclude, with an error probability of α, that a sample is non-compliant. For substances with no permitted limit α is 1%, whereas for all other substances α is 5%. Thus, the result of an analysis shall be considered non-compliant if the CCα of the confirmatory method for the analyte is exceeded.

The detection capability is the smallest content of the substance that may be detected, identified and/or quantified in a sample with an error probability of β (β is 5%). Procedures to determine the $CC\alpha$ and $CC\beta$ are given in Decision 2002/657/EC and its corresponding guidelines document (SANCO/2004/2726-rev-4-December-2008).

Decision 2002/657/EC also introduces the concept of *identification point (IP)*. A minimum of three IPs is required to confirm the identity of a compound that has a permitted limit, whereas at least four IPs are required for a banned compound. The number of IPs provided by the analytical method depends on the technique used. For instance, with low-resolution MS each ion earns 1 point, and with low-resolution MS^n each precursor ion earns 1 point, and each transition product, 1.5 points. More details on IPs for the different techniques can be found in Decision 2002/657/EC. This IP system has made MS an essential technique for laboratories that analyze residues and contaminants in foodstuffs.

In addition to the performance criteria requirements for screening and confirmatory methods, Decision 2002/657/EC also provides guidelines for the validation of analytical methods. Validation should demonstrate that the method complies with its performance criteria. Therefore, depending on the method category (*e.g.* qualitative or quantitative; screening or confirmatory), different performance characteristics must be determined. Table 2 shows an overview of EU legislation on analytical methods for environmental and food samples

Directive 98/83/EC	Quality of water intended for human consumption
Directive 2008/50/EC	Ambient air quality and cleaner air for Europe
Directive 2009/90/EC	Technical specifications for chemical analysis and monitoring of water status
Regulation (EC) 333/2007	Methods of sampling and analysis of lead, cadmium, mercury, inorganic tin, 3-MCPD and benzo(a)pyrene in foodstuffs.
Decision 2002/657/EC	Performance of analytical methods and interpretation of results
Regulation (EC) 401/2006	Methods of sampling and analysis of mycotoxins in foodstuffs

Table 2. Overview of EU legislation on analytical methods for environmental and food samples

5. QC management

At the LPHAB QC activities are managed by the Quality Assurance Unit (QAU), in close cooperation with the head of the Chemical Analysis Service (CAS) and the senior technicians responsible for each analytical technique or methodology.

The QAU comprises two senior technicians and one mid-level technician. Its functions include:

- Coordinating implementation and maintenance of the Quality Management System (QMS)
- Cooperating with the LPHAB's top management in the annual system review and in preparation of the annual staff training program
- Preparing and conducting an annual internal audit
- Managing any complaints received from customers or third parties

- Defining corrective and preventive actions, supervising their implementation and verifying their efficacy
- Managing documentation (Quality Manual, general procedures and SOPs, etc.), distributing and maintaining documents, and preparing lists for flexible-scope accreditation
- Approving the auxiliary equipment program control
- Advising technicians on method validation and QC activities
- Managing the LIMS

Moreover, the LPHAB's QC activities are described in several documents of its QMS. Table 3 shows these documents in a hierarchical order.

Document	Scope
Quality Manual (Section 14)	Whole laboratory
General procedure: "Assessment of the Quality of Analytical Results"	Whole laboratory
General procedure: "Management of Complaints, Non-conforming Work, and Corrective and Preventive actions"	Whole laboratory
General procedure "Management of Flexible-Scope Accreditation"	Whole laboratory
Standard Operating Procedure (SOP): "Application of the General Quality Criteria to the Chemical Analysis Service"	CAS
SOP: "Management of Standards"	CAS
Annual QC Plan	CAS
Specific SOPs (per method)	CAS
Records	CAS

Table 3. Major QC documents from the LPHAB's QMS (CAS: Chemical Analysis Service).

One of the chapters in LPHAB's Quality Manual defines the basis of QC in accordance with the requirements of the ISO/IEC 17025 standard (ISO, 2005b). General procedures, which are applicable either to the whole laboratory, or to the Microbiology Analysis Service or the CAS, outline the LPHAB's general QC activities. Standard operating procedures provide detailed specifications on the CAS's QC activities (both internal and external QC). Internal QC activities are either performed within each analytical run or are scheduled. The within-run activities are done in accordance with the specific SOP for regular samples received by the LPHAB; these encompass analysis of reagent blanks, blank samples, spiked samples and verification of instrument sensitivity. They are employed to prevent releasing of any erroneous results to customers. Scheduled activities are used to check the efficacy of within-run controls. External QC (EQC) relies on the regular and frequent participation of the LPHAB in proficiency tests organized by competent bodies (whenever possible, accredited as Proficiency Test Providers). All these activities are agreed upon by the head of the CAS, the director of the QAU and the senior technicians and are reflected in the Annual QC Plan. Table 4 shows a sample page from the CAS's Annual QC Plan.

All of the QC activities are described in the SOP entitled "Application of General Quality Criteria to the Chemical Analysis Service", as are the procedures for handling of all scheduled internal and external QC samples (in terms of ordering, analysis and evaluation). These activities are summarized in Table 8.

Technical Unit: Instrumental Organic Chemistry		
Technician: XXXXXX		

SOP	External Quality Control	Periodicity:	Internal Quality Control
MA/2/19510	FAPAS	**Blank**	✓ Every run
Determination of Chloramphenicol	Progetto Trieste	**Reference Material**	✓ At least once a year
by LC-MS/MS	CNA	**Spiked samples**	✓ At CCα level- every run
		Others	✓ Matrix matched surrogates/ Internal standard in samples checking
MA/2/19560	FAPAS	**Periodicity:**	
Determination of Nitrofuran metabolites	Progetto Trieste	**Blank**	✓ Every run/ Internal standard checking
by LC-MS/MS	Anfaco	**Reference Material**	✓ At least once a year
		Spiked samples	✓ At CCα level-every run
		Others	✓ Calibration curve from processed samples, including MRPL (1 µg/kg). Internal Standard in samples checking
MA/2/19560	FAPAS	**Periodicity:**	
Determination of corticosteroids	Progetto Trieste	**Blank**	✓ Every run/ Internal standard checking
by LC-MS/MS		**Reference Material**	✓ At least once a year
		Spiked samples	✓ At CCα level- every run
		Others	✓ Calibration curve from processed samples, including MRL (2 µg/kg). Internal Standard in samples checking

Table 4. Sample page from the CAS's Annual QC Plan.

5.1 Management of external QC

External QC is managed through proficiency tests. Participation in each test is scheduled according to proposals by the technician responsible for each analytical procedure. Each procedure is to be tested in at least one exercise per year, if possible. In parallel, certain samples are requested in duplicate for use in scheduled internal QC.

The LPHAB tends to be extremely active in this area, since it considers external QC among the strongest point of its QC system. In the CAS, in 2010, 458 samples were analyzed in proficiency tests that encompassed 1,915 assays, 420 different analytes and 89 analytical procedures (SOPs).

Given that the market lacks universal exercises for all types of matrices and assays, the CAS aims to assess all families of analytes and all instruments. Usually, matrices included in the accreditation scope are used. Importantly, for assays included in the flexible-scope accreditation, different matrices that represent the entire assay category should be employed whenever possible. To evaluate some of the procedures for which no exercises are currently available, the CAS, together with other laboratories, has organized specific activities.

It is extremely important that any organization that aims to organize these types of evaluations be accredited according to ISO/IEC 17043 (ISO, 2010). For non-accredited entities, the quality of their exercises will be assessed.

In accordance with the aforementioned principles, CAS actively participates in the programs FAPAS® (for food) and LEAP (for water), both of which are accredited by the United Kingdom Accreditation Service (UKAS).

For each exercise, a technician is assigned to handle and follow the sample, which must be analyzed using the typical procedures and which must not be treated differently because of its interlaboratory status. Once the organizer's report has been received, an internal evaluation report is written up, which includes the results found by CAS, the mean result assigned by the organizer, and the calculated z-score for each analyte.

Upon receiving the report, each manager performs a complementary evaluation of the results obtained, considering all of the documentation referring to the analysis performed, in order to confirm that all of the QC criteria have been met. Another very important and highly utile aspect to consider is the information on the methods applied by different laboratories, which can help the CAS to improve its methods.

If the evaluation is unsatisfactory, then a report on corrective actions is written up. The results of proficiency tests are generally evaluated based on the z-scores. Nonetheless, other criteria (*e.g.* compatibility index) may also be used; these are described in the final evaluation report for the exercise.

One of the critical points for evaluating z-scores is the standard deviation used in the calculations. The standard deviation used is generally that which is documented by the organizer, which tends to the value obtained from the Horwitz equation. Nevertheless, another value can be used, as deemed necessary by the technician responsible for the evaluation, as long as it is justified in the internal evaluation report for the exercise. Fig. 5 shows a sample evaluation form for external QC samples.

The results obtained are introduced into a database, which enables tracking of any possible trends as well as confirmation of validation data over time.

The figure below illustrates moisture analysis results for various types of samples from the FAPAS® exercises in which LPHAB has participated over the past few years.

PROFICIENCY TEST REPORT

Sample No. : 11_00237	Date of entry :	11/01/2011
Proficiency Test : LEAP - Series: CHEM 72	Round:	C1316
Sample : Potable water	Laboratory id No.:	
Area : 2/0103 Chromatography	Responsible: Dr. Laura Pineda	

ANALYTES:	RESULTS LIMS / Sent result / Consensus value / z-score				SOP / Equipment used	
Toluene	3,9	3,87	3,2	1,7	MA/2/30470	2-170
Ethylbenzene	0,9	0,89	0,75	1,5	MA/2/30470	2-170
Benzene	1,7	1,65	1,37	1,6	MA/2/30470	2-170
Xylene (mixture of isomers)	6,1	6,13	5	1,8	MA/2/30470	2-170

ASSESSMENT OF RESULTS:

Successful results: ☒

Questionable results: Punctual Repetitive Preventive action PA number

Unsuccessful results: Non-conforming work No.   Date.

OTHER METHODS USED IN THE PROFICIENCY TEST:

After reviewing the methods used by the other participants, an improving opportunity has been detected:

Yes ☐   No ☒   Preventive action No

GENERAL COMMENTS ON RESULTS:

Date:
Signature:

| Final assessment of the head of the Chemical Analysis Section | Revision QAU | Date |
| Date: Signature: | | |

Fig. 5. Sample evaluation form for external QC samples (completed by CAS based on the organizer's report).

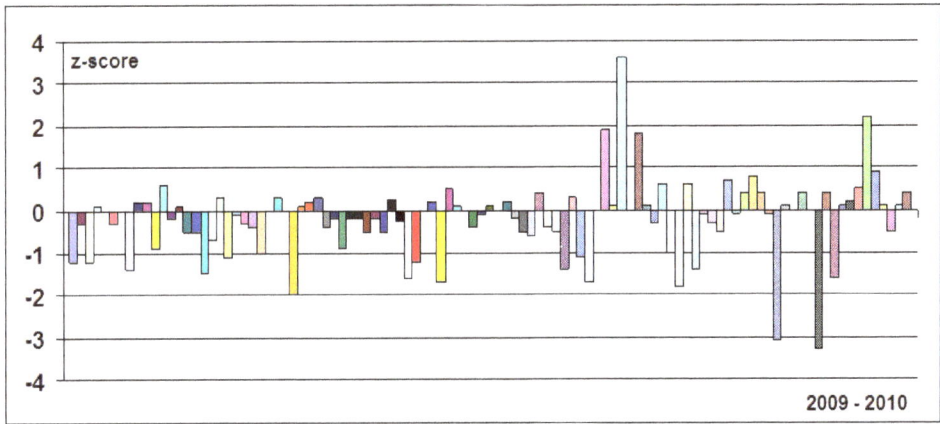

Fig. 6. Proficiency testing: results from moisture analysis of different samples performed for FAPAS® exercises.

Another important factor concerning the results obtained from proficiency tests is their utility for systematic expansion of validation data. The CAS has established a dynamic validation system in which the overall validity of the uncertainty of a procedure is checked against the different sample types and different concentration levels analyzed.

5.2 Management of internal QC

The scheduled internal QC samples generally correspond to duplicates of samples from proficiency tests (the duplicates are purchased annually at the time the test is performed).
In 2010 the CAS analyzed 99 samples for internal QC, encompassing 371 assays, 209 analytes and 77 SOPs.
Once the samples arrive at the LPHAB, their information is entered into the reference materials database, and they are carefully handled, taking their particular storage needs and expiration dates into account. The new sample is added to the sample registry in the LIMS according to the schedule. The results are analyzed using an internal evaluation form in which the z-score (accuracy) is re-calculated, and the reproducibility is calculated based on the results from the external QC and from the internal QC test. This approach enables evaluation of both accuracy and precision. Fig. 7 shows a sample evaluation form for internal QC.

5.3 Handling of any inconsistencies detected in the QC activities

Results obtained in both internal and external QC activities are suitably recorded. Out of control situations can be categorized as incidences and deviations. Incidences are sporadic events that usually do not occur in subsequent applications of the analytical method. Contrariwise, deviations are non-conforming work that must be managed through corrective actions. Detection of these events, and subsequent causal analysis, sometimes leads to proposal of preventive actions. Fig. 8 shows a general schematic of QC management.

INTERNAL QUALITY CONTROL REPORT

Sample No. : 11_00065	Date of entry :	04/01/2011
Sample reference : 2-R0525	Original sample:	10_28944
Sample : Meat		
Area : 2/0401 Food	Responsible:	José Antonio León

ANALYTES:	RESULTS								SOP
	Result / Certif. value / ± U / Recovery / z-score / Reprod. / Correct								
Nitrate (E-251, E-252)	205	226	55	90.7	-0.76	12	77	☒	MA/2/30102
Nitrite (E-249, E-250)	121	150	36,6	80,6	-1,58	24	51	☒	MA/2/30101

ASSESSMENT OF RESULTS:

PARAMETER	ACCEPTANCE CRITERIA	COMMENTS
Uncertainty		
Recovery		
Z-scores		
Reproducibility		
Cause analysis:		
Non-conforming work No.:	Date:	

GENERAL COMMENTS ON RESULTS:

Date:
Signature:

Final assessment of the head of the Chemical Analysis Section	Revision QAU	Date
Date: Signature:		

Fig. 7. Sample evaluation form for internal QC samples.

Figures 9 and 10 show the number of scheduled internal QC and external QC samples in absolute values and as percentages of the total number of samples analyzed, respectively, in the CAS. These figures are testament to the LPHAB's major efforts to ensure the reliability of its results and demonstrate its commitment to quality. Moreover, this approach also implies sizeable financial investment: participation in proficiency testing costs the LPHAB roughly €60,000 per year.

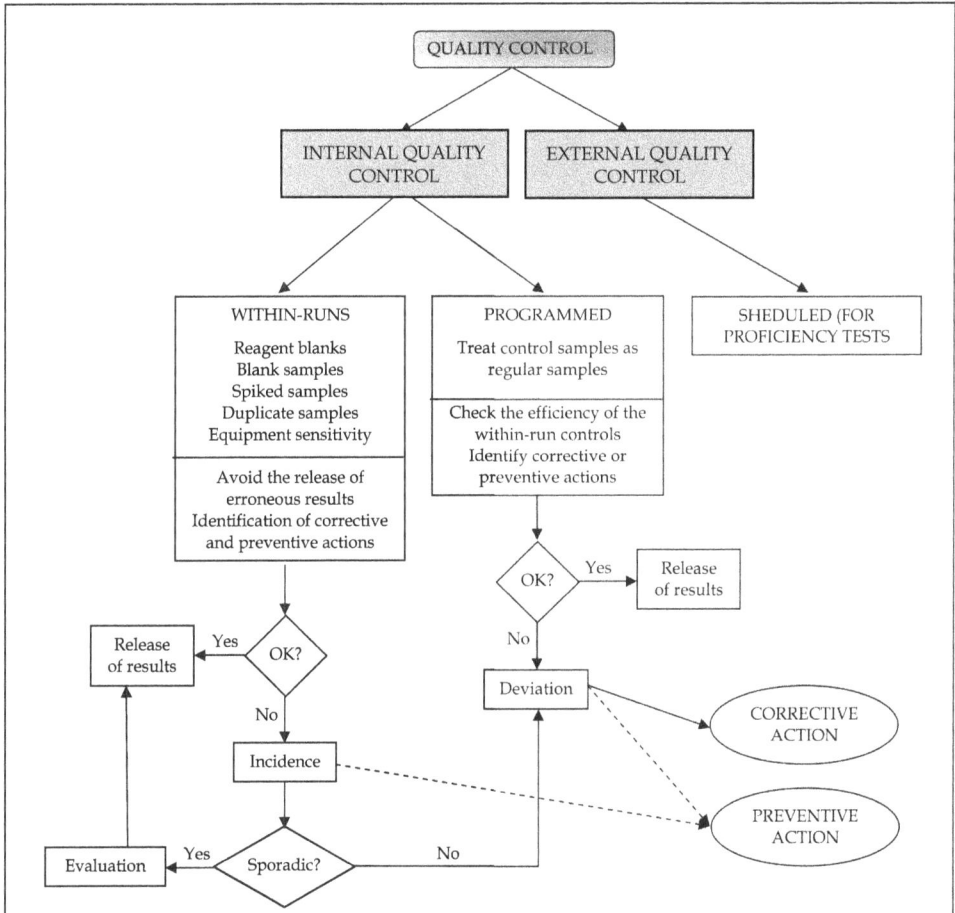

Fig. 8. General schematic of QC management.

The reliability of QC activities is greatly based on the suitability of the criteria applied. Depending on whether the limits established are too strict or too lax, α or β errors, respectively, may be committed. Over the past few recent years, the CAS has adapted the criteria applied in its internal QC to the values obtained during method validation. Improving the frequency and quality of internal QC has enabled improved detection of non-conforming results, and therefore, has enabled optimization of external QC activities.

5.4 QC in the framework of flexible scope accreditation
Accreditation of a laboratory is usually based on a concrete definition of the laboratory's scope. Thus, the technical annexes for accreditation certificates comprise detailed lists of the tests for which the laboratory has been accredited. The lists clearly specify matrices, analytes, ranges of concentration, and methods. This scheme is known as *fixed-scope accreditation*.

However, in recent years, in order to meet the needs of customers, laboratories have had to quickly expand their accreditation scope without compromising their technical competence or altering definition of the scope. Thus, highly experienced laboratories with a long history of accreditation can now adopt a new scheme, known as *flexible-scope accreditation*, whereby they perform analyses using appropriate validated methods, and then report the results as being accredited, without prior evaluation by the accreditation body. This may entail incorporation of new matrices or analytes, or inclusion of new tests within a generic method. Thus, the flexibility of the accreditation scope implies sufficient technical competence and operational capacity, which places more of the responsibility on the laboratory. This in turn means that the laboratory must endeavor to increase its QC operations in order to guarantee the quality of the results of the expanded scope. In any case, the bounds within which a scope is flexible must be precisely stated.

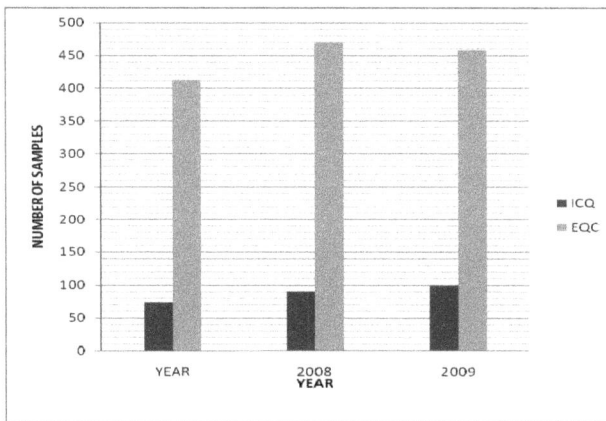

Fig. 9. Scheduled internal and external QC samples (ICQ and ECQ, respectively), expressed as number of samples.

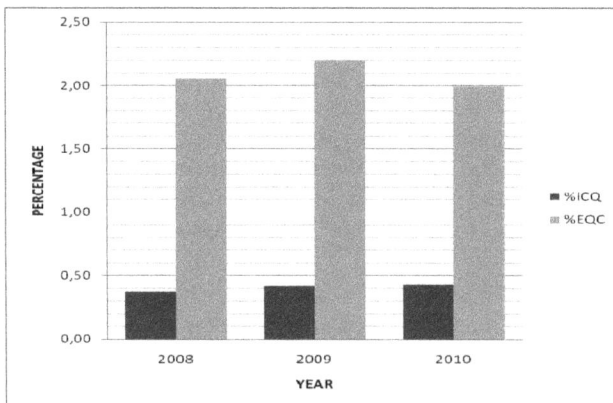

Fig. 10. Scheduled internal and external QC samples (ICQ and ECQ, respectively), expressed as percentage of total samples.

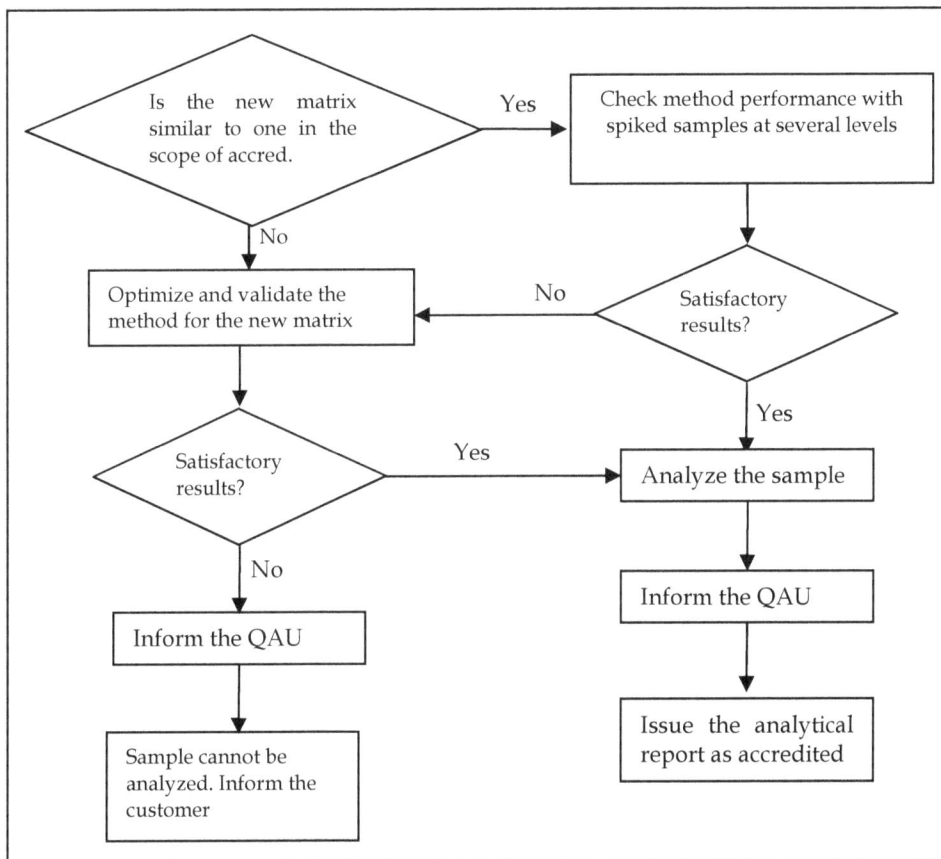

Fig. 11. Procedure for analysis of an established analyte in a new matrix.

In this context, once a laboratory receives a request for an analysis that falls within the bounds of a flexible scope, it must do the following:

• Inform the customer that the analysis will be performed in the framework of a flexible scope, and therefore, prior validation studies will be required; this will involve some delay in the delivery of results; and, if the results of the validation studies are unsatisfactory, then the report cannot be issued as being accredited.

• Perform validation studies. A scheme of this process for analysis of an established analyte in a new material is illustrated in Fig. 11. An analogous process would be employed for the opposite case (*i.e.* analysis of a new analyte in an established matrix).

Flexible-scope accreditation was initiated in 2004 for pesticide analysis and was later extended to other analyte families. The LPHAB defines these families according to the type of analyte studied and the analytical technique used. Therefore, these vary from very broad (organic compounds studied by chromatographic techniques) to rather narrow (ions studied by liquid chromatography). The CAS's current fixed-scope and flexible-scope of accreditation are summarized in Table 5.

PARAMETER MATRIX	ANALYTE OR FAMILY OF ANALYTES (NUMBER OF ANALYTES)	FIXED SCOPE	FLEXIBLE SCOPE
Additives by LC Food	Phenolic antioxidants (7), artificial dyes (14), Sudan dyes (10), preservatives (7), sweeteners (3) caffeine		X
Organic pollutants by LC and GC Food / Water Air sampling supports	Acrylamide, BFRs (8), BTEX (5), 3-MCPD, 3-MCPD ester, VOCs (15), ethyl carbamate, drugs (3), furan, PAHs (17), BADGEs (11), melamine, PBCs (7), PCNs (5) pesticides (44) , nicotine,		X
Ions by LC or segmented continuous flow analysis Food / Water	BrO_3^-, BrO^-, Cl^-, F^-, P, NO_3^-, NO_2^-, SO_4^{-2}, NH_4^+, Ca^{+2}, Mg^{+2}, K^+, Na^+, ClO^{3-}, ClO_2^-, SO_2	X	X
Heavy metals by atomic spectroscopy Food /Water Air sampling supports	As, As inorganic, Ba, B, Cd, Ca, Co, Cu, Cr, Sn, Fe, Mg, Mn, Hg, CH_3Hg, Ni, Ag, Pb, K, Se, Na, Ti, V, Zn		X
Mycotoxins by LC Food	aflatoxin M1, aflatoxins B i G (4), fusarium mycotoxins (6), ochratoxin A, patulin		X
Residues of veterinary drugs by LC Food of animal origin	Antibiotics: β-lactams, macrolides, quinolones, sulphonamides, tetracyclines (41), coccidiostats (10), antihelmintics (1), tireostatics (6), benzodiazepines (3), chloramphenicol, dyes (5), corticosteroids (6), synthetic hormones (9), metabolites of nitrofurans (4), nitroimidazole, β-agonists (18)		X
Residues of pesticides by GC and LC Food	Pesticides (200)		X
Physicochemical parameters Food	Mass, volume, density, acidity, pH, fat (3), alcohol degree, peroxide index, moisture, ash	X	
ELISA determinations Food	Gluten	X	
Ions by flow techniques Food / Water	P_{total}, $P_{soluble}$, N_{total}, NO_3^-, NO_2^-, Cl^-, NH_4^+, SO_2, volatile nitrogenous bases, trimethylamine	X	
Several compounds by chromatographic techniques Food	Biogenic amines (6), sugars (6), methanol, ethanol, other alcohols (5), ethyl acetate, fatty acids, steroids and estradiols	X	
Physicochemical parameters Water	Colour, turbidity, conductivity, suspended solids, dissolved matter, pH, chemical oxygen demand, alkalinity, cyanide, oxidisability	X	

Table 5. The CAS's current fixed-scope and flexible-scope of accreditation.

Managing the flexible scope implies a significant amount of extra documentation that must be completely updated. Indeed, in 2010 alone six new analytical methods were added, together with numerous matrices and analytes. The flexible scope is currently in its 22nd edition (an average of three editions are created per year).

6. QC in the analytical method SOPs: examples of general and specific QC activities

This section provides examples of the some of the QC activities summarized in Table 8, as well as the corresponding documentation for recording and evaluating the data. Generally, each analytical procedure (SOP) features a section describing internal QC activities that are performed within each run and the corresponding criteria for accepting the results, which must be evaluated by the technician responsible for the procedure before they are communicated to the customer. Several concrete examples are presented below.

6.1 Spiked samples

An example of control analysis of spiked samples is illustrated in Fig. 12, which shows a plot of arsenic analysis in food samples by inductively coupled plasma mass spectrometry (ICP-MS). The results are evaluated based on the recovery (% Rec) of samples spiked at different concentrations and with different matrices, such that the entire scope of the flexible-scope accreditation can be addressed.

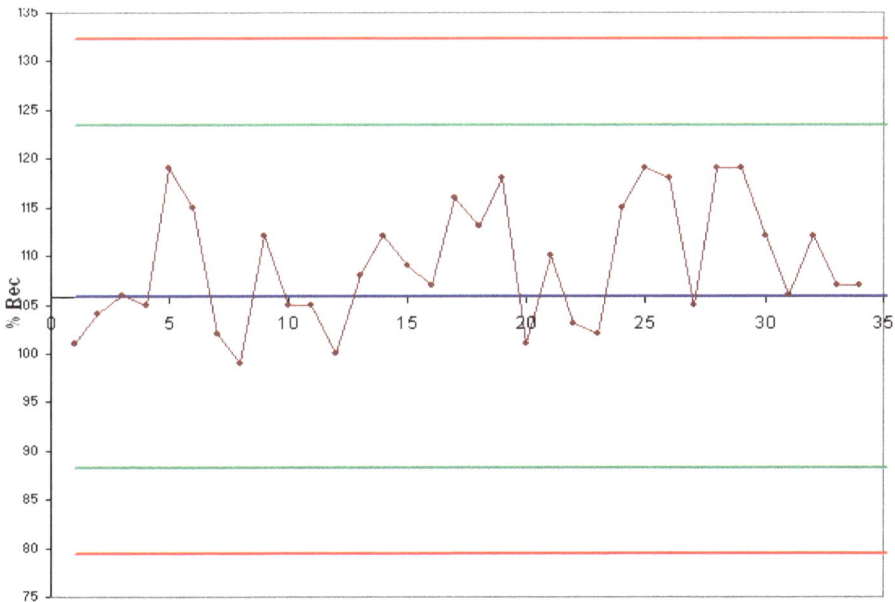

Fig. 12. Plot of arsenic recovery levels from spiked samples of different food types, as determined by ICP-MS.

6.2 Use of QC records for an LC-MS/MS procedure (detection of antibiotics)

Table 6 shows an example of a QC records for a procedure in which 44 antibiotics are analyzed in samples of products of animal origin by LC-MS/MS. The following data are recorded for representative analytes (in the case of Table 6, two antibiotics): the area of the peak corresponding to the standard used for verifying the instrument; retention time (TR) and the ratio of transitions (ion ratio [IR]) at CCα level, which are the data used for identifying and confirming the two compounds. The peak area value is checked against the minimum peak area that guarantees response at the lowest level of validation, which also verifies the confirmation. Lastly, the analytical sequence and the user's initials are also recorded.

SOP	MA/2/19210	Analysis of antibiotics by CL / MS-MS							
Instrument	2-201	CL MS-MS							
Compound	SULFAMETHAZINE				ENROFLOXACIN				
Average	4,50				4,30				
Maximum value	4,75				4,55				**Antibiotics**
Minimum value	4,25	100000	0,8		4,05	1000	1,0		
DATE	TR	Area	I.R. (CC α)	OK	TR	Area	I.R.(CC α)	OK	Run/User
15/02/2011	4,36	135483	0,9		4,45	1022	1,1		110215ant/AR
22/02/2011	4,60	175563	0,8		4,20	1156	1,0		110222ant/AR
01/03/2011	4,55	115632	0,8		4,15	1325	1,2		110321ant/AR
08/03/2011	4,34	134856	0,8		4,35	1278	1,2		110308ant/AR
15/03/2011	4,57	142569	0,9		4,10	1486	1,2		110315ant/AR

Table 6. QC records from analysis of antibiotics in products of animal origin by LC-MS/MS.

In similar QC records, the responses of the internal standards (which are typically deuterated or C^{13}-labeled analogs of the test compounds) from analysis of various types of samples are recorded. This control step can also be used to broaden the validation data by incorporating new matrices (*i.e. online validation*). Based on the values of the responses of the internal standards, one can deduce the validity of the matrix-matched surrogate quantifications in the different sample types that can be incorporated into the analytical sequence.

6.3 QC records for verification of the instrument, its calibration levels, and the blank in the turbidity analysis procedure

The format of the QC records used for turbidity analysis of water samples is illustrated in Table 7 as a representative example of a physicochemical assay.

The upper and lower limits traceable to the values obtained in the validation are shown. In this case, the experimental readings obtained are recorded for each certified standard and are used to verify calibration of the instrument and to confirm the response of the blank (in this case, ASTM type I purified water).

SOP	MA/2/30504	Determination of turbidity in water					
Instrument	2-001	Turbidimeter					
STABCAL	0,5 NTU	3 NTU	6 NTU	18 NTU	30 NTU	Blank	**Turbidity**
Upper val.	0,48	3,11	6,43	19,1	32,5	0,094	
Lower val	0,32	2,81	5,81	17,3	29,4		
Date	Result (NTU)						User
10/03/2011	0,38	2,85	5,93	17,7	30,1	0,079	NG
11/03/2011	0,37	2,92	5,98	17,8	30,3	0,07	NG

Table 7. QC records from turbidity analysis of water samples.

QC factor	Action	Objective	Calculations and tolerance limits	Frequency at the LASPB
General internal QC				
Reagent blank	The analytical procedure is performed using only the reagents.	Enables monitoring for any contamination in materials, reagents, the environment, etc.	LOD: limit of detection	Within-run (the type of blank depends on the method) See Table 7
Matrix blank	The analytical procedure is performed using a blank sample.	Enables monitoring for any contamination, and confirmation that the matrix is not responsible for any interference	Evaluation: Blank < LOD	
Duplicate samples (intermediate precision)	Full analysis of duplicate samples on different dates	Enables monitoring of the reproducibility (R) relative to the standard deviation of the validation (s_R)	$R = 2 * \sqrt{2} * s_R$ Evaluation: $x_1 - x_2 \leq R$ x_1 and x_2 are the results of duplicate samples	Scheduled: usually, once per year, using a reference material (normally, a duplicate sample from a proficiency test)
Spiked samples	The analytical procedure is performed on a sample that has been spiked with the analyte (whenever possible, previously analyzed samples containing the analyte at levels lower than the limit of detection).	Enables monitoring of the bias or the trueness based on the recovery (% Rec), and compared with the recovery (%Rec$_{val}$) and the standard deviation (s) obtained in the validation	$Re\,c(\%) = \dfrac{(x_{lab})}{x_{spiked}} \times 100$ X_{lab}: obtained value X_{spiked}: spiked value Evaluation: $Re\,c(\%) = Re\,c_{val}(\%) \pm 2s$	Within-run See Fig. 12
Reference materials	The analytical procedure is performed on a sample which has been prepared under concrete specifications and which contains the analyte in question at a known value.	Enables monitoring of the accuracy of the results based on the compatibility index (CI), which is calculated from the reference value (x_{ref}) and the obtained value (x_{lab})	$CI = \dfrac{\left| X_{ref} - X_{lab} \right|}{\sqrt{(U_{ref}^2 + U_{lab}^2)}}$ U_{ref}: expanded uncertainty of reference material U_{lab}: expanded uncertainty laboratory Evaluation: CI≤1	Scheduled

Table 8. Part I

QC factor	Action	Objective	Calculations and tolerance limits	Frequency at the LASPB
External QC				
Proficiency test	The analytical procedure is performed on a sample which has been part of an interlaboratory comparison scheme.	Enables monitoring of the accuracy of the results based on the z-score (z), which is calculated from the assigned value, the obtained value (x_{lab}) and the standard deviation of the participants (σ_P).	$z = \dfrac{\lvert x_a - x_{lab} \rvert}{\sigma_p}$ Evaluation: $z \le 2$ satisfactory result $2 < z \le 3$ questionable result $z > 3$ unsatisfactory result	Scheduled See Fig. 5
Internal QC to verify equipment or reagents				
Verification of instrument at the beginning of the run and monitoring of instrument drift	A standard is injected under the instrumental conditions established in the analytical procedure.	Enables verification of proper instrument performance before the sequence is started, and at every n samples, based on confirmation that the response of the standard (A) falls within a pre-established range of acceptable values (x %) that guarantee the limit of quantification (LOQ)	A : response of the standard Evaluation: $A \le \pm x\%$	Within-run See Table 7
Calibration of the instruments associated with the analytical method	The standards used to generate the calibration curve are injected.	Enables monitoring of the **quality of the fit of the calibration curve**, based on at least two different criteria: for example, the coefficient of correlation (r) and the residual error of the standard (Er %), which is the ratio of the value of the concentration of the standard in the curve (V_{curve}) to the nominal concentration value ($V_{nominal}$)	$r \ge$ (see specific SOP) $E_r(\%) = \dfrac{V_{curve}}{V_{no\,min\,al}} * 100$ Evaluation: $Er\,\% \le r \ge$ (see specific SOP)	Upon generation of a new calibration curve
Verification of a new lot of standards	Two different samples of the same standard are injected: one from a regularly used lot, and one from a newly prepared lot.	Enables confirmation that a standard has been correctly prepared, based on verification that the ratio of the response of the new sample (A) to the response of the sample from a previously used lot (B) falls within a pre-established range of acceptable values (x %)	A/B : response ratio Evaluation: $A/B \le \pm x\%$	Upon preparation of new lots of standards

Table. 8. Part II

QC factor	Action	Objective	Calculations and tolerance limits	Frequency at the LASPB
Specific internal QC procedures: chromatographic methods				
Verification of the response of the internal standard	Addition of the internal standard to all samples, spiked samples, and other standards (matrix-matched surrogate) at the beginning of the procedure	Verification of the procedure for each sample: extraction, and performance of different matrices	Signal traceable to previous analyses Quantification based on internal standard	Within-run
Identification of the chromatographic peak	Retention time of each compound relative to that of the internal standard	Verification of the criteria described in the chromatographic method	According to chromatographic system; TR ± % tolerance limit	Performed for each chromatographic peak identified that corresponds to a standard See Table 6
Confirmation of the identified compounds	DAD, FLD, etc.: The compound spectra are compared to the internal standard spectra	Verification of the criteria described in the chromatographic method	Spectral match	Performed for each chromatographic peak identified See Table 6
	MS (SIM): Mass spectra ion ratios		According to analysis; generally ± 20%	
	MS/MS: Transition ratios		According to regulations, analysis type, concentration, intensity of the transitions, etc.	

Part III

Table 8. QC activities at the LPHAB's Chemical Analysis Service (CAS).

7. References

Shewhart (1931). *Economic Control of Quality of Manufactured Products,* Van Nostrand, New York, USA. ISBN 0-87389-076-0.

ISO (2005a). *ISO 9000:2005 standard. Quality management systems. Fundamentals and vocabulary,* ISO, Geneva, Switzerland.

ISO (2005b). *ISO 17025:2005 standard. General requirements for the competence of testing and calibration laboratories.* ISO, Geneva, Switzerland.

ISO (2010). *ISO/IEC 17043:2010 standard. Conformity assessment. General requirements for proficiency testing.* ISO, Geneva, Switzerland.

IUPAC (1998). *Compendium of analytical nomenclature: Definitive rules 1997*, 3rd ed. Blackwell Science, ISBN 978-84-7283-870-3, Oxford, United Kingdom.

Thompson, M. & Lowthian, P. J. (1993). Effectiveness of Analytical Quality Control is Related to the Subsequent Performance of Laboratories in Proficiency Tests. *Analyst*, Vol.118, No.12, (December 1993), pp. 1495-1500.

14

Comparison Sequential Test for Mean Times Between Failures

Yefim Haim Michlin[1] and Genady Grabarnik[2]

[1]Technion - Israel Institute of Technology
[2]St' Johns University
Israel
USA

1. Introduction

The present study deals with the planning methodology of tests in which the parameters of two exponentially-distributed random variables are compared. The largest application field of such tests is reliability checking of electronics equipment. They are highly cost-intensive, and the requirements as to their resolution capability become stricter all the time. Hence the topicality and importance of an optimal plan permitting decisions at a given risk level on the basis of a minimal sample size.

Such comparison tests are required for example in assessing the desirability of replacing a "basic" object whose reliability is unknown, by a "new" one; or when the influence of test conditions on the results has to be eliminated.

This is the case when an electronics manufacturing process is transferred to another site and the product undergoes accelerated testing.

Recently, equipment and methods were developed for accelerated product testing through continuous observation of a population of copies and replacement of failed objects without interrupting the test. For such a procedure, the sequential approach is a feasible and efficacious solution with substantial shortening – on the average – of the test duration (see e.g. Chandramouli et al. 1998; Chien et al. 2007).

In these circumstances there is high uncertainty in the acceleration factor, with the same effect on the estimated reliability parameters of the product. This drawback can be remedied by recourse to comparison testing. The latter serves also for reliability matching in objects of the same design and different origins, or a redesigned product versus its earlier counterpart, or different products with the same function (see e.g. Chien & Yang, 2007; Kececioglu, 2002).

The exponential nature of the Time Between Failures (TBF) of repairable objects, or the time to failure of non-repairable ones – is noted in the extensive literature on the reliability of electronic equipment (Kececioglu, 2002; Chandramouli et al, 1998; Drenick, 1960; Sr-332, 2001; MIL-HDBK-781A, 1996). For brevity, the TBF acronym is used in the sequel for both these notations.

Mace (1974, Sec. 6.12) proposed, for this purpose, the so-called fixed sample size test with the number of failures of each object fixed in advance – which is highly inconvenient from the practical viewpoint. For example, when the "basic" object has "accumulated" the specified number of failures, one has to wait until the "new" one has done the same, and if the latter is substantially more reliable, the waiting time may be very long.

The international standard IEC 61650 (1997) deals with two constant failure rates, which is equivalent to the problem just described. However, this standard, which forms part of an international system of techniques for reliability data analysis, does not refer to the planning aspect of the tests.

A solution to our problem was outlined in (Michlin & Grabarnik, 2007), where it was converted into binomial form, for which Wald's sequential probability ratio test (SPRT) is suitable (Wald, 1947, chap. 5). Wald and Wolfowitz (1948) also proved that this test is the most efficacious at two points of its characteristic, but it has one drawback – the sample size up to a decision can be many times larger than the average. This is usually remedied by resorting to truncation (see e.g. Wald, 1947; Siegmund, 1985).

A methodology is available for exact determination of the characteristics of such a truncated test with known decision boundaries. It was proposed by Barnard (1946) and developed by Aroian (1968). It served as basis for an algorithm and computer programmes (Michlin et al. 2007, 2009) used in examining its properties.

Hare we consider the inverse problem – determination of the test boundaries from specified characteristics.

In the absence of analytical dependences between the boundary parameters and characteristics, the search is hampered by the following circumstances:

- The number of parameter-value combinations may be very large.
- While shortening of the step makes for more combinations, it cannot be guaranteed that combinations with optimal characteristics are not missed.
- The standard optimum-search programmes are unsuitable for some of the discrete data of the type in question.

The theme of this chapter is the planning methodology for comparison truncated SPRT's. Formulae derived on its basis are presented for calculation of the test boundary parameters.

The rest of the chapter is organised as follows: In Section 2 is given a description of the test and its conversion to SPRT form. In Section 3 are described the quality indices for a truncated test and criteria for the optimal test search. In Section 4 are discussed the discrete nature of the test boundaries and its characteristics; a search algorithm is presented for the oblique boundaries. Section 5 describes the planning methodology, and approximative dependences are presented for calculation of the boundary parameters. Section 6 deals with planning of group tests. Section 7 presents a planning example and applications. Section 8 – the conclusion.

2. Description of test and its SPRT presentation

2.1 Description of test procedure in time domain. Checked hypothesis

In the proposed test two objects are compared – one "basic" (subscript "b") and the other "new" (subscript "n"). In the course of such tests, the "null" hypothesis is checked, that the ratio of the mean TBF (MTBF) of these objects exceeds or equals a prescribed value Φ_0, versus the alternative of it being smaller than the latter. The compared objects work concurrently (Figure 1). When one of them fails, it is immediately repaired or replaced. The unfailed object is not replaced but allowed to continue working until it fails in turn (in which case it is neither replaced nor repaired), or until the test terminates. A situation may occur in which there has been no failure in one object and it kept working throughout the whole test, as against several failures in the other object. The total work times T are equal for both objects.

Fig. 1. Scheme of test course (Upward marks – failures of basic item; downward marks – those of new item; T – time, common to both systems) (Michlin et al., 2011).

The probability density of the TBF for each of the compared objects has the form:

$$f_{TBF}(t)=(1/\theta)*\exp(-t/\theta)$$

where θ is the MTBF for the "new" (θ_n) and "basic" (θ_b) objects respectively. At each failure, a decision is taken – continuing the test versus stopping and accepting the null hypothesis, or rejecting it in favour of the alternative (Michlin & Migdali, 2002; Michlin & Grabarnik, 2007):

$$\begin{aligned} H_0 &: \quad \Phi \geq \Phi_0 \quad \left(P_a(\Phi_0)=1-\alpha\right) \\ H_1 &: \quad \Phi < \Phi_0 \quad \left(P_a(\Phi_1)=\beta\right) \end{aligned} \tag{1}$$

where

$$\Phi = \theta_n / \theta_b \tag{2}$$

a and β are the probabilities of I- and II-type errors; in the sequel, their target values will be denoted by the subscript "tg", and their actual values – by the subscript "$real$".
$P_a(\Phi)$ is the probability of acceptance of H_0, which is the Operating Characteristic (OC) of the test;

$$\Phi_1 = \Phi_0 / d \tag{3}$$

$d>1$ being the discrimination ratio.

Mace (1974 , Sec. 6.12) presents the following estimate $\hat{\Phi}$ for Φ, obtained with the aid of the maximum likelihood function (for the proof, see Kapur & Lamberson 1977, Sec. 10.C):

$$\hat{\Phi} = (T_n / r_n)/(T_b / r_b)$$

where r_n and r_b – the accumulated number of failures over times T_n and T_b.
As in this test $T_n=T_b=T$, we have:

$$\hat{\Phi} = r_b / r_n \tag{4}$$

Figure 2 shows an example of the test field. In the course of the test, it can reside at a point of this field characterised by an integer number of failures of each of the objects. When one of them fails, the test "jumps" to a neighbouring point located above (failure of "n") or to the right (failure of "b"). With the test course thus described, shifts from point to point occur only on failures in one of the objects, i.e. the time factor is eliminated from the analysis. When the Accept boundary is crossed, the test stops at the Accept Decision Point (ADT), and when its Reject counterpart is crossed – at the RDP. The boundaries consist of two parallel oblique straight lines (accept line (AL) and reject line (RL)) and the truncation lines parallel to the coordinate axes and intersecting at the Truncation Apex (TA).

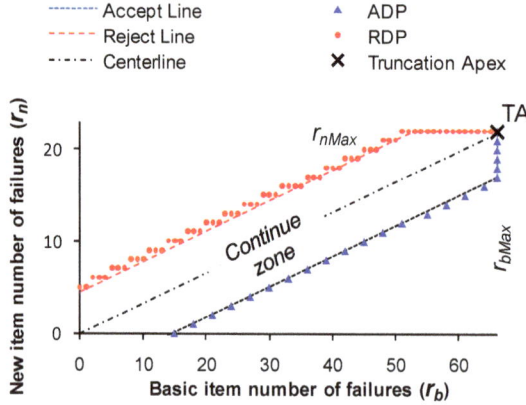

Fig. 2. Truncated test field for Φ_0=4.3, d=2, a_{real}=0.098, β_{real}=0.099, R_{ASN} =9.2%, (Michlin et al., 2011).

2.2 Binomial presentation of test and SPRT solution

For all points of the test field, the probability of the next failure occurring in the new object (i.e. of a step upwards) is constant and given by the following expression (for proof see Michlin & Grabarnik, 2007):

$$P_R(\Phi) = 1/(1+\Phi) \tag{5}$$

A binomial SPRT is available for such a test (Wald, 1947, chap. 5), whose oblique boundaries are:

$$\text{Accept line (AL):} \qquad r_b = r_n/s + h'_b \tag{6}$$

$$\text{Reject line (RL):} \qquad r_n = r_b \cdot s + h_n \tag{7}$$

where s is their slope, uniquely determined by the SPRT theory depending on a, β, Φ_0, d (Wald,1947; Michlin & Grabarnik, 2007), and given by:

$$s = -\ln q/(\ln q + \ln d) \tag{8}$$

where

$$q = (1+\Phi_0)/(d+\Phi_0) \tag{9}$$

The absolute terms of (6) and (7) are given by:

$$h_a = \ln\left(\beta^*/(1-\alpha^*)\right)/(\ln q + \ln d) \tag{10}$$

$$h_n = \ln\left(\left(1 - \beta^*\right)/\alpha^*\right)/\left(\ln q + \ln d\right) \tag{11}$$

$$h_b' = -h_a / s \tag{12}$$

The expressions (10)-(12) have one drawback: the parameters a^* and β^* are unknown. Their dependence on a_0, β, Φ_0, d, and on the TA coordinates is available only in the form of the limits between which the parameters lie (Michlin et al., 2009). Still, these limits suffice for determining – from the above expressions – corresponding search limits for h_b' and h_n. A search methodology, within these limits, for exact values ensuring the target characteristics – is, basically, the goal of this work.

2.3 Calculation of test characteristics acc. to given boundaries

The probability of hitting a given point of the test is given by (Barnard, 1946; Michlin & Grabarnik, 2007):

$$P_{r_b, r_n}(\Phi) = P_{r_b, r_n - 1}(\Phi) \cdot P_R(\Phi) + P_{r_b - 1, r_n}(\Phi) \cdot \left[1 - P_R(\Phi)\right] \tag{13}$$

while that of hitting the given ADP is:

$$P_{ADP}(r_n, \Phi) = P_{r_b - 1, r_n}(\Phi) \cdot \left[1 - P_R(\Phi)\right] \tag{14}$$

and that for the given RDP is:

$$P_{RDP}(r_b, \Phi) = P_{r_b, r_n - 1}(\Phi) \cdot P_R(\Phi) \tag{15}$$

$P_a(\Phi)$ is the sum of all the probabilities $P_{ADP}(r_n, \Phi)$ of hitting all ADP, hence the actual values of a and β, namely a_{real} and β_{real}, are given by:

$$\alpha_{real} = 1 - P_a(\Phi_0); \quad \beta_{real} = P_a(\Phi_1) \tag{16}$$

The Average Sample Number (ASN) of a truncated test is calculated as:

$$ASN(\Phi) = \sum_{r_b = 0}^{TA_b} \left[r_b + r_{nRDP}(r_b)\right] P_{RDP}(r_b, \Phi) + \sum_{r_n = 0}^{TA_n} \left[r_n + r_{bADP}(r_n)\right] P_{ADP}(r_n, \Phi) \tag{17}$$

where $r_{nRDP}(r_b)$ is the r_n-coordinate of the RDP with given r_b.
The Average Test Duration (ATD) for each object is:

$$ATD(\Phi) = \theta_b \cdot ASN(\Phi)/\left(1 + 1/\Phi\right) \tag{18}$$

3. Comparative characteristics and optimality of test

In this Section the optimality criteria for the test, on which the comparison- and selection algorithm is based, are substantiated, and the problems of the study are clarified.
In (Michlin & Grabarnik, 2007) were presented three optimality criteria which can be calculated for the specified boundaries:

- Closeness of the test OC to the prescribed one. For given d, the measure of this closeness is R_D:

$$R_D = \sqrt{\left[\left(\alpha_{real} - \alpha_{tg}\right)\Big/\alpha_{tg}\right]^2 + \left[\left(\beta_{real} - \beta_{tg}\right)\Big/\beta_{tg}\right]^2} \tag{19}$$

  with α_{real} and β_{real} as per (16).
- The degree of truncation, which characterises the maximum test duration whose measure can be, for example, the sum of the TA coordinates.
- The efficacy of the test according to Wald (1947) and to Eisenberg & Ghosh (1991), as the measure of which R_{ASN} was adopted (Michlin et al., 2009) – the relative excess of the function $ASN(\Phi)$ of the truncated test over $ASN_{nTR}(\Phi)$, its non-truncated counterpart which can be taken as ideal:

$$R_{ASN} = \left\{\sum_{i=1}^{5}\left[ASN(\Phi_i) - ASN_{nTr}(\Phi_i)\right]\right\}\Big/\sum_{i=1}^{5}ASN_{nTr}(\Phi_i) \tag{20}$$

  where Φ_i - values of Φ in geometric progression:

$$\Phi_0 \cdot \left(\sqrt{d}\right)^{i-4} \quad \text{for } i = 1...5 \tag{21}$$

$ASN(\Phi)$ – calculated as per the recursive formulae (17), (13...15) $ASN_{nTr}(\Phi)$ – calculated by Wald's formulae (1947, chap. 3) obtained for a non-truncated test of the type in question:

$$ASN_{nTr}(\Phi) = \frac{\left(1+\Phi(\eta)\right)\left[P_a(\eta)\ln B + \left(1-P_a(\eta)\right)\ln A\right]}{\left(1+\Phi(\eta)\right)\ln\left[\left(1+\Phi_0\right)\Big/\left(d+\Phi_0\right)\right] + \ln d} \tag{22}$$

where

$$\Phi(\eta) = \frac{d^\eta\left(1+\Phi_0\right)^\eta - \left(d+\Phi_0\right)^\eta}{\left(d+\Phi_0\right)^\eta - \left(1+\Phi_0\right)^\eta} \tag{23}$$

$$P_a(\eta) = \left(A^\eta - 1\right)\Big/\left(A^\eta - B^\eta\right) \tag{24}$$

$$A = \left(1-\beta_{real}\right)\Big/\alpha_{real} \tag{25}$$

$$B = \beta_{real}\Big/\left(1-\alpha_{real}\right) \tag{26}$$

η – an auxiliary parameter calculated by (23) for Φ values as per the progression (21). The choice criterion for the optimal test is:

$$\min(TA_n + TA_b) \tag{27}$$

subject to:

$$(\min R_d \text{ at given } TA)\&(R_d < R_{d0})\&(R_{ASN} < R_{ASN0}) \tag{28}$$

where R_{d0} and R_{ASN0} – threshold values of R_d and R_{ASN}.
The TA of such a test called Optimal TA (OTA). Section 5 presents approximative formulae for determination of those OTA coordinates which permit reduction of the search field to 2 to 6 points. A particular problem in this context is: for a given TA, find h'_b and h_n (eqs. (6), (7)) ensuring min R_d.

4. Discreteness of test boundaries and their search at given TA

This Section deals with the interrelationships between the boundary parameters of the test on the one hand, and the characteristics of the test itself (namely, a_{real} and β_{real}) and those of its quality (introduced in the preceding Section, R_d and R_{ASN}) – on the other. These interrelationships lack analytical expression and are further complicated by the discreteness of the test. Thus one had to make do with typical examples of their behaviour in the vicinity of the optimum. With this behaviour clarified, an efficacious search algorithm could be developed for the optimum in the discrete space in question. Clarity of the picture is essential both for the developer of the planning methodology and for the practitioner planning the binomial test in the field.

4.1 Discreteness of test boundaries
As the slope s of the oblique test boundaries, described by eqs. (6) and (7), is unrelated to a and β (see eq. (8)), the search for them under the min R_d stipulation reduced to finding the absolute terms in the describing equation, namely the intercepts h'_b and h_n on the coordinate axes (Figure 3).

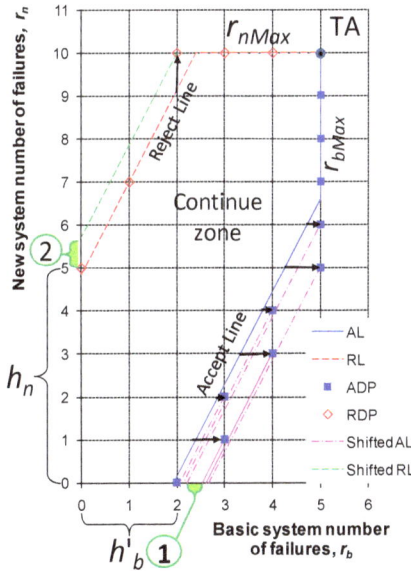

Fig. 3. Test Plane (Michlin & Grabarnik, 2010). 1 – Example of interval of h'_b values over which the test ADP's do not change. 2 – Ditto for h_n and RDP.

Stopping of the test occurs not on the decision lines, but at points with integer coordinates, with ADP to the right of the AL and RDP above the RL. If the AL is shifted from an initial position (solid line in Figure 3) to the right, the test characteristics remain unchanged until it crosses an ADP, which in turn is then shifted in the same direction by one failure. The AL positions at these crossings are shown as dot-dashed lines, and its shifts are marked with arrows. Projecting the termini of these arrows, parallel to the AL, on the r_b axis, we obtain the values of h'_b at which the changes occur. An analogous process takes place when the RL is shifted upwards.

The intervals of h'_b and h_n over which the test characteristics remain unchanged are marked in Figure 3 by the circled numbers 1 and 2 respectively.

When the AL is shifted to the right (h'_b increased) $P_a(\Phi)$ is reduced, i.e. a_{real} increases and β_{real} decreases. When the RL is shifted upwards, the effects are interchanged. These relationships are monotonic and stepwise, and differ in that change of h'_b is reflected more strongly in β_{real} and more weakly in a_{real}. With h_n the pattern is reversed.

4.2 Basic dependences between oblique boundaries and test characteristics

In (Michlin et al., 2009, 2011; Michlin & Kaplunov, 2007) were found the limits within which a^* and β^* of the optimal tests should be sought. These limits can also serve for determining the search limits of h'_b and h_n, as per (10) – (12).

Figure 4 shows an example of the above, with the limits for h'_b and h_n calculated, according to the data of (Michlin et al., 2009), for $d=2$, $\Phi_0=1$, $a_{tg}=\beta_{tg}=0.1$, $TA_b=27$, $TA_n=38$, $R_{ASN}\leq12\%$. In the figure, the points mark the centres of rectangles within which the characteristics remain unchanged. The resulting picture is fairly regular, even though the spacings of the columns and rows are variable. In space, the R_d points form a cone-shaped surface.

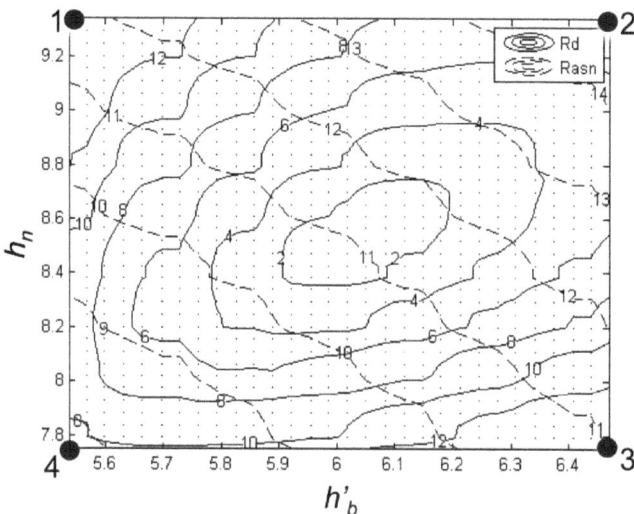

Fig. 4. Contours of R_{ASN} (dashed lines) and R_D (solid lines) vs. h'_b and h_n. (Michlin & Grabarnik, 2010). The dots mark the centres of rectangles within which the test characteristics do not change. 1 – 4 are the corner points at which the test characteristics are calculated in the search for the optimum (Subsection 4.3, stage ‹1.1›).

The figure also contains the contours (isopleths) of R_d (solid lines) and R_{ASN} (dashed lines), given as percentages. In macro the R_d contours can be described as oval-shaped, whereas in micro they are quite uneven, so that derivatives calculated from a small set of points would show large jumps, which would hamper the search for the minimum R_d. It is seen that in the vicinity of that minimum, $R_{ASN}\approx11\%$.

Figure 5 shows two projections representing α_{real} and β_{real}, calculated according to the coordinates of Figure 4, so that to each point of the latter corresponds one of α_{real} and β_{real}. These points form intersecting almost-plane surfaces. In the upper figure the coordinate axes are oriented so that the intersection zone (α_{real} - β_{real}) is perpendicular to the page; in the lower figure. the orientation is such that the rows of β_{real} points reduce in projection to a single point – in other words, they form parallel or almost-parallel lines.

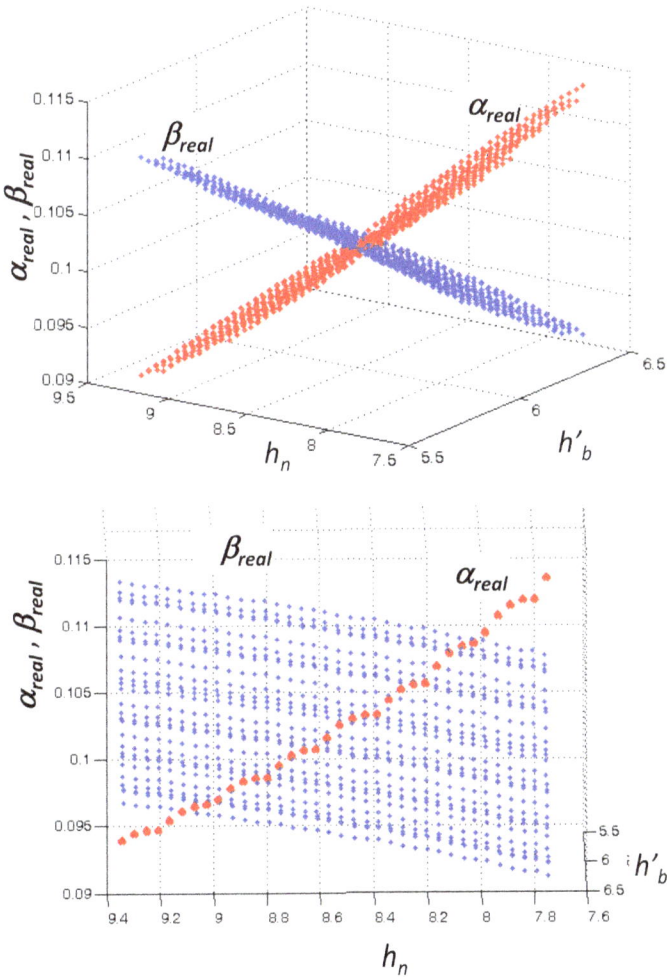

Fig. 5. Two projections of α_{real} and β_{real} "planes". (Michlin & Grabarnik, 2010).

Figure 6 shows analogous projections for R_{ASN}, and we again have an almost-plane surface, monotonic and uneven in micro.

The provided examples show that the described patterns characterise the dependences of a_{real}, β_{real} and R_{ASN} on h'_b and h_n within the limits determined in Subsection 5.3 (Michlin et al., 2009, 2011; Michlin & Kaplunov, 2007). Over small intervals these dependences are stepwise, and the lines through the step midpoints are uneven as well.

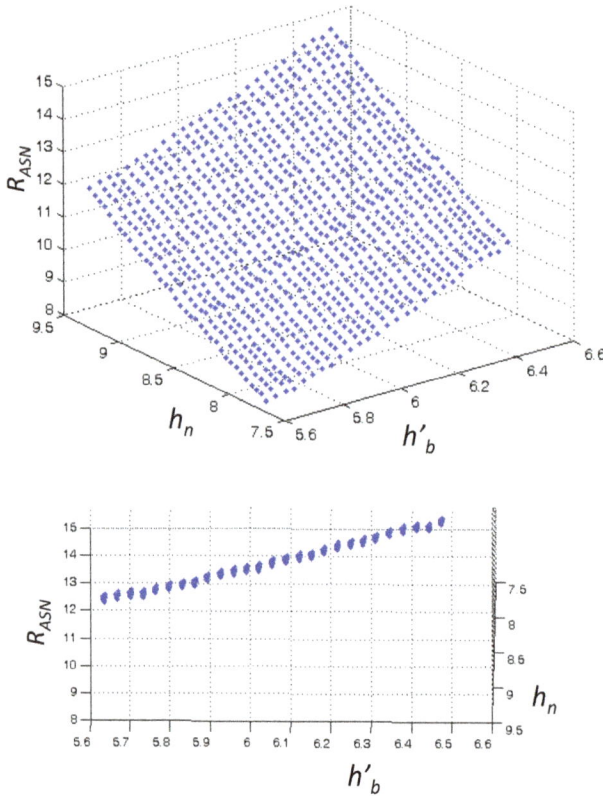

Fig. 6. Two projections of R_{ASN} "plane". (Michlin & Grabarnik, 2010).

4.3 Search algorithm for oblique test boundaries

Standard search programmes for minima (such as those in Matlab) operate poorly, or not at all, with discrete data of the type in question. Availability of known regularities in the behaviour of the functions a_{real}, β_{real}, R_{ASN}, R_D makes it possible to construct a fast and efficacious algorithm.

These known regularities are:

* The values of h'_b and h_n at which the test characteristics change.
* The limits of h'_b and h_n, yielding tests with the specified characteristics.
* Almost-plane monotonic dependences of a_{real}, β_{real} and R_{ASN} within the above limits, stepwise and unstable but also monotonic in narrower intervals.

- Stronger dependence of a_{real} on h_n than on h'_b; the reverse – for β_{real}.

In expanded form, the search algorithm for min R_d consists in the following:

1st stage.

‹1.1› Calculation of the test characteristics at the four vertices of a rectangle (Figure 4) whose coordinates are obtained from the relationships presented in Subsection 5.3.

‹1.2› Approximation of $a_{real}(h'_b, h_n)$ and $\beta_{real}(h'_b, h_n)$ as planes, and determination of the first estimate h'_{b1}, h_{n1} yielding min R_D (point 5, Figure 7). Checking for $R_D \leq R_{D0}$. If satisfied, stopping of search.

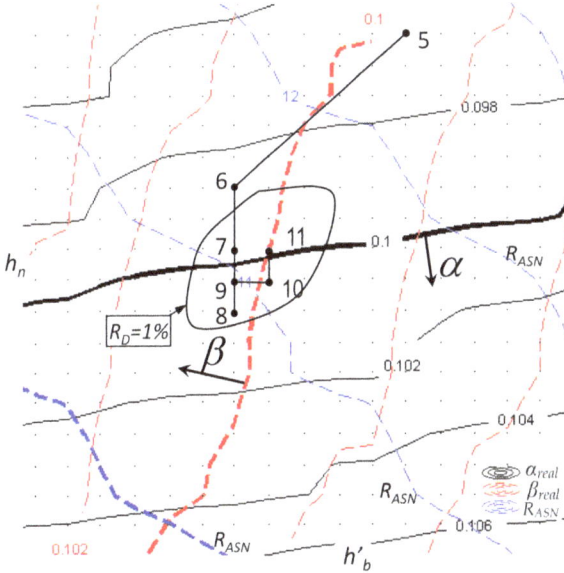

Fig. 7. Example of search scheme for min(R_D). (Michlin & Grabarnik, 2010). 5 – 11 are points of test characteristics calculation.

2nd stage.

Determination of point 6 – from a_{real5}, β_{real5} and the slopes of the a-, β-planes as per ‹1.2›. Re-checking for $R_D \leq R_{D0}$.

3rd stage.

Alternating advance parallel to the h'_b- and h_n-axes. In view of the discreteness and complexity of the R_D function, the search for its minimum was reduced to one for the points h'_b and h_n where Δa and $\Delta \beta$ change sign:

$$\Delta a = a_{real} - a_{tg}; \quad \Delta \beta = \beta_{real} - \beta_{tg}$$

This problem is easier to solve, as both Δa and $\Delta \beta$ are monotonic functions of h'_b and h_n. The search can be stopped at every step, subject to $R_D \leq R_{D0}$.

‹3.1› If at point 6 (‹2› above) $\Delta a_6 > \Delta \beta_6$, a path parallel to the h_n-axis is taken in uniform steps Δh_n, until Δa changes its sign (points 6,7,8 on Figure 7), $\Delta h_n = \Delta a_6 / a_3$, where a_3 is the coefficient in the equation of the a-plane as per ‹1.2›. Beyond that point, the root $\Delta a(h_n)$ is searched for by the modified Regula Falsi method (point 9). (The term "root" refers here to one of a pair of adjoining points at which the function changes its sign and has the smaller absolute value). The special feature of this procedure is accounting for the discreteness of the solution.

‹3.2› At the point of the root Δa, a right-angled turn is executed and a path parallel to the h'_b-axis is taken, searching for the $\Delta\beta$ root (point 10).

‹3.3› The alternating procedure is continued until a situation is reached where two consecutive turns involve only movement to an adjoining point. This point 10 corresponds to $\min(R_D)$. If in ‹3.1› $\Delta a_6 < \Delta\beta_6$, we begin from ‹3.2›.

4.4 Efficacy of algorithm

With a view to assessing the efficacy of the proposed algorithm, a search for the h'_b and h_n values yielding min R_d was conducted with the aid of a Matlab programme which realised this algorithm, and alternatively with the Matlab fminsearch command, with the same function WAS (Michlin & Grabarnik, 2007) referred to in both cases. This function determines the test characteristics according to its specified boundaries. The run covered different tests with $R_{ASN0}=5$ and 10%.

The calculation results are shown in Figure 8.

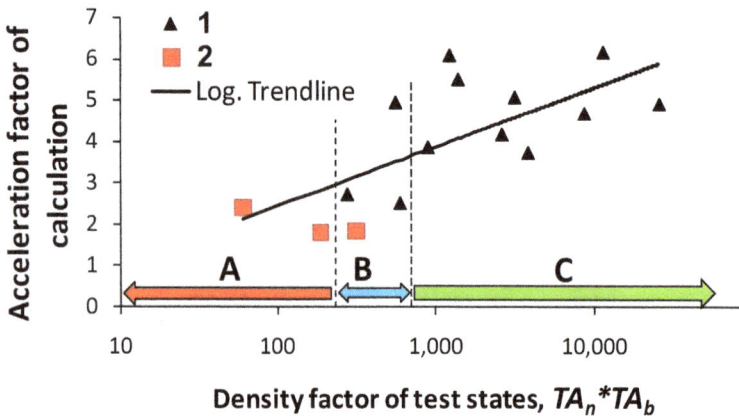

Fig. 8. Comparative efficacy of proposed algorithm. (Michlin & Grabarnik, 2010).
1 – fminsearch (Matlab) found min R_D or stopped close to it;
2 – fminsearch failed to find R_D;
A, B, C – short, medium and long tests, respectively.

In it, the abscissa axis represents the product TA_b*TA_n, which we term "density factor of test states". The higher the latter, the denser the disposition of the test points in the search zone (see Figure 4), the smaller the changes in the test characteristics from point to point, and the closer the search to one over a continuous smooth surface. A small value of the product is associated with a short test, due to be completed at small sample size and moderate computation times for the characteristics; a large value – with long tests, completed on the average at large sample sizes and long computation times.

The ordinate axis represents the "acceleration factor of calculation", which is the ratio of references to the WAS-function by fminsearch and the proposed algorithm respectively. The larger the ratio, the faster the algorithm compared with the standard Matlab function.

The diagram shows that at low densities (short tests, zone A) fminsearch fails to find min R_d. In zone C (long tests) the command finds it or stops close to it, but with 3 to 6 times more references to WAS. In zone B (medium tests) the minimum is either not found, or found with 2.5 to 5 times more references to WAS. By contrast, the programme based on the proposed algorithm found the minimum in all cases.

Accordingly, for the present task – searching for the optimum in a discrete space – the proposed algorithm accomplishes it much faster than the Matlab standard fminsearch command, thus saving computation time in long tests. Moreover, it guarantees a solution – a critical aspect in short tests, where fminsearch usually fails to find one.

5. Estimates for boundary parameters

5.1 Search methodology for optimal test boundaries

In (Michlin & Grabarnik, 2007) it was established that for $\Phi_0=1$ and $a=\beta$, the OTA lie on the centreline (which runs through the origin parallel to the AL/RL), so that

$$r_n = s \cdot r_b \tag{29}$$

This was checked for different Φ_0. With given $a_{tg}=\beta_{tg}$, d, and $R_D \leq 1\%$, a search was conducted for three location zones of the TA – namely, with $R_{ASN} \leq 5\%$, $5\% < R_{ASN} \leq 10\%$, and $R_{ASN} > 10\%$, the last-named being restricted by the above requirement on R_D, i.e. achievability of a_{tg} and β_{tg}.

A typical example of such zones for $a_{tg}=\beta_{tg}=0.05$, $d=2$, and $\Phi_0=1, 2, 3$ is shown in Figure 9. The fan-shaped zones have their apices on the corresponding centrelines. These apices are the OTA locations, as with the imposed limits satisfied they are closest to the origin (heaviest truncation). In these circumstances the search zone is narrowed, the location problem being converted from two- to one-dimensional.

To study the relationships between the sought boundary parameters (TA, a^*, β^*) and the specified test characteristics (Φ_0, d, $a_{tg}=\beta_{tg}$, $R_{ASN\ max}$), a search was run over a large population of optimal tests with the characteristics given in the Table below.

	Lower limit	Upper limit	Number of levels
Φ_0	0.3	5	9
d	1.5	5	12
$a_{tg}=\beta_{tg}$	0.05	0.25	5
$R_{ASN\ max}$	5%	10%	2

Table 1. Regions of characteristics covered by search

5.2 Search results for OTA and their curve fitting

The dots in Figure 10a mark the OTA for $a_{tg}=\beta_{tg}=0.05$, $R_{ASN} \approx 10\%$, and wide intervals of d and Φ_0. Figure 10b is a zoom on the domain in 3a representing the "short" tests, namely those with small ASN and – correspondingly – low TA coordinates. It is seen that all curves smooth out as the distance from the origin increases (the tests become longer), the reason

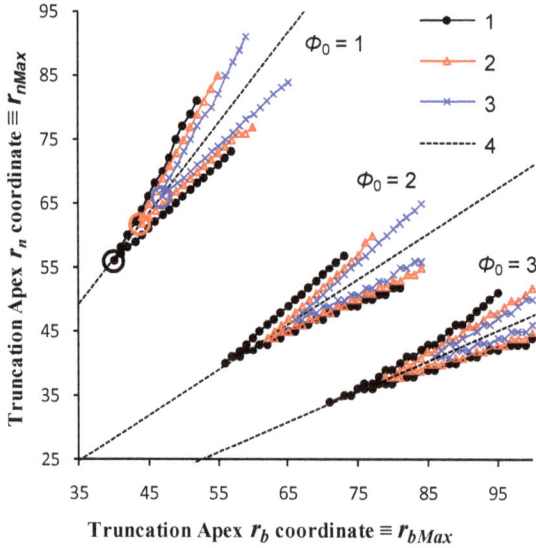

Fig. 9. TA zone boundaries for three Φ_0 values and three R_{ASN} zones (Michlin et al., 2011): 1 = Boundary beyond which $R_D \leq 1\%$ is unachievable at any R_{ASN}; 2 = Boundary for $R_{ASN} \leq 10\%$; 3 = Boundary for $R_{ASN} \leq 5\%$; 4 = Centreline. Remark 1. $\Phi_0 = 1$ subgraph: OTA for each R_{ASN} zone circled. Remark 2. For this figure: $d=2$, $a_{tg}=\beta_{tg}=0.05$, $R_D =1\%$.

being the weakening influence of discreteness of the test characteristics   (Michlin et al., 2009).

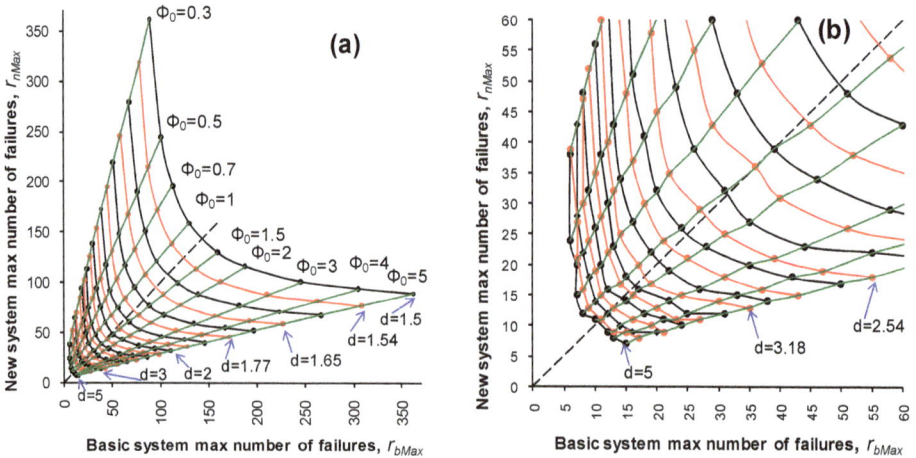

Fig. 10. (a) OTA locations for different d and Φ_0, and for $a_{tg}=\beta_{tg}=0.05$, $R_{ASN} \approx 10\%$. (b) Zoom on short test zone. (Michlin et al., 2011).

The Φ_0-isopleths in the figures are broken radial lines, whereas their d-counterparts are symmetrical about the $r_n = r_b$ line and approximate neatly to a hyperbola:

$$r_n(r_b) = \left\{ \left[k(d) \cdot q(x, R_{ASN}) \right]^{-1} - r_b^{-1} \right\}^{-1} \tag{30}$$

where

$$k(d) = \exp\left[5.58(d-1)^{-1/4} \right] - 1 ; \tag{31}$$

$$q(x, R_{ASN}) = -\frac{1}{20}(1 + 1.10\ln x + 0.41 R_{ASN} - 1.03 R_{ASN} \ln x) \tag{32}$$

x – common target value for a and β, $x = a_{tg} = \beta_{tg}$;
R_{ASN} – in relative units rather than in percent.
The formulae indicate that the approximate curves differ only in the scale factor $k(d)$, common to both axes – it remains the same for any pair (x, R_{ASN}).
As the formulae do not contain Φ_0, the OTA is searched for through its required adherence to the centreline, whose expression (29) is uniquely determined by d and Φ_0. Accordingly, the sought OTA is the integer point closest to the intersection of the curve (30) and the centreline (29) (Figure 11).

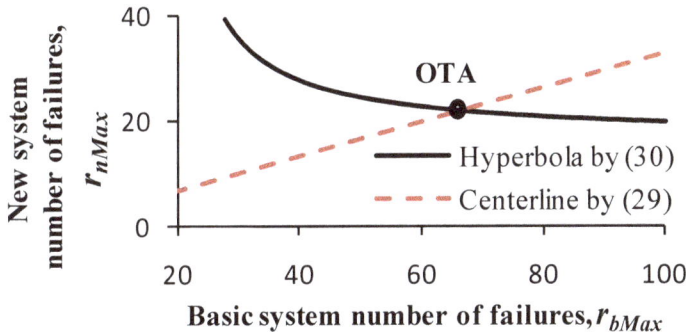

Fig. 11. Determination of OTA. (Michlin et al., 2011).
The coefficients in (31) and (32) were found through the requirement of minimal root mean square error ($RMSE$) – the difference between the OTA's found as per eqs. (29) and (30). For the data in the Table, $RMSE = 0.88$, indicating high estimation accuracy for such a broad domain.

5.3 Estimates for a^* and β^*

As already mentioned, the problem of finding the oblique boundaries reduces to that of finding a^* and β^*. This Subsection presents regressional dependences of the latter on the test characteristics Φ_0, d, $x = a_{tg} = \beta_{tg}$, and $R_{ASN\ max}$, as well as their counterparts for the upper and lower limits (a_U^* and a_L^*, β_U^* and β_L^*) of these parameters. These dependences, determined on the basis of the total data on optimal tests with the characteristics in the Table, were sought in the form:

$$\alpha_M^* = c_\alpha \cdot x;$$
$$\beta_M^* = c_\beta \cdot x. \tag{33}$$

The Matlab tool for stepwise regression yielded the coefficients for the above:

$$c_\alpha = 1.10 - 0.021(\ln x)^2 - 0.0081\Phi_0^2 + 0.036\Phi_0 d + 1.07 R_{ASN}\ln x \tag{34}$$

where $RMSE$=0.061 and R^2=0.83, the latter being the coefficient of determination, and

$$c_\beta = 1.09 + 0.096\ln x + 0.14d - 0.018\Phi_0 d + 1.11 R_{ASN}\ln x \tag{35}$$

with $RMSE$=0.069 and R^2=0.80.
The limit formulae read

$$\left.\begin{matrix}\alpha_U^*\\\alpha_L^*\end{matrix}\right\} = (1 \pm c_{\alpha B})\alpha_M^* \tag{36}$$

$$\left.\begin{matrix}\beta_U^*\\\beta_L^*\end{matrix}\right\} = (1 \pm c_{\beta B})\beta_M^* \tag{37}$$

where

$$c_{\alpha B} = -0.045 + 0.14\ln d - 0.031\ln x \tag{38}$$

$$c_{\beta B} = -0.059 + 0.16\ln d - 0.048\ln x \tag{39}$$

and such that all a^* and β^* obtained for the Table are included.
Figure 12 shows example dependences for the regressional value α_M^* and the upper and lower limits, versus $x=a_{tg}$ for Φ_0=3 and d=1.5, 3. Also included are the actual values of a^*. (The graphs for β^* are analogues). The bounded zone becomes narrower as d and a_{tg} decrease. It is seen that at low d, α_M^* and β_M^* can serve as the calculation values without undue deviation of a_{real} and β_{real} from their targets.
The search methodology for a^* and β^* of the optimal test, described in detail in Section 4, is based on knowledge of the limits (36), (37), which is one of the reasons for its high efficacy.

5.4 Accuracy assessment of proposed planning
The accuracy of the proposed planning, using eqs. (30) – (32) and (33) – (35) – was assessed by applying them in calculating the test boundaries for all characteristic values in the Table. This was followed by calculation of a_{real}, β_{real} and R_{ASN} for these tests and their deviation from the targets. The $RMSE$'s of a_{real} and β_{real} decrease with decreasing d and R_{ASN}. For $d\leq2$ they do not exceed 3 to 4% of the target value and for large d they reach 8 and 10% at R_{ASN}=5 and 10% respectively. In the former case this is very satisfactory accuracy, while in the latter case it may become necessary to find more accurate values of the boundary parameters – for which the methodology outlined in Section 4 is recommended, using eqs. (30) – (32) and (34) – (39) for the search limits.

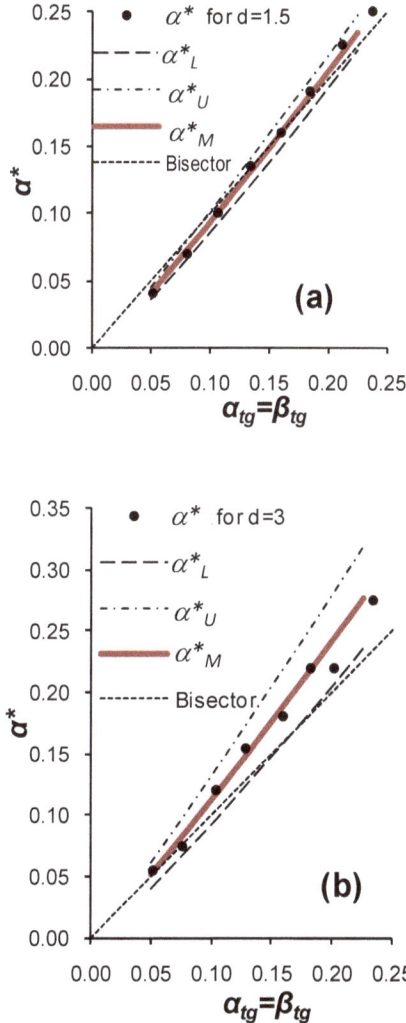

Fig. 12. Actual a^*, regressional dependence, and upper and lower search limits. $R_{ASN}=5\%$. (a) $d=1.5$. (b) $d=3$. (Michlin et al., 2011).

6. Group tests

In this case the items are compared groupwise, which makes for economy in the time to a decision. The items of the respective subgroups, N_b and N_n in number, are drawn at random from their respective populations with exponential TBF's, and tested simultaneously. On failing, they are immediately replaced or repaired – just as in the two-item tests. The subgroup can be treated as a single item with an N-times shorter MTBF (Epstein & Sobel, 1955). The planning procedure remains the same, except that Φ in the calculations is replaced by Φ_g:

$$\Phi_g = \Phi \cdot N_b / N_n \tag{40}$$

Thus when $N_b = N_n = N$, the test boundaries remain as in the two-item case, except that the test duration is also N times shorter (see (18)). When $N_b \neq N_n$, it is recommended to check the efficacy of larger groups, e.g. in terms of a shorter average test duration $ATD_g(\Phi)$. By (18) and (40) we obtain:

$$ATD_g\left(\Phi_g\right) = \left(\theta_b / N_b\right) \cdot ASN_g\left(\Phi_g\right) \Big/ \left(1 + 1/\Phi_g\right) \tag{41}$$

where $ASN_g(\Phi_g)$ is the ASN of the group test as per (17) or (22), except for Φ_g replacing Φ of (40).

The planning example covers also the problem of choice of N_b and N_n, while ensuring min ATD_g and satisfying additional essential test-planning conditions.

7. Example of test planning

A large organisation operates a correspondingly large body of mobile electronic apparatus whose MTBF is substantially shortened under the stressful exploitation conditions. The manufacturer offers to modify this equipment, thereby significantly improving its resistance to external impacts, albeit at increased weight and cost.

In a fast laboratory test the modified (hereinafter "new") apparatus exhibits high reliability, but so does the original ("basic") one. Accordingly, it is decided to check the MTBF increase under field conditions on an experimental batch.

The requirements regarding the test OC are established as follows. If the MTBF of the new product is 5 times that of the basic ($\Phi=5$), replacement is beneficial; at $\Phi=2.5$ it does no harm; but at $\Phi=1.5$ it is unacceptable. These findings follow from the OC_{nTr} of a non-truncated SPRT with $a=\beta=0.1$, $d=2$, $\Phi_0=5$ (Figure 13), constructed as per (23) – (24).

The apparatus are operated in sets of 28 items, so that conditions within a set are practically uniform. Each set comprises both new and basic items, so as to offset the influence of fluctuating conditions.

A "failure" in this context is defined as any event that necessitates repair or re-tuning of the item, with enforced idleness for more than 20 seconds. The failed item is either treated in situ – or replaced by a spare, repaired and stored with the spares. Thus the size of the operative set remains 28.

The assignment is – planning a truncated test with the proportions of new and basic items in the test group chosen so as to ensure a minimal ATD. Below is the planning procedure:

a. As the OC's are practically the same for truncated and non-truncated tests when their Φ_0, d, a_{real} and β_{real} coincide (Michlin & Grabarnik, 2007) – we chose the initial parameters given above:

$$\Phi_0=5, \ d=2, \ a_{tg}=\beta_{tg}=0.1 \tag{42}$$

and specified

$$R_{ASN \ max} = 10\% \tag{43}$$

Thus the test has an ASN and ATD close to that of the non-truncated SPRT, and at the same time its maximal duration is heavily restricted, a fact of practical importance for the organisation.

b. Eqs. (41) and (18) yielded the approximate dependences of $ATD_g(\Phi)/\theta_b$ on N_n for different Φ, given $N_n + N_b = 28$. A minimum was found at $N_n \approx 18$. Figure 14 shows examples of these dependences at $\Phi = \Phi_0$ and $\Phi = \Phi_1$, which are seem to be almost flat over a wide interval around the minimum, and $N_n = 15$ was chosen accordingly. With this choice, $ATD_g(\Phi)$ only slightly exceeds the minimum, while the number of new items is lower, with the attendant saving in preparing the experimental batch. By (40) we have

$$\Phi_{0g} = \Phi_0 \cdot 13 / 15 = 4\tfrac{1}{3} \tag{44}$$

The values of $ASN_g(\Phi)$ and $ATD(\Phi)$, obtained by (41) and (18) with allowance for (44) – confirmed the practicability of the test.

c. Eq. (8) yielded $s=0.330$. Simultaneous solution of (29) and (30) yielded, after rounding-off, the TA coordinates: $r_{bMax}=66$, $r_{nMax}=22$.
Eqs. (33) through (35) yielded $a^*=0.0909$, $\beta^*=0.074$, which in turn, by (11) and (12), yielded $h'_b=4.804$, $h_n=4.453$.
The decision boundaries for a test planed on the basis of these parameters are shown in Figure 2.
Figure 13 shows the exact values of the functions $OC(\Phi)$ and $ASN(\Phi)$ as per eqs. (13) – (18), which in turn yield the test's real characteristics: $\Phi_0=5$, $d=2$, $a_{real}=0.098$, $\beta_{real}=0.099$, $R_{ASN}=9.2\%$, in very close agreement with the given (42) and (43) – evidence of the high accuracy of eqs. (30) – (35).

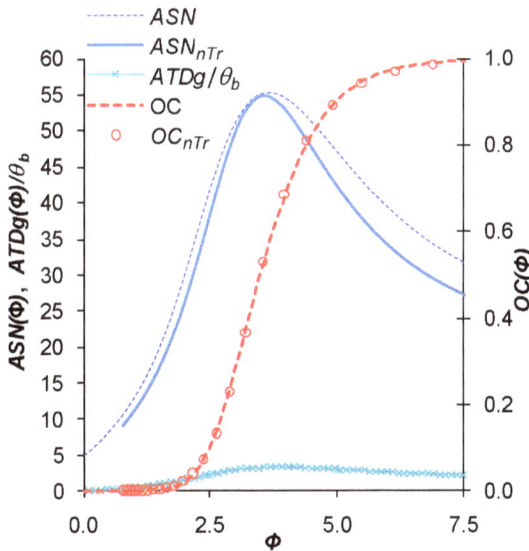

Fig. 13. OC and ASN of truncated group test and of non-truncated theoretical (subscript nTr) test; normalised expected duration of group test $ATDg(\Phi)/\theta_b$ for $\Phi_0=5$, $\Phi_{0g}=4^1/_3$, $d=2$, $a_{real}=0.098$, $\beta_{real}=0.099$, $r_{bMax}=66$, $r_{nMax}=22$. (Michlin et al., 2011).

The $OC(\Phi)$ of the planned test (Figure 13) practically coincides with that of the non-truncated test $OC_{nTr}(\Phi)$ with the same a_{real} and β_{real}. The ASN of the former is higher than that of the latter, in accordance with R_{ASN}=9.2%. The diagram also shows the estimate for the normalised ATD, i.e. the ratio $ATD_g(\Phi)/\theta_b$. Assuming $\hat{\theta}_b = 10$ hr , the time requirement of the test should be reasonable. In practice, it ended with acceptance of the null hypothesis in 16 hr, following the twenty-first failure in the basic subgroup, by which time a total of 2 failures in the new subgroup had been observed.

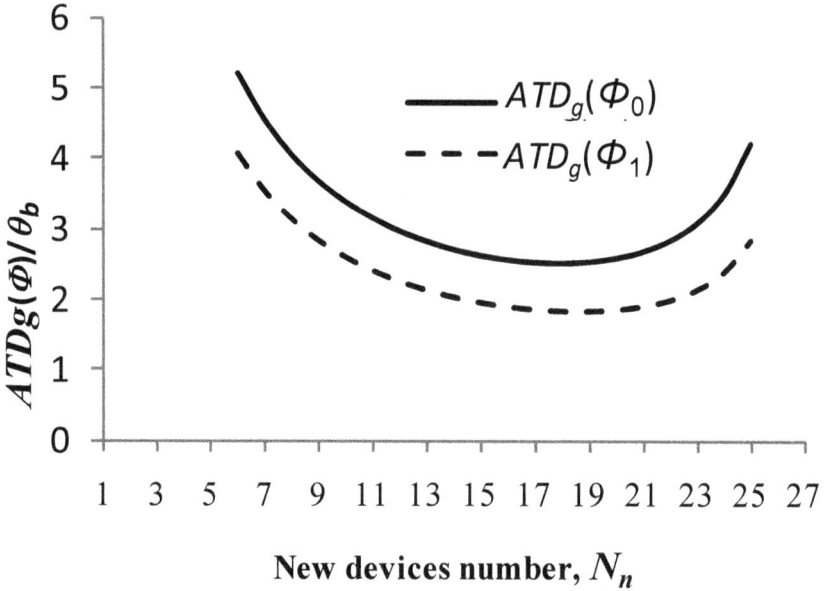

Fig. 14. Normalised expected group test duration vs. number of new devices, for $\Phi=\Phi_0$, and $\Phi=\Phi_1$. (Michlin et al., 2011).

8. Conclusion

The example in Section 7 demonstrated the potential of the proposed planning methodology for a truncated discrete SPRT. An innovative feature in it are the test-quality characteristics R_{ASN} and R_D – which represent, respectively, increase of the ASN on truncation and closeness of the test OC to the non-truncated one. This innovation permitted comparison of different SPRT and automatisation of the optimum-choice process. It was found that over a large domain about the solution, the R_{ASN} and boundary parameters are linked monotonically and almost linearly. This implies sound choice of this characteristic and simplifies the planning. An efficacious search algorithm was developed for the optimal test boundaries, incorporating the obtained interrelationships.

The findings can be summed up as follows:

- A truncated SPRT was studied with a view to checking the hypothesis on the ratio of the MTBF of two objects with exponential distribution of TBF.

- It was established that the basic test characteristics a_{real}, β_{real}, R_{ASN} depend monotonically on the absolute terms in the equations of the oblique test boundaries.
- At the search limits for these absolute terms, determined in Section 5, these dependences are almost plane.
- a_{real} and β_{real} change stepwise with the smooth changes in the absolute terms of the oblique boundaries; expressions are derived for the minimal intervals of these terms, over which a_{real} and β_{real} remain unchanged.
- These and other established regularities yielded an efficacious algorithm and programme for determining the optimal location of the test boundaries.
- The found links between the input and output characteristics of the test, and the fast-working algorithm for its planning, permit improvement of the planning methodology and its extension to all binomial truncated SPRT.
- On the basis of the above body of information, regressional relationships were derived for determining the TA coordinates and oblique-boundary parameters of the optimal tests. Also derived were formulae for the limits of the latter parameters. These are very close at low d and R_{ASN} and draw apart as the characteristics increase; the reason being increasing influence of the test's discreteness. The regressional relationships and boundary-parameter limits permit quick determination of these boundaries for the optimal test with specified characteristics.
- The methodology is also applicable in group tests, with the attendant time economy; moreover, it permits optimisation of the respective group sizes.
- A planning and implementation example of this test is presented.

9. Acknowledgements

The authors are indebted to Mr. E. Goldberg for editorial assistance, and to MSc students of the "Quality Assurance and Reliability" Division of the Technion: Mrs. E. Leshchenko, and Messrs. Y. Dayan, D. Grinberg, Y. Shai and V. Kaplunov, who participated in different stages of this project.

The project was supported by the Israel Ministry of Absorption and the Planning and Budgeting Committee of the Israel Council for Higher Education.

10. Acronyms

ADP	accept decision point
AL	accept line
ASN	average sample number
ATD	average test duration
MTBF	mean TBF
OC$\equiv P_a(\Phi)$	operating characteristic
OTA	truncation apex of the optimal test
RDP	reject decision point
RL	reject line
RMSE	root mean square error
SPRT	sequential probability ratio test

TA	truncation apex
TBF	time between failures or time to failure
WAS	program name

11. Notations

$ASN(\Phi)$	exact value of ASN for a truncated test, obtained recursively (17)
$ASN_{nTr}(\Phi)$	ASN calculated via an analytical formula (22) for a non-truncated test
$ATD(\Phi)$	ATD function for given Φ
c	with the appropriate subscripts, coefficients in the approximative equations
$d = \Phi_0 / \Phi_1$	discrimination ratio
h'_b, h_n	absolute terms of Accept, and Reject oblique boundaries, respectively
N_b, N_n	item numbers of "basic" and "new" subgroups in group test
$P_a(\Phi) \equiv OC$	acceptance probability of H_0 at given Φ
$P_{ADP}(r_n, \Phi), P_{RDP}(r_b, \Phi)$	probabilities of reaching the given points ADP, RDP
$P_R(\Phi)$	probability of new system failing next during test
r_b, r_n	system number of failures observed up to time T
$r_{bADP}(r_n)$	r_b-coordinates of ADP for given r_n
$r_{nRDP}(r_b)$	r_n-coordinates of RDP for given r_b
R^2	coefficient of determination
R_{ASN}	relative excess of the ASN of the truncated test over its non-truncated counterpart
R_D	relative deviation a_{real} and β_{real} from their targets
R_{d0} and R_{ASN0}	threshold values of R_d and R_{ASN}
s	slope of oblique boundaries
T	current test time
TA_b, TA_n	r_b- and r_n-coordinates of TA, respectively
x	common target value for a and β, $x = a_{tg} = \beta_{tg}$
a, β	probabilities of I- and II-type errors in test
a_{real}, β_{real}	exact real values of α and β computed for prescribed stopping boundaries
a_{tg}, β_{tg}	target values of a, β
a^*, β^*	parameters determining the constant terms of initial boundary lines
$\alpha_M^*, \alpha_U^*, \alpha_L^*, \beta_M^*, \beta_U^*, \beta_L^*$	regressional value, upper and lower search limits of a^* and β^*
$\theta, \theta_b, \theta_n$	MTBF, same for the basic system θ_b, and for the new system θ_n respectively
$\Phi = \theta_n / \theta_b$	true MTBF ratio
Φ_0	Φ value for which the null hypothesis is rejected with probability a
Φ_1	Φ value for which the null hypothesis is rejected with probability $1-\beta$
Φ_g	Φ for group test

12. References

Aroian L. A. (1968). Sequential analysis-direct method. *Technometrics*. Vol. 10, pp. 125-132.

Barnard, G. A. (1946). Sequential test in industrial statistics, *Journal of the Royal Statistical Society.* Suppl., Vol. 8, pp. 1-21.

Chandramouli, R.; Vijaykrishnan N. & Ranganathan, N. (1998). Sequential Tests for Integrated-Circuit Failures, *IEEE Transactions on Reliability.* Vol. 47, No. 4, pp. 463–471.

Chien, W. T. K. & Yang, S. F. (2007). A New Method to Determine the Reliability Comparability for Products, Components, and Systems in Reliability Testing. *IEEE Transactions on Reliability*, Vol. 56, No. 1, pp. 69–76.

Drenick, R. F. (1960). The failure law of complex equipment. *The Journal of the Society for Industrial Applications of Mathematics*, Vol. 8, No. 4, pp. 680-689.

Eisenberg, B., & Ghosh, B. K. (1991). The sequential probability ratio test. In: *Handbook of Sequential Analysis*, B. K. Ghosh, Sen P.K (Ed.), pp. 47-66, Marcel Dekker, NY.

Epstein, B. & Sobel, M. (1955). Sequential life test in the exponential case. *The Annals of Mathematical Statistics*, Vol. 26, pp. 82-93.

IEC 61650 (1997) *Reliability Data Analysis Techniques – Procedures for Comparison of Two Constant Failure Rates and Two Constant Failure (Event) Intensities.*

Kapur, K. C. & Lamberson, L. R. (1977). *Reliability in Engineering Design.* Wiley, NY, pp. 342-363.

Kececioglu, D. (1993). *Reliability & Life Testing: Handbook.* Vol. 1, Prentice Hall, NJ, pp. 133-156.

Mace, A. E. (1974). *Sample Size Determination.* Robert E. Krieger Pub. Co., NY, pp. 110-114.

Michlin, Y. H. & Grabarnik, G. (2007). Sequential testing for comparison of the mean time between failures for two systems. *IEEE Transactions on Reliability*, Vol. 56, No. 2, pp. 321-331.

Michlin, Y. H.; Grabarnik, G., & Leshchenko, L. (2009). Comparison of the mean time between failures for two systems under short tests. *IEEE Transactions on Reliability*, Vol. 58, No. 4, pp. 589-596.

Michlin, Y. H. & Grabarnik, G. (2010). Search boundaries of truncated discrete sequential test. *Journal of Applied Statistics.* Vol. 37, No. 05, pp. 707-724.

Michlin, Y. H.; Ingman, D. & Dayan, Y. (2011). Sequential test for arbitrary ratio of mean times between failures. *Int. J. of Operations Research and Information Systems*, Vol. 2, No. 1, pp. 66-81.

Michlin, Y. H. & Kaplunov, V. (2007). Optimal truncation of comparison reliability tests under unequal types I and II error probabilities, *Proceedings of the 9th Conference of Israel Society for Quality*, Tel-Aviv, Nov. 2007, 6 pp.

Michlin, Y. H. & Migdali, R. (2004). Test duration in choice of helicopter maintenance policy. *Reliability Engineering & System Safety*, Vol. 86, No. 3, pp. 317-321.

MIL-HDBK-781A (1996). *Reliability test methods, plans, and environments for engineering, development, qualification, and production.* US DOD, pp. 32-42.

Siegmund, D. (1985). *Sequential Analysis: Tests and Confidence Intervals*, Springer, NY, pp. 34-63.

Sr-332. (2001). *Reliability prediction procedure for electronic equipment.*, Telcordia Technologies Inc.,. Red Bank, NJ, Section 2.4.

Wald, A. (1947). *Sequential Analysis*, John Wiley & Sons, NY.

Wald, A. & Wolfowitz, J. (1948). Optimum character of the sequential probability ratio test. *The Annals of Mathematical Statistics*, Vol. 19, No. 3, pp. 326-339.

15

Dependence of Determination Quality on Performance Capacity of Researching Technique, Exemplified by the Electron Probe X-Ray Microanalysis

Liudmila Pavlova

Vinogradov Institute of Geochemistry, Siberian Branch of Russian Academy of Sciences,
Irkutsk
Russia

1. Introduction

The quality of results obtained by any analytical method depends on every stage of data acquisition: representativeness of study object; appropriate sample preparation; optimum conditions for analytical signal excitation and registration; availability of reference materials for comparison; procedure to process acquired values referred to the content to be determined. These characteristics vary in different analytical methods. In some cases, the sample preparation represents the major source for analytical errors, in some others, the complexities arise from the incorrect selection of calibrating plot, and thus availability of reference materials for comparison is essential. The technique specifications pose the requirements to every stage of analysis.

The problem of quality has ever been critical in analytical work, and every time it depends on the level of progress in the theory and application of selected technique.

Quality has emerged and remained the dominant theme in management thinking since the mid-twentieth century (Beckford, 2010).

2. Quality of the electron probe X-ray microanalysis at each stage

The electron probe X-ray microanalysis is a fairly young technique. The first papers describing the basics of electron probe microanalysis (EPMA) (Castaing, 1951; Borovskii, 1953) and original designs and constructions of microanalyzers (Castaing & Guinier, 1953; Borovskii & Il'in, 1956) were published in 1951-1956 in France and the USSR. The technique was rapidly progressing. In 1973 Borovskii (Borovskii, 1973), the founder of EPMA in Russia, admitted that one could hardly identify the fields of science and engineering, where the EPMA had not been successfully used. The instruments and theory were developing simultaneously.

At present it is difficult to overestimate the significance and application of the method. It is one of the leading methods in mineralogy, and similarly in metallurgy and biology. The investigations of the micro-level are required at approbation of technological processes in all fields of science and engineering.

The quality of results is of prime importance for researchers in all fields of science. The specifics of the method related to obtaining information suggest it to be the research technique, thus intensifying the problem of determination quality, rather than the analytical method. The studies on developing the theoretical foundation of the method and upgrading the instruments, reported in numerous publications, and partly mentioned in the articles reviewed by authors of articles (Szaloki et al., 2004; Pavlova et al., 2000), allow the way of improving quality of the electron probe X-ray microanalysis to be observed. The topical studies undertaken by the author are included into this chapter.

2.1 Representativeness of materials
The representativeness of the sample substance is initially defined when posing problem, and it depends on the requirements of particular analytical technique. These requirements imply the subsample weight, homogeneity of components, and particle size in the subsample, solubility and miscibility of subsample substance with a binding substance or a solvent, and others.

In case of EPMA the representativeness of determinant, namely the quantity of samples and size of the surface, prepared for examination, depend on the frequency of component occurrence in material, probability of their occurrence on the prepared surface and their phase distribution over the investigated surface. In electron probe microanalysis, when we study the inclusions rarely occurring in the groundmass, it is essential to have a sufficient amount of the geological substance to be examined.

In EPMA a correct solution of the problem posed is dependent on the frequency of determinant occurrence within the observation zone. With EPMA, an absence on the studied surface of the element to be defined is not suggestive of its complete lack in the sample. It might be assumed, that the sought element has skipped from the study zone.

It can be exemplified by searching for the invisible gold in lithochemical stream sediments of the Dukat gold-silver deposit in northeastern Russia. Initially, the studies of rock in thin and polished sections did not provide wanted results - fine gold inclusions have not been detected. The probability of gold inclusion occurrence on the studied surface was negligibly small. Only having extracted the heavy fraction and prepared the briquette thin sections and after locating grains on the surface and polishing thin sections and thoroughly studied numerous grains in the thin section we managed to obtain positive results.

If the sought inclusion had not been found on the surface, a thin layer of substance was removed, and researched through entire sample. The process was being repeated until the sought inclusion was found (Fig. 1). If the grains of heavy fraction were completely polished down a new briquette thin section of the same sample was prepared and searching was continued. Figure 1 presents the gold inclusion found in the mineral only after the third try of surface polishing. The size of the inclusions is 10 μm, while the area of thin section is about 25mm². In such cases the quality of investigations and conclusion correctness result from both correctly selected and prepared material and thorough search.

2.2 Sample preparation
In any technique sample preparation is truly important, that is decisive for analytical procedures. In different analytical methods the laboriousness of sample preparation varies. Because the EPMA, in effect, is the analysis of a surface, the sample surface is required to be flat, well polished, smooth and clean and to have a good conductivity. Sample preparation

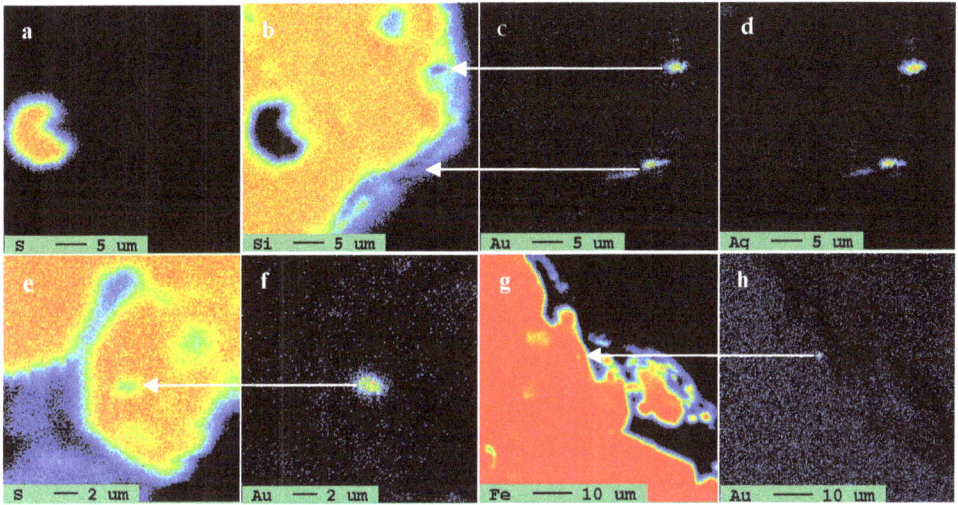

Fig. 1. The gold inclusion found in the mineral only after the third try of surface polishing.

for the electron probe microanalysis is in many ways still as much of an art as a science (Goldstein et al., 1992). Different procedures are applied in EPMA for sample preparation. The solid samples for EPMA may be thin sections, thick micro sections, briquette sections and isolated particles. The preparation of polished thin sections and thick micro sections consists in selecting sized solid material, its cutting, polishing selected surface with diamond pastes.

2.2.1 Particle preparation

When examining the grinded substance the particles should be fixed so that they are not scattered in air and retain the representativity of material when investigated in the vacuum microanalyzer. Various preparation procedures are used to analyse different solid particles (Pavlova et al., 2001): (1) fixing particles on substrate with colloid; (2) pasting particles on the carbon double-faced adhesive tape; (3) preparing briquette sections. In the first case particles are fixed on the polished surface of the substrate with collodion. The particles are distributed as a thin layer on the surface of the substrate, which is preliminarily covered by a thin layer of the liquid collodion. While drying the collodion fixes the particles on the surface. Gluing particles on the carbonic adhesive tape is commonly applied when studying conductive materials. In this case the glued particles cannot be covered by the conductive layer.

Two techniques to prepare briquette thin sections: 1) Particles of any solid material are mixed with epoxy resin, and after the surface is hardened it is polished. 2) Grains of any solid material are glued on the adhesive tape and coated with epoxy resin. After the resin is hardened, the sample is removed from the adhesive tape and the surface with particles is polished to make it mirror-smooth. Due to this procedure plenty of grains are included in the same puck. The briquette thin sections with particles are often used for quantitative determinations of particle composition using the wave spectrometers applied for studying horizontal well polished surfaces. Table 1 presents the data on the comparison of two garnet particles composition processed by different methods of sample preparation.

Preparation		Non-polished particles				Polished particles			
Analytical method		EDS		WDS		EDS		WDS	
Sample	Oxide	Certified concentration, wt.%	Concentration, wt.%	Relative error, %	Concentration, wt.%	Relative error, %	Concentration, wt.%	Relative error, %	Concentration, wt.%	Relative error, %
Garnet O-145	MgO	20	20,59	2,95	21,325	6,63	20,91	4,55	19,93	-0,35
	Al_2O_3	23,4	23,62	0,94	21,286	-9,03	23,19	-0,90	23,36	-0,17
	SiO_2	42,3	42,55	0,59	43,527	2,9	42,07	-0,54	42,56	0,61
	CaO	4,03	4,14	2,73	3,06	-24,02	4,34	7,69	4,03	0,00
	MnO	0,17	0,07	-58,82	0,12	-29,41	0,16	-5,88	0,16	-5,88
	FeO	10,1	9,3	-7,92	9,63	-4,65	9,36	-7,33	10,04	-0,59
Garnet C-153	MgO	21,09	22,61	7,21	23,46	11,24	21,25	0,76	20,93	-0,76
	Al_2O_3	18,09	17,73	-1,99	19,97	10,39	18,86	4,26	17,84	-1,38
	SiO_2	41,52	42,24	1,73	42,98	3,52	42,27	1,81	41,94	1,01
	CaO	3,4	3,7	8,82	3,94	15,88	3,62	6,47	3,38	-0,59
	Cr_2O_3	7,41	6,59	-11,07	6,89	-7,02	7,09	-4,32	7,3	-1,48
	MnO	0,32	0,21	-34,38	0,27	-15,63	0,534	66,88	0,34	6,25
	FeO	7,59	6,85	-9,75	6,57	-13,44	6,163	-18,8	7,41	-2,37

Table 1. Compared compositions of two garnet particles processed by different methods of sample preparation. Relative error=100*(C-Ccer)/Ccer; C- is concentration; Ccer – is certified concentration.

The composition of non-polished particles is not determined with wave spectrometers splitting the X-ray spectrum by the wave length. The quality of determinations is low and the error can reach as high as tens percent. When the particles are glued on the substratum and adhesive tape it is feasible to study the shape and size of particles; when using the energy-dispersive spectrometer it is possible to identify (i) what elements compose grains, (ii) how elements are distributed over the surface, (iii) element contents.

2.2.2 Preparation of biological samples

The surface suitable for the analysis is hard to receive in examining porous samples, which are often biological samples. Preparation of biological materials is differently approached. For example, sponges are first rinsed in distilled water, then dehydrated in alcohol, freed from alcohol with acetone and impregnated with epoxy resin (Pavlova et al., 2004). The samples obtained are fit in one or some briquette sections. The briquette sections with specimens are polished with diamond paste to get the surface flat and mirror-smooth. The Figure 1 displays the sponge image. The solid part of the sponge, its skeleton consists of spicules. This kind of preparation of fragile biological specimens ensures intact solid part of sponge. It avoids distortions due to destructions when polishing. With this procedure we determine silicon concentration in the center of spicule cross-section and on its margins: they are higher in the center than in the margins.

2.2.3 The influence of surface on the quality of results

The distortion of horizontal position of surface (effect of absence of flat horizontal surface) is the cause of false conclusions and considerable deterioration of analytical results. If the surface is either not flat or not horizontal the analytical signal is distorted.

Figure 3 presents the pattern of x-ray radiation distribution of manganese in Mn-pure (a) and gold in Au-pure (b). Samples of pure Mn and Au have flat polished, even if not horizontal surfaces, displaying the x-ray intensity distortion. Manganese shows decrease in intensity to the right and to the left of center (Fig. 3a), because the sample surface is tilted relative to the center: the right part is higher, and the left one is lower. In Fig. 3b the gold particle bottom is in focus, but the top part is elevated toward horizon, therefore AuL_α – intensity is deformed in the top markedly lower, though the sample is homogeneous, but some inclusions. The gold image in backscattered electrons does not exhibit the surface inclination (Fig. 3c). Both gold images in Fig. 3b and Fig. 3c were simultaneously produced.

Because the grains of majority of natural samples are dielectrics, a layer of carbon (20-30 nm thick) is vacuum-sprayed onto the polished surface of all tablets to make it conductive and to remove the accumulative charge.

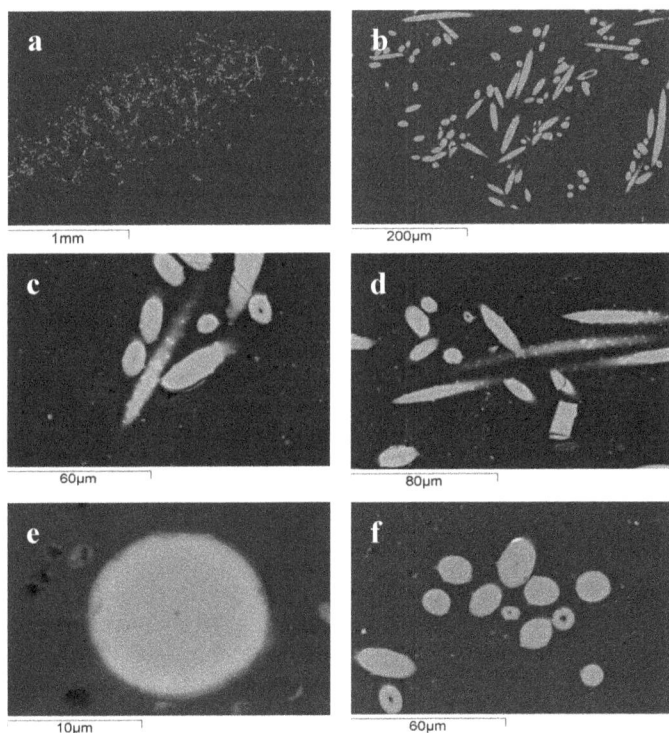

Fig. 2. The sponge surface prepared for EPMA studies. The image is given in back- scattered electrons. Section of whole sponge in epoxy resin (a). Section of sponge part (b). Sections several separate sponge spicules (c, d). Cross-sections of sponge spicules (e, f).

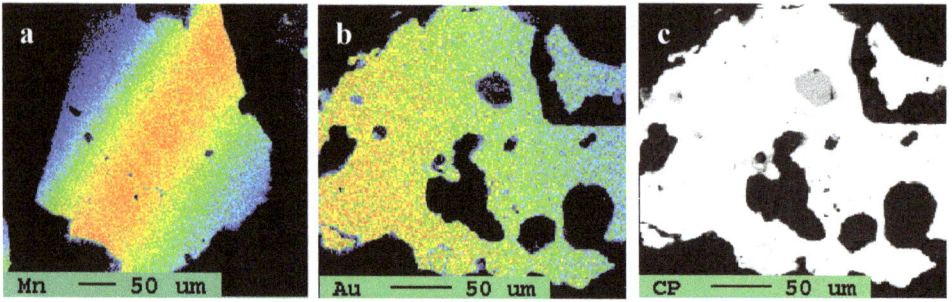

Fig. 3. Patterns of x-ray radiation distribution of manganese (a) and gold (b) in Mn- and Au-pure samples. The samples have flat polished but not horizontal surfaces.

The mass absorption coefficients are basic quantities used in calculations of the penetration and the energy deposition by photons of x-ray in biological, shielding and other materials. The different authors offer special absorption coefficients of x-ray radiation in the same element, specifically in carbon. Varying mass absorption coefficients defined in about twenty articles are discussed in the work (Hubbell et al., 2011). In Table 2 the x-ray absorption cross sections, determined by different authors for energy 0,277 keV in carbon have been compared with the experimental data contained in the National Bureau of Standards collection of measured x-ray attenuation data (Saloman et al., 1988)

The x-ray attenuation coefficients have been approximated by Finkelshtein and Farkov (Finkelshtein & Farkov, 2002). The article by Farkov with co-autors (Farkov et al., 2005) reports the data on compared absorption coefficients of nitrogen and oxygen radiation in carbon from different literature sources (Table 2). Table 3 presents the absorption coefficients of carbon, nitrogen and oxygen radiation in carbon, as well as relative deviations in the data of different authors.

Authors (Reference from the article of Saloman et al., 1988)	Fomichev V.A. & Zhukova I.I. (68FO1)	Henke et al. (67HE1)	Denne D.R. (70DE2)	Denne D.R. (70DE1)	Weiswe-iler W. (68WE1)	Messner R.H. (33ME1)	Duncumb P. & Melford D.A. (65DU1)	Kurtz H. (28KU1)	Dershem, E. & Schein, M. (31DE1)
XACS, barns/atom $*10^4$	5,600	4,569	3,850	4,09	5,056	3,43	4,587	4,200	4,400
Relative error, %	3,7	-17	-40	-31,8	-6,6	-57,2	-17,5	-28,4	-22,5

Table 2. X-ray absorption cross sections, determinated by different authors (Saloman et al., 1988). XACS - is X-ray absorption cross sections of authors; Relative error= -EXPS - XACS)/ EXPS*100; EXPS - is experimental data contained in the National Bureau of Standards collection of measured x-ray attenuation data.

Emitter	CK_α			NK_α			OK_α		
Authors	(Heinrich, 1986)	Henke*, 1967	(Henke, 1993)	(Heinrich, 1986)	Henke*, 1967	(Henke, 1993)	(Heinrich, 1986)	Henke*, 1967	(Henke, 1993)
Carbon absorber	2147	2292	1960	23585	21784	25000	11575	9079	12100
Athor-Heinrich, 1986	-	6,3	-9,5	-	-8,3	5,7	-	-27,5	4,3
Athor – Henke*, 1967	-6,7	-	-16,9	7,6	-	12,9	21,6	-	25,0
Athor – Henke, 1993	8,7	14,5	-	-6,0	-14,8	-	-4,5	-33,3	-

Table 3. X-ray absorption coefficients (centimeter^2/g) in the carbon and relative deviations
in the data of different authors. *Henke's data from work of Saloman (Saloman et al., 1988).
Relative deviation=(Author-Comp)/Author*100%. Author - is data of author; Comp – is
comparison data.

It might be assumed how the quality light elements (carbon, nitrogen and oxigen)
determination depends on choice of absorption coefficients.
The thickness of carbonic film on the studied surface and on the sample for comparison
should be the same not to deteriorate the quality of results.

2.3 Optimum conditions to excite and register analytical signals
Ensuring a high quality of results depends on uniformity of analytical signal through the
time of observation. The conditions of exciting and registering analytical signal are to
provide such uniformity. Selection of optimum conditions to excite and register analytical
signals is the major condition to obtain correct information on a sample. It is one of the main
conditions for any method of analytical chemistry.
With EPMA applied the surface of different samples: glassy materials, sponges, bones,
argentiferous samples and others can be destroyed during excitation with x-radiation, when
electron probe falls on the sample surface, and as this takes place, the analytical signals
become heavily contorted. Selection of optimum conditions of measurement can be based on
the criteria of (a) minimum detection limit of sought elements, (b) uniformity of analytical
signal during measurement, (c) variations of element intensity depending on the electron
beam density or (d) sample stability while performing measurements (Buseck @ Goldstein,
1969).

2.3.1 The criterion of minimum detection limit
The detection limit variations depending on measurement conditions were identified for
scandium, strontium and barium in basalt glasses. Figure 4 illustrates the plotted variations
of detection limits for scandium, strontium and barium in basalt glasses depending on
measurement conditions (Paradina @ Pavlova, 1999).
Radiation of SrL_α -, BaL_β - and ScK_α - lines is measured with the PET crystal for different
accelerating voltages, counting times and probe currents. The curves show that the optimum
conditions for electron probe microanalysis with wavelength spectrometers (WD EPMA) of

Ba, *Sc* and *Sr* are accelerating voltage 25 kV, sample current 80-100 nA and counting time 20 s. With the other measurement conditions the detection limits of above-indicated elements are higher, thus the determination quality of these elements in glasses is worse.

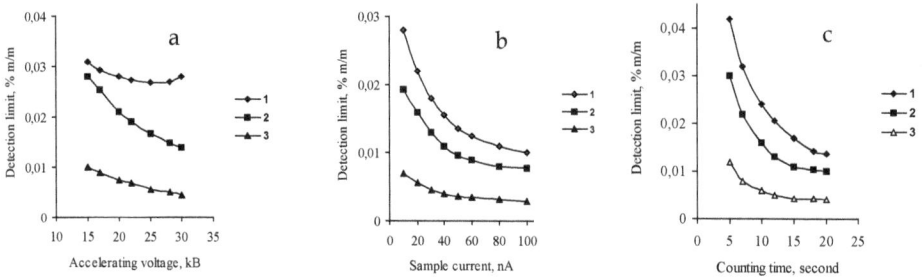

Fig. 4. Detection limits for SrL_α - (1), BaL_β - (2), and ScK_α - (3) as the functions of: (a) accelerating voltage (sample current 15 nA, counting time 10 s); (b) sample current (accelerating voltage 25 kV, counting time 10 s); (c) counting time (probe current 15 nA, accelerating voltage 25 kV). Measurements were made by Camebax-micro and Camebax SX-50 microprobes (Paradina @ Pavlova, 1999).

2.3.2 Variations of x-ray intensity versus counting time

Intensity variations of the elements present in the bones versus counting time are examined by microprobe Superprobe-733 (Pavlova et al., 2001). Intensity variations of elements present in bone tissue are shown in the fig. 5.

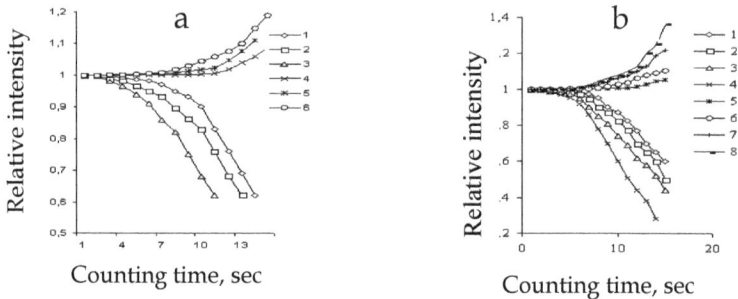

Fig. 5. Intensity variations of elements present in bone tissue versus counting time: (a) NaK_α (curves 1–3 correspond to accelerating voltage 10, 15 and 20 kV, respectively, and probe current 15 nA) and MgK_α (curves 4–6 correspond to accelerating voltage 10, 15 and 20 kV, respectively, and probe current 15 nA); (b) NaK_α (curves 1–4 correspond to sample current 10, 15, 20 and 25 nA, respectively, and accelerating voltage 15 kV) and MgK_α (curves 5–8 correspond to sample current 10, 15, 20 and 25 nA, respectively, and accelerating voltage 15 kV). Measurements were made by Superprobe-733 microprobe (Pavlova et al., 2001).

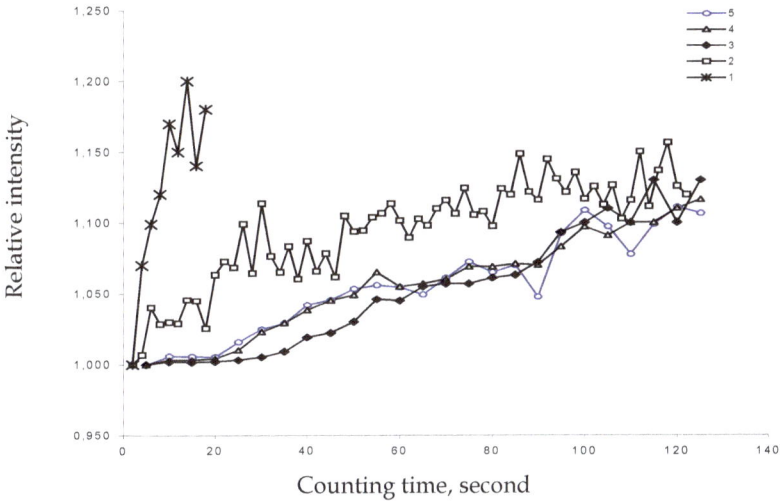

Fig. 6. The change of CaK_α - line intensity in accordance with the electron beam power
densities ($\mu W / \mu m^2$): 1 – 191,00; 2 – 3,82; 3 – 2,54; 4 – 2,86; 5 – 3,06.

The X-ray intensity of calcium versus counting time (Fig. 6) with varying electron beam
power density assessed for omul fish otoliths (Pavlova et al., 2003). This change of CaK_α -
line intensity versus the electron beam power densities has been chosen as the criterion for
selecting optimum conditions for analyzing the elements in omul fish otoliths. As seen on
the plot the CaK_α intensity of otoliths remains constant during twenty seconds, if the beam
power density is below 2.8 $\mu W / \mu m^2$. The increase of measuring time causes signal
distortion and deterioration of result quality.

2.3.3 Criterion of sample stability while performing measurements
The stability of argentiferous samples can be studied quantitatively by Buseck's technique
(Buseck @ Goldstein, 1969), and it can also be used as the criterion for selecting optimum
conditions for EPMA and thereby refinement of substance test (Pavlova @ Kravtsova, 2006).
According to the recommendations (Buseck and Goldstein, 1969) the sample is stable
(analytical signal is constant for counting time), if it obeys the relationship

$$S_c < \sigma_c \qquad (1)$$

where the repeatability of standard deviation of the measured x-ray counts for the major
elements Sc is given by:

$$S_c = \sqrt{\sum_{i+1}^{n}(N_i - \overline{N})^2 / (n-1)} \qquad (2)$$

where N_i is the number of x-ray counts in each measurement i and n is the number of
individual measurements, and average number of x-ray counts is

$$\overline{N} = \sum_{i+1}^{n}(N_i / n) \qquad (3)$$

the population standard deviation stipulated by the Poisson counting statistics σ_c is:

$$\sigma_c = \sqrt{\overline{N}} . \qquad (4)$$

The stability characteristics of argentiferous sample were determined according to the recommendations (Buseck and Goldstein, 1969). Table 4 provides stability characteristics of argentiferous samples.

It was found that the response from argentiferous sample was stable through period from 1 to 1.5 min depending on the beam power densities.

The beam power density <2.55 $\mu W / \mu m^2$ is admissible for the analysis of argentiferous samples. The best compromise conditions for exciting the x-ray radiation and recording analytical signal for WD EPMA of argentiferous samples are: accelerating voltage of 15-20 kV, beam current of 10 nA, probe diameter of 10 μm and counting time of 10 s. Hence, so as to obtain the correct results of analysis it is necessary to select optimum conditions for exciting and registering analytical signals.

Counting time, second				10		20		30		40		50		60	
Conditions of measurements															
E_0 kV	i nA	d_b μm	P μW /	S_c	σ_c	S_c	σ_c	S_c	σ_c	S_c	σ_c	S_c	σ_c	S_c	σ_c
2	1	2	0,64	15	33,	16,	33,	15,	34,	16,	33,	16,	34,	15,	33,
2	1	2	0,96	16	35,	15,	35,	16,	35,	16,	35,	16,	35,	16,	34,
2	2	2	1,27	15	37,	16,	37,	16,	37,	15,	37,	15,	37,	42,	38,
2	5	1	1,27	16	37,	16,	37,	16,	37,	16,	37,	16,	37,	42,	38,
2	8	1	2,04	15	39,	16,	39,	16,	38,	18,	39,	50,	39,	49,	39,
2	1	1	2,55	16	40,	16,	40,	16,	39,	49,	41,	52,	41,	62,	42,
2	1	1	3,06	16	41,	48,	41,	51,	42,	53,	42,	59,	43,	63,	43,
2	1	1	3,82	47	45,	69,	44,	69,	43,	59,	44,	63,	44,	65,	44,
2	5	1	127,	49	48,	96,	48,	98,	49,	97,	50,	98,	56,	99,	49,

Table 4. The stability characteristics of argentiferous sample. d_b – beam diameter; E_0 – accelerating voltage; i – probe current; P – beam power densities; S_c is repeatability standard deviation of the measured X-ray counts; σ_c is population standard deviation, stipulated by Poisson counting statistics.

2.4 Reference samples and standard materials for electron probe microanalysis

The accuracy of results of any analytical technique depends on the application of a sufficient number of required reference and control samples of top quality. An important condition to acquire appropriate analytical results of required quality is the application of a sufficient number of required reference samples meeting all requirements of analytical method employed. It is desirable to use certified standards as control and reference samples.

A specific feature of electron probe X-ray microanalysis is its locality, which is 10^{-13} g of the substance, causing toughening the requirements for standard and reference materials claimed for EPMA. In the case of EPMA, reference and connrol samples should correspond with the following requirements simultaneously: (1) have known chemical composition; (2) be uniform at macro and micro levels of spatial resolution; (3) remain stable under action of electron probe; (4) do not decay in vacuum (up to 10^{-16} mm Hg); (5) must be well polished. Evidently, only the samples being standard samples of structure and properties at the same time may become standard and reference samples for EPMA. The standard samples widely utilized in the other analytical techniques are often inapplicable in the case of EPMA, because they are not homogeneous both at macro and micro levels.

2.4.1 Homogeneity and stability of samples

The homogeneity (at macro- and micro-level) of space resolution and substance stability resulting from the effect of electron beam is of particular interest for the EPMA. There are different approaches to assess the substance homogeneity (Borkhodoev, 2010; Buseck @ Goldstein, 1969; MI, 1988). Two methods to test the homogeneity of elements distribution at macro and micro levels (Buseck @ Goldstein, 1969; MI, 1988) are described for the copper-rich alloys and basaltic glass.

According to the recommendations of Buseck and Goldstein (Buseck @ Goldstein, 1969) the sample is homogeneous on micrometer- to millimeter scale if

$$S_c / 2\sigma_c < 1 \tag{5}$$

the homogeneity is doubtful if

$$1 < S_c / 2\sigma_c < 2 , \tag{6}$$

the sample is inhomogeneous if

$$2 < S_c / 2\sigma_c , \tag{7}$$

S_c and σ_c values are determined from (2)-(4) formulae.

The macro homogeneity at the micrometer to millimeter scale was evaluated by recording EPMA profiles across a few different profile lines of sample.

The beam diameter ranged from 1 to 10 μm, the profile length was about 1000 μm, the point spacing was 10 μm and the line spacing was 20 μm. $S_c / 2\sigma_c$ values in Table 5 demonstrate that at micron-mm level in alloys CA-2 and CA-6 all elements are distributed irregularly and the tin homogeneity in CA-4 alloy is doubtful.

According to the National Standard for homogeneity assessment (MI, 1988) the material is uniform on a micrometer scale, if

$$\sigma_i = \sqrt{\sigma_b^2 + \sigma_l^2} < RSD , \tag{8}$$

Sample		Element	C_{cer}, wt.%	Stability (for 2 min) σ_c / S_c	homogeneity $S_c / 2\sigma$	σ_b	σ_l	σ_i	$\sigma / 8$
Coppery alloys	CA-1	Sn	1,62	1,38	0,77	0,03	0,71	0,71	0,71
		Ni	0,10	n.d	0,76	0,08	1,94	1,94	2,10
		Fe	0,05	n.d	0,88	0,09	2,05	2,05	2,50
	CA -2	Cu	68,74	1,48	2,96	n.d	n.d	n.d	n.d
		Sn	0,97	1,75	2,95	n.d	n.d	n.d	n.d
		Zn	30,00	1,13	2,27	n.d	n.d	n.d	n.d
	CA -3	Zn	0,71	n.d	0,80	0,12	1,12	1,12	0,26
		Cu	98,53	1,49	0,98	0,14	1,12	1,13	0,31
		Sn	0,19	n.d	0,79	0,00	1,55	1,55	1,56
		Ni	0,26	n.d	0,79	0,07	1,13	1,13	1,20
		Fe	0,30	n.d	0,76	0,06	1,16	1,16	0,81
	CA -4	Zn	1,56	1,36	0,73	0,04	0,68	0,68	0,85
		Sn	0,09	n.d	1,48	n.d	n.d	n.d	n.d
	CA -5	Zn	1,00	1,46	0,92	0,11	1,12	1,12	1,13
		Sn	0,12	n.d	0,83	0,00	1,55	1,55	1,56
		Ni	0,51	n.d	0,89	0,07	1,13	1,13	1,20
		Fe	3,63	1,45	0,87	0,14	0,79	0,80	0,81
	CA -6	Cu	63,00	1,23	2,46	n.d	n.d	n.d	0,31
		Zn	33,80	1,43	2,87	n.d	n.d	n.d	n.d
Basalt glass		Ba	2,30	1.94	0,47	0,00	1,73	1,73	3,80
		Sr	2,01	1.75	0,375	0,00	1,56	1,56	2,80
		Si	21,84	1.84	0,42	0,02	0,13	0,13	0,13
		Al	9,50	1.72	0,36	0,20	0,34	0,40	0,44
		Ca	7,34	1.68	0,34	0,13	0,38	0,40	0,40
		Mg	3,49	1.88	0,44	0,17	0,49	0,52	0,58
		Fe	8,94	1.66	0,33	0,12	0,32	0,35	0,35
		Na	2,02	1.58	0,29	0,02	0,96	0,96	1,00
		Ti	0,87	n.d	0,39	0,17	0,81	0,83	0,88

Table 5. Characteristics of stability and uniformity of copper alloys and basalt glass. n.d. means that the value was not determined. CA stands for coppery alloys; S_c is repeatability standard deviation of the measured X-ray counts; σ_c is population standard deviation, stipulated by Poisson counting statistics; σ_i is total root mean square deviation; σ_b is the root mean square deviation of random components of error for some parts of the specimen; σ_l is the root mean square deviation of random components of error for the analytical areas; RSD is the relative standard deviation (Thompson et al., 1997).

Dependence of Determination Quality on Performance Capacity of Researching Technique, Exemplified by
the Electron Probe X-Ray Microanalysis

269

where σ_b is the root mean square deviation of random components of error for some parts of the specimen; σ_l is the root mean square deviation of random components of error for the analytical areas; RSD is the relative standard deviation used for certification (Thompson et al., 1997).

The total X-ray counts were recorded at each point of the sample by the static beam to detect possible micro heterogeneity at $1–10\ \mu m$ scale. The X-ray intensities of analytical lines were measured from 20 to 30 points of a sample. The pattern of these points was chosen from the table of random numbers. At each point two measurements were made using the optimum conditions. The homogeneity of the material in the surface layers was estimated by repeating measurements after the repeated polishing. The results for each element were processed by a dispersion analysis with a three-step grouping of material. Evidently the samples are homogeneous on micrometer scale excepting alloys CA-2, CA-6 and CA-4.

At EPMA the stability of reference samples and standards under the microprobe is significant both during measuring the analytical signal and using the sample for studies. The data on stability of alloys and glasses, given in Table 5, are obtained by the following method (Pavlova et al., 2001). The X-ray radiation of the most intensive lines was excited by the electron beam at 5 arbitrary points of the selected sample. This was the x-ray radiation of copper Kα-line for alloys and calcium Kα-line for basalt glass recorded by the WD EPMA technique. The measurements were made 30 times and lasted for 5 minutes per point. Relationship (1) was applicable for some elements of all copper alloys and basalt glass. Table 5 indicates that all copper alloys and basalt glass are resistant to electron action during 2 minutes.

2.4.2 Valuation of laboratory reference samples

The necessity to have certified standard samples is obvious; they help to achieve the requisite quality of results. But in some cases the certified standard samples are absent and researchers are forced to apply laboratory reference materials after having the control study fulfilled and these materials assessed (Wilson & Taggart, 2000).

One of the techniques to assess quality of the EPMA results obtained using laboratory reference samples instead of standards is described below for the silicate mineral analysis (Pavlova et al., 2003). Measurements were performed on the Jeol Superprobe-733 electron microprobe using accelerating voltage of 15 kV and probe current of 20 nA with an electron beam diameter of 1 and 10 μm. Assessment was done by analyzing 26 control samples (glass K-412, garnet IGEM, Ti - glass, Mn-glass, Cr –glass, basalt glasses BHVO-2G, BCR-2G, BIR-1G, diopside, ilmenite, garnet O-145, olivine, spinel, garnet C-153, albite, garnet UD-92, orthoclase, chromite UV-126, oxides MgO, MnO, Fe_2O_3, Cr_2O_3, Al_2O_3, TiO_2, SiO_2 and $CaSiO_3$). The glass K-412 was supplied by the National Institute of Standards and Technology of the USA (NIST, 1990). Garnet IGEM, Ti - glass, Mn-glass and Cr -glass were prepared as laboratory reference samples at the Institute of Geology and Mineralogy, Moscow, Russia. Three basalt glasses BHVO-2G, BCR-2G andBIR-1G were produced by the US Geological Survey (Denver Federal Center) (Wilson & Taggart, 2000). The diopside, ilmenite, olivine, spinel, garnets C-153 and UD-92, albite and orthoclase were certified as the laboratory reference samples at the Institute of Geology and Geophysics, Novosibirsk,

Russia. The oxides MgO, MnO, Fe_2O_3, Cr_2O_3, Al_2O_3, TiO_2, SiO_2 and $CaSiO_3$ were supplied by JEOL Ltd., Tokyo, Japan. Each reference sample producer has reported that all samples meet the requirements needed for the samples for comparison at EPMA.

These control samples comprised the following. (I) Seven glasses, in which the contents of the elements analyzed varied over the following ranges (wt.%): SiO_2, 45.35-54.11; MgO, 3.48-19.33; Al_2O_3, 1.40-16.68; FeO, 8.46-12.04; CaO, 6.87-23.38; TiO_2, 0.81-9.11; MnO, 0.17-8.48; and Cr_2O_3, 10.20. (II) Eight oxides: MgO, SiO_2, Al_2O_3, TiO_2, Fe_2O_3, MnO, Cr_2O_3 and $CaSiO_3$. (III) Twelve minerals with concentrations varying as follows (wt.%): MgO, 1.02-49.20; Al_2O_3, 0.50-26.10; SiO_2, 0.30-55.50; CaO, 2.24-53.80; TiO_2, 0.33-50.00; Cr_2O_3, 0.08-44.80; MnO, 0.17-30.76; and FeO, 0.05-62.40.

All samples were at first used as control samples before being accepted as reference samples. The available reference samples were divided into three groups (Table 6). Group I comprised the basic components (Si, Al, Fe, Mg and Ca) which were defined using glass K-412, while Ti, Mn and Cr were defined from Ti-glass, Mn-glass and Cr-glass. The second complete set (II) consisted of simple minerals: diopside (for Si and Ca); ilmenite GF-55 (for Ti); olivine (for Mg); spinel $MnFe_2O_4$ (for Fe); garnets C-153 (for Al), UD-92 (for Cr) and IGEM (for Mn). Simple oxides MgO, MnO, Fe_2O_3, Cr_2O_3, Al_2O_3, TiO_2, SiO_2 and $CaSiO_3$ (Ca) represent the third complete set of reference samples (III). In all three cases the sodium content was determined from albite and that of potassium from the orthoclase standard sample. The analyzed values were corrected for matrix effects using the PAP method (Pouchou @ Pichoir, 1984) through the MARCHELL program (Kanakin @ Karmanov, 2006) adapted for the Superprobe-733 operating system. When applying three complete sets of calibration samples the three series of concentration data (C_I, C_{II} and C_{III}) have been received. Table 6 shows the analytical results for USGS TB-1 glass measured by EPMA. The same analytical results have been acquired for every control sample.

The results obtained in this study (Table 6) showed that deviations from the recommended/certified value varied in value and sign, however, they did not depend on the group of the reference samples selected for calibration. The largest deviations were observed in the elements with concentrations close to the detection limit. The relative standard deviations do not exceed the target precision (σ_r) in all cases. The values of z-scores for all elements determined lie within permissible limits ($-2<z<2$) for the elements ranging in concentration from 0.1 to 100%. The relative standard deviation for each element assessed in all control samples depends on the concentration and varies as (%): Na_2O, 0.30-2.89; MgO, 0.42-1.76; Al_2O_3, 0.29-2.4; SiO_2, 0.11-2.32; K_2O, 0.43-2.00; CaO, 0.37-1.91; TiO_2, 0.84-2.16; Cr_2O_3, 0.71-2.25; MnO, 0.72-2.59; and FeO, 0.45-2.80. The relative standard deviations for each element in all control samples were not higher than the admissible relative standard deviations (σ_r), defined for 'applied geochemistry' category of analysis (category 2) in the GeoPT proficiency testing program (Thompson et al., 1997). Fig. 7 shows the correlation of concentrations for MgO and Cr_2O_3 analyzed using the different sets of reference samples vs. their recommended or certified values.

Each trend of data plotted in Figure 7 is well described as a straight line. In all cases the correlation coefficients (R^2), describing the reliability of the linear dependence, is close or equal to 1 (Table 7). This confirms the absence of systematic differences and confirms the reliability of each set of reference samples in the calibration of the EPMA instrument.

Dependence of Determination Quality on Performance Capacity of Researching Technique, Exemplified by
the Electron Probe X-Ray Microanalysis

271

Group of reference samples	Metrological performance	Na_2O	MgO	Al_2O_3	SiO_2	P_2O_5	K_2O	CaO	TiO_2	MnO	FeO
	C_{cer}, wt. %	3,3	3,5	13,6	54,1	0,36	1,7	7	2,2	0,18	12
	n	17									
I	C_{av}, wt.%	3,3	3,5	13,6	54	0,35	1,7	7	2,2	0,18	12
	s	0,1	0,1	0,17	0,6	0,01	0	0,1	0	0,01	0,2
	s_r, %	1,4	1,4	1,25	1,1	2,31	1,8	1,5	1,7	2,53	1,3
	σ_r, %	1,7	1,7	1,35	1,1	2,33	1,9	1,5	1,8	2,59	1,4
	ΔC, wt. %	0	0	0,08	0,3	0	0	0,1	0	0	0,1
	z	-0	0,4	-0,5	-0,2	-1,2	0,3	0,4	-1	-0,4	-0,1
	n	8									
II	C_{av}, wt.%	3,3	3,5	13,7	54	0,36	1,7	6,9	2,3	0,18	12
	s	0,1	0,1	0,16	0,6	0,01	0	0,1	0	0,01	0,2
	s_r, %	1,7	1,6	1,17	1	2,31	1,8	1,5	1,7	2,54	1,3
	σ_r, %	1,7	1,7	1,35	1,1	2,33	1,9	1,5	1,8	2,59	1,4
	ΔC, wt. %	0	0	0,11	0,4	0,01	0	0,1	0	0	0,1
	z	0,7	-0	0,38	0,2	-0,1	-1	-1	0,5	-0,7	-0,7
	n	7									
III	C_{av}, wt.%	3,3	3,6	13,7	54	0,37	1,7	7,1	2,2	0,18	12
	s	0,1	0	0,09	0,4	0,01	0	0,1	0	0,01	0,1
	s_r, %	1,5	1,1	0,64	0,7	2,27	1,8	1,4	1,7	2,5	1,2
	σ_r, %	1,7	1,7	1,35	1,1	2,33	1,9	1,5	1,8	2,59	1,4
	ΔC, wt. %	0,1	0,1	0,1	0,4	0,01	0	0,1	0	0,01	0,2
	z	1	1,8	0,46	0,4	1,19	-1	1,7	-1	0	1,2

Table 6. Analytical results for USGS TB-1 glass measured by EPMA with the calibrations obtained from three separate groups of reference materials for each of samples: C_{cer} is the certified concentration; n is number of measurements; $\Delta C = t * s / \sqrt{n}$ is confidence intervals; $s_r = s / C_{av}$ is standard deviations, where $s = \sqrt{\{[n\sum_{i=1}^{n}(C_i)^2 - (\sum_{i=1}^{n}C_i^2)]/[n(n-1)]\}}$; $C_{av} = \frac{1}{n}\sum_{i=1}^{n}C_i$ is average concentrations; $z = (C_{av} - C_{cer})/\sigma_r$ is z-scores (Thompson et al., 1997); $\sigma_r = 100 * 0.02 * C^{0.8495}/C$ is acceptable standard deviation (Thompson et al., 1997), where C is the concentration expressed as fraction.

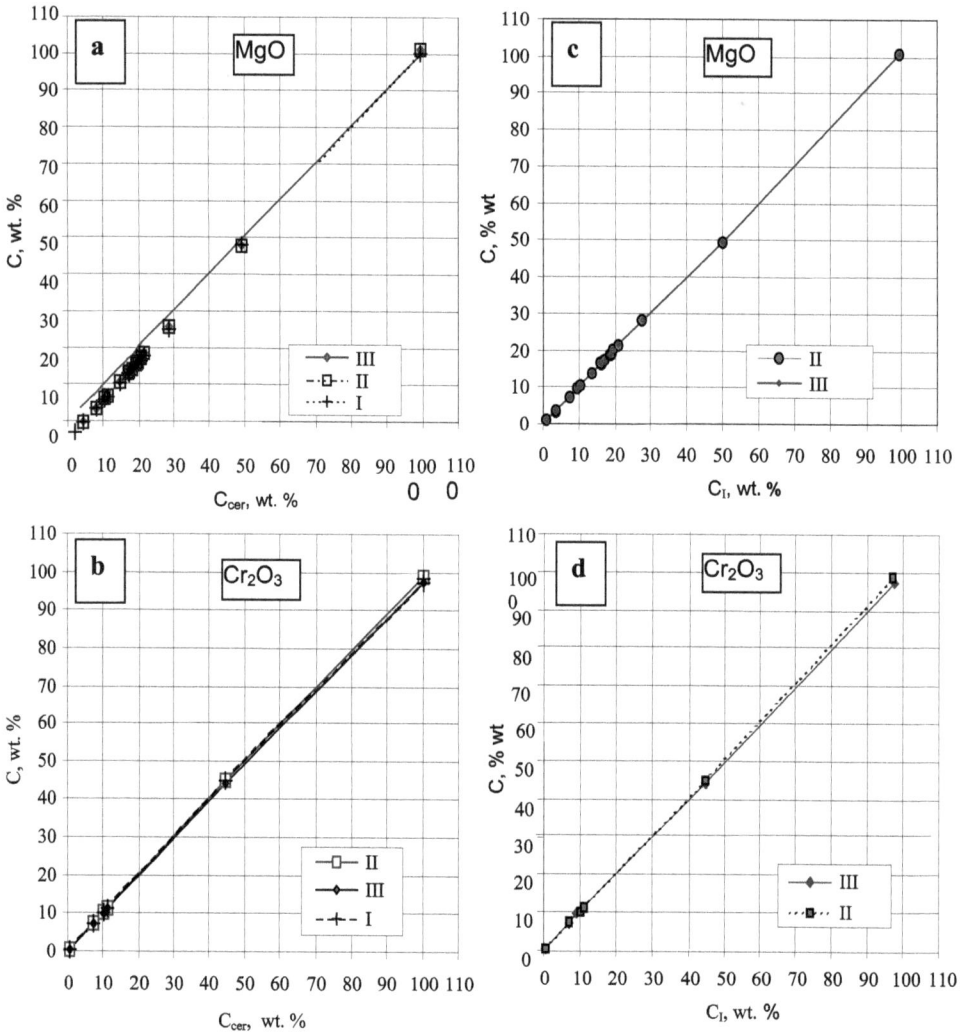

Fig. 7. The graphic correlation of concentrations for MgO and Cr_2O_3. a, b – the graphic correlation of concentrations (C) received according to three different reference sample sets (I, II, III), to their certified values (C_{cer}). Graphic representation of the ratio between the concentrations determined using laboratory reference samples II (C_{II}) and III (C_{III}) and C_I determined using standard sample glasses I. Graphs correspond to the concentrations, determined using: fist set of reference samples (I) – glass K-412, Ti-glass, Mn-glass, Cr-glass, albite and orthoclase; the second set (II) – diopside, ilmenite GF-55, olivine, spinel, garnets C-153, UD-92, IGEM albite and orthoclase; for the third set (III) – MgO, MnO, Fe_2O_3, Cr_2O_3, Al_2O_3, TiO_2, SiO_2, $CaSiO_3$, albite and orthoclase.

Dependence of Determination Quality on Performance Capacity of Researching Technique, Exemplified by
the Electron Probe X-Ray Microanalysis

273

Oxides	C_{III} (C_{cer})	C_{II} (C_{cer})	C_I (C_{cer})	C_{III} (C_I)	C_{II} (C_I)
Al_2O_3	$C_{III} = 1.003\,C_{cer}$ - 0.0985; R^2 =0.9999	$C_{II} = 1.0020\,C_{cer}$ - 0.0782; R^2 = 0.9999	$C_I = 1.0001\,C_{cer}$ + 0.0674; R^2 = 0.9999	$C_{III} = 1,0396\,C_I$ - 0,0105; R^2 = 0,9998	$C_{II} = 1,0414\,C_I$ - 0,0104; R^2 = 0,9997
MgO	$C_{III} = 1.0081\,C_{cer}$ - 0.1342; R^2 =0.9999	$C_{II} = 1.0068\,C_{cer}$ - 0.0978; R^2 = 0.9999	$C_I = 0.9977\,C_{cer}$ - 0.1246; R^2 = 0.9998	$C_{III} = 1,0101\,C_I$ - 0,0048; R^2 = 0,9998	$C_{II} = 1,0089\,C_I$ + 0,0280; R^2 = 0,9998
SiO_2	$C_{III} = 0.9943\,C_{cer}$ + 0.234; R^2 =0.9997	$C_{II} = 0.9966\,C_{cer}$ + 0.1482; R^2 = 0.9997	$C_I = 0.9964\,C_{cer}$ + 0.0369; R^2 = 0.9991	$C_{III} = 1,0044\,C_I$ - 0,4210; R^2 = 0,9991	$C_{II} = 1,0034\,C_I$ - 0,3896; R^2 =0,9991;
Na_2O	$C_{III} = 1.005\,C_{cer}$ - 0.0361; R^2 =0.9987	$C_{II} = 1.0001\,C_{cer}$ + 0.0193; R^2 = 0.9997	$C_I = 1.0042\,C_{cer}$ - 0.0415; R^2 = 0.9996	$C_{III} = 1,0016\,C_I$ + 0,0838; R^2 = 0,9999	$C_{II} = 0,9991\,C_I$ + 0,2071; R^2 =0,9989;
K_2O	$C_{III} = 0.9974\,C_{cer}$ - 0.0532; R^2 =0.9998	$C_{II} = 0.9992\,C_{cer}$ + 0.0359; м = 0.9990	$C_I = 0.9821\,C_{cer}$ + 0.0581; R^2 = 0.9999	$C_{III} = 1,0148\,C_I$ - 0,0107; R^2 = 0,9998	$C_{II} = 1,0157\,C_I$ - 0,0126; R^2 = 0,9990
CaO	$C_{III} = 0.9925\,C_{cer}$ + 0.0026; R^2 = 0.9999	$C_{II} = 0.9946\,C_{cer}$ + 0.0351; R^2 = 0.9999	$C_I = 1.0057\,C_{cer}$ - 0.0203; R^2 = 0.9999	$C_{III} = 0,9855\,C_I$ + 0,0584; R^2 = 0,9998	$C_{II} = 0,9889\,C_I$ + 0,0720; R^2 = 0,9999
TiO_2	$C_{III} = 1.0140\,C_{cer}$ - 0.0105; R^2 = 1.0000	$C_{II} = 1.0058\,C_{cer}$ + 0.0202; R^2 = 0.9999	$C_I = 1.0051\,C_{cer}$ + 0.0339; R^2 = 1.0000	$C_{III} = 1,0093\,C_I$ - 0,0389; R^2 = 1,0000	$C_{II} = 1,0005\,C_I$ + 0,0036; R^2 = 1,0000
Cr_2O_3	$C_{III} = 0.9747\,C_{cer}$ + 0.0672; R^2 = 0.9999	$C_{II} = 0.9872\,C_{cer}$ + 0.1281; R^2 = 0.9999	$C_I = 0.9780\,C_{cer}$ + 0.2218; R^2 = 0.9998	$C_{III} = 0,9950\,C_I$ + 0,1126; R^2 = 0,9999	$C_{II} = 1,0144\,C_I$ - 0,1526; R^2 = 0,9999
MnO	$C_{III} = 1.0130\,C_{cer}$ - 0.0512; R^2 = 0.9999	$C_{II} = 1.0075\,C_{cer}$ + 0.0027; R^2 = 1.0000	$C_I = 1.0140\,C_{cer}$ - 0.0066; R^2 = 1.0000	$C_{III} = 1,0013\,C_I$ - 0,0551; R^2 = 0,9990	$C_{II} = 0,9935\,C_I$ + 0,0099; R^2 = 0,9999
FeO	$C_{III} = 1.0031\,C_{cer}$ - 0.0332; R^2 = 0.9999	$C_{II} = 1.0013\,C_{cer}$ - 0.0112; R^2 = 0.9999	$C_I = 0.9880\,C_{cer}$ + 0.1251; R^2 = 0.9999	$C_{III} = 1,0107\,C_I$ - 0,1443; R^2 = 0,9999	$C_{II} = 1,0090\,C_I$ - 0,1295; R^2 = 0,9999

Table 7. Evaluation of the linear function and correlation coefficients (R^2) between determinations made against the three sets of calibration samples (C_I, C_{II}, C_{III}) and the certified concentrations of control samples (C_{cer}) as well as between C_{II}, C_{III} and C_I. Concentrations C_I, C_{II}, C_{III} are determined using the respective sets of reference samples: (I) – glass K-412, Ti -glass, Mn-glass, Cr-glass, albite and orthoclase; (II) – diopside, ilmenite GF-55, olivine, spinel, garnets C-153, UD-92, IGEM, albite and orthoclase; (III) – MgO, MnO, Fe_2O_3, Cr_2O_3, Al_2O_3, TiO_2, SiO_2, $CaSiO_3$, albite and orthoclase.

Figure 7 includes the data for elements in which the maximum differences were observed between the results. It is evident that in all cases there are no essential systematic differences between the data sets. This confirms the absence of systematic differences and confirms the reliability of each set of reference samples in the calibration of the EPMA instrument.

Table 8 gives the results of calculations designed to check the hypothesis that there are no differences between sets of results obtained from any of these three sets of calibration samples.

The series of concentrations determined using calibrations established using three sets of reference samples have been compared with certified values using a two-tailed selective Student's t-test. Numerical values of probabilities for each pair of the series are significant, showing that the populations of results from all three series are statistically indistinguishable certified values.

The closeness in value of the significance data listed in Table 8 demonstrates the absence of systematic differences between the certified concentrations and the analyzed data for every series. Thus, the sets of reference samples tested in such a way can be successfully applied in EPMA for obtaining high-quality results.

Oxides	$C_{cer} - C_I$	$C_{cer} - C_{II}$	$C_{cer} - C_{III}$	$C_I - C_{II}$	$C_I - C_{III}$
MgO	0.312	0.355	0.299	0.359	0.268
AL$_2$O$_3$	0.244	0.237	0.215	0.251	0.267
SiO$_2$	0.310	0.363	0.328	0.265	0.368
CaO	0.455	0.444	-	0.292	-
MnO	-	0.503	0.490	-	-
FeO	0.359	0.390	0.346	0.374	0.328

Table 8. Results from checking the hypothesis that there is no significant difference between determinations made by any of the three sets of calibration and certified/recommended values using a 2-pair selective Students t-test. C_{cer} – certified concentrations of control samples; C_I, C_{II}, C_{III} – concentrations, determined using the following sets of reference samples: (I) – glass K-412, Ti-glass, Mn-glass, Cr-glass, albite and orthoclase; (II) – diopside, ilmenite GF-55, olivine, spinel, garnets C-153, UD-92, IGEM, albite and orthclase; (III) – MgO, MnO, Fe$_2$O$_3$, Cr$_2$O$_3$, Al$_2$O$_3$, TiO$_2$, SiO$_2$, CaSiO$_3$, albite and orthoclase.

2.5 The dependence of EPMA quality on homogeneity of reference samples

The influence of inhomogeneity of reference samples on the EPMA quality has been exemplified by copper-containing alloys (Pavlova, 2009). Ten copper-rich alloys, the standards for chemical, optical and x-ray fluorescence analysis have been quantitatively evaluated as the reference materials to be employed in the electron probe microanalysis.

The optimum conditions for measurements were selected considering the dependence of intensity and detection limit on conditions of the x-ray radiation excitation and analytical signal recording.

The reference samples were divided into the three groups: the control group 1 consisted of the certified glass (Fe), simple minerals (Zn, Sn), metals (Cu, Ni); two groups of samples for comparison included the assessed alloys: group 2 comprised the alloys MC76 (Fe), MC44 (Sn, Cu), MC104 (Ni), MC153 (Zn); and group 3 included MC71 (Zn), MC74 (Fe, Ni), MC42 (Sn) and metallic copper (Cu).

Three data sets (1, 2 and 3 - in agreement with the groups of reference samples) comprising the average concentrations, standard deviations, relative standard deviations, confidence interval and the z-score of data quality were calculated for 10 copper-rich alloys.

The average concentrations for all elements of every control sample were being defined from 8 to 18 times. The measured values were corrected for matrix effects using the PAP method (Pouchou and Pichoir, 1984) and applying the MARCHELL program (Kanakin & Karmanov, 2006) adapted for the microanalyzer Superprobe-733 operating system. Table 9 presents the data for one alloy.

The relative standard deviations obtained for each element were lower than the target values for all determinations in all cases except for set 2, where group 2 of reference samples was used. In two sets of data the z-score values for all elements determined lie within acceptable limits (-2<z<2) for concentrations ranging from 0.1 to 100%.

Figure 8 shows the graphic dependence of certified Ni, Fe, Sn, Zn concentrations and values obtained from different (1, 2 and 3) groups of reference samples. Obtained sets of concentrations were compared between each other and with the certificated/recommended values using 2-pair selective Student's t-test. The table 9 gives the results of calculations to

Group of reference samples	Metrological performance	Zn	Cu	Sn	Ni	Fe
1	C_{cer} , wt. %	1,96	88,8	4,89	1,09	2,46
	n	14				
	C_{av} , wt.%	1,94	88,5	4,85	1,11	2,48
	s	0,07	1,78	0,14	0,04	0,08
	s_r , %	3,56	2,01	2,98	3,89	3,41
	σ_r , %	3,61	2,04	3,15	3,95	3,49
	ΔC , wt. %	0,04	1,05	0,09	0,03	0,05
	z	-0,3	-0,2	-0,3	0,46	0,24
2	n	17				
	C_{av} , wt.%	1,75	85,7	4,56	1,34	2,95
	s	0,07	2,25	0,16	0,06	0,11
	s_r , %	3,78	2,63	3,45	4,22	3,76
	σ_r , %	3,61	2,04	3,15	3,95	3,49
	ΔC , wt. %	0,05	1,67	0,12	0,04	0,08
	z	-3,2	-1,4	-2,1	4,42	4,34
3	n	12				
	C_{av} , wt.%	1,98	89,3	4,92	1,08	2,45
	s	0,07	1,81	0,15	0,04	0,08
	s_r , %	3,54	2,03	3,12	3,92	3,46
	σ_r , %	3,61	2,04	3,15	3,95	3,49
	ΔC , wt. %	0,04	1,03	0,09	0,02	0,05
	z	0,29	0,26	0,2	-0,2	-0,1

Table 9. Analytical results for copper alloys measured by EPMA using calibrations obtained from three separate groups (1, 2 and 3) of reference materials for each of samples: C_{cer} is the certified concentration; n is number of measurements; $s_r = s / C_{av}$ is standard deviations,

where $s = \sqrt{\{[n\sum_{i=1}^{n}(C_i)^2 - (\sum_{i=1}^{n}C_i^2)] / [n(n-1)]\}}$; $C_{av} = \frac{1}{n}\sum_{i=1}^{n}C_i$ is average concentrations;

$\Delta C = t * s / \sqrt{n}$ is confidence intervals; $z = (C_{av} - C_{cer}) / \sigma_r$ is z-scores (Thompson et al., 1997); $\sigma_r = 100 * 0.02 * C^{0.8495} / C$ is acceptable standard deviation (Thompson et al., 1997), where C is the concentration expressed as fraction.

check the hypothesis that there are no differences between the series of data obtained from any of the three groups of reference samples. We can see, the numerical values of probabilities in columns II, III and VI are not significant, thus indicating that the population of results from series C_2 is statistically different from the recommended values (column II), and the values C_1 and C_3 obtained with groups 1 and 3 of reference samples (columns III and VI).

No systematic divergence was the case between the concentrations obtained from set 1 and set 2, when analyzed results were compared with the certified compositions. The set (2) of reference samples for copper-rich alloys yields erratic data because this set contained the inhomogeneous reference samples.

Lack of close values of probabilities (columns II, III and VI) demonstrates the presence of systematic differences between the concentration series C_2 and recommended concentrations (column II) as well as between the results of the second series C_2 and the data on the series C_1 and C_3 (columns III and VI). This confirms incorrectness of reference samples in group 2. It was previously shown that alloys CA-6 and CA-2 are not homogeneous (table 5).

The numerical values of probabilities are significant in columns I, IV, V and exhibit that the concentrations from selected sets are statistically indistinguishable. Similar values of probabilities listed in these columns for alloys demonstrate the absence of systematic differences of the first and third concentration series both between each other and the recommended concentrations.

Samples	Elements	Numerical values of probabilities for each pair of the series					
		I	II	III	IV	V	VI
		C_{cer} & C_1	C_{cer} & C_2	C_1 & C_2	C_{cer} & C_3	C_1 & C_3	C_2 & C_3
Alloys	Fe	0.341	0.031	0.013	0.421	0.296	0.021
	Ni	0.324	0.012	0.009	0.367	0.311	0.008
	Cu	0.336	0.003	0.026	0.383	0.382	0.017
	Sn	0.445	0.0154	0.0142	0.435	0.278	0.007
	Zn	0.425	0.009	0.001	0.398	0.318	0.011

Table 10. Comparison of concentrations using a coupled selective Student's t-test; C_{cer} is the certified concentration; C_1, C_2, C_3 are three series of concentration data obtained for each of copper-rich alloys using three groups of reference samples.

As a result, the systematic error is very small in the case when the compositions of control samples are calculated using groups 1 and 3 of reference samples. These experiments show that the quality of data obtained from alloy reference samples of group 3 is not inferior to that from the certified reference samples. The quality of results obtained from group 3 of reference samples corresponds to the 'applied geochemistry' type of analysis (category 2) as defined in the GeoPT proficiency testing program (Thompson et al., 1997)

2.6 Method to recalculate experimental values into concentrations

The choice of method of processing experimental values, when the concentration to be determined is concerned, influences determination of a true composition and thus the quality of analysis. Almost every analytical method of recalculating of experimental values into concentrations is developed for certain samples and conditions. Different methods of analytical chemistry have several ways of recalculating experimentally measured values in the concentration. Every way has its own advantages and disadvantages.

Dependence of Determination Quality on Performance Capacity of Researching Technique, Exemplified by
the Electron Probe X-Ray Microanalysis

277

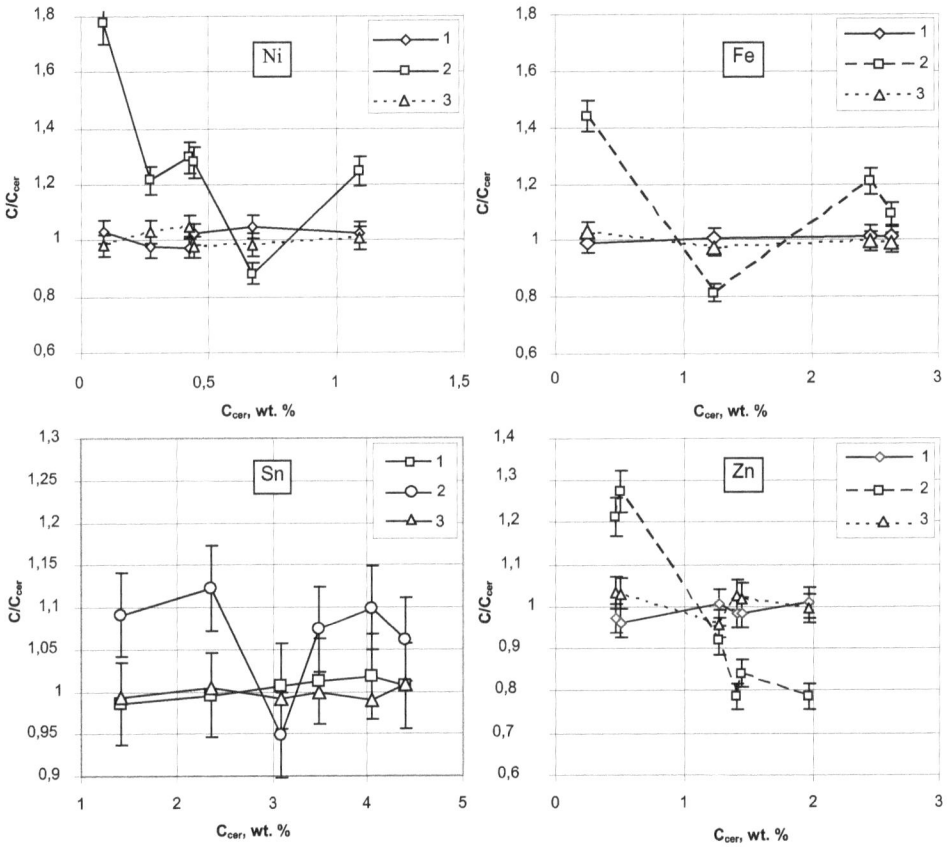

Fig. 8. Dependence of certified Ni, Fe, Sn, Zn concentrations and values obtained from different (1, 2 and 3) groups of reference samples.

When applying EPMA, the recalculation of x-ray radiation intensity in the concentration of silver group elements depends on the correct choice of x-ray radiation absorption coefficients and ability to consider the matrix effects. The determination of the composition of silver-containing compounds can exemplify how the counting procedure influences the quality of composition determination. The results on the composition of silver minerals, differently obtained by calculating absorption coefficients, are compared in Table 10 (Pavlova @ Kravtsova, 2006). The measured intensities, acquired by microprobes JXA8200 and JCXA-733 (Japan), were corrected for matrix effects using programs of JXA8200 Unix platform and the PAP method (Pouchou & Pichoir, 1984) applying the original controlling computer program MARshell (Kanakin & Karmanov, 2006) adapted for JCXA-733 microprobe.

One can observe how available results differ in various methods of calculating matrix effects and how the quality of results depends on the correct calculation of absorption coefficients. It is evident, that the best results are gained by the PAP method (Pouchou & Pichoir, 1984), when the Marenkov's absorption coefficients are applied (Marenkov, 1982), as well as program of JXA8200 Unix platform by PPX method (Pouchou & Pichoir, 1991) from software of microprobe JXA8200 (Japan).

The change of the subject under study often requires the change in the method of recalculating of experimental intensities into concentrations. Thus, a correct determination of the composition of particles comparable in size with the area of generation of x-ray radiation is dependent on the particle size. The quality of obtained results for such particles depends on a correct consideration of size factor (Table 12). The dependence of quality of particle composition versus the method of calculation of their size is exemplified in the article (Belozerova, 2003). Here it is seen that the results obtained for particles with size 3–5 µm are closer to the stoichiometrical composition than those obtained for particles sized 1–2 µm size. One of the best commonly used methods of calculating element content in bulk samples (PAP-method) provides a relative error of determining the composition of one micron particles ranging from 0.5 to 45 %. Using the exponent model of particle composition calculation lowers this inaccuracy.

Method to calculate matrix effects	Authors of absorption coefficients	Sample 1		Sample 2		Sample 3	
		Concent-ration, wt. %	Relative error, %	Concent-ration, wt. %	Relative error, %	Concent-ration, wt. %	Relative error, %
(Pouchou & Pichoir, 1984)	(Marenkov, 1982)	65.56	-0.23	62.69	0.11	77.52	-0.51
	(Pouchou @ Pichoir, 1984)	67.16	-2.67	62.91	-0.24	79.51	-3.08
	(Heinrich, 1986)	66.03	-0.95	61.70	1.69	77.72	-0.77
(Lavrentiev et al., 1980)	(Marenkov, 1982)	73.20	-11.91	68.54	-9.20	86.65	-12.35
	(Pouchou @ Pichoir, 1984)	73.78	-12.79	68.95	-9.87	87.32	-13.21
	(Heinrich, 1986)	68.45	-4.64	62.49	0.42	78.82	-2.19
(Brizuela & Riveros,1990)	(Marenkov, 1982)	67.33	-2.94	63.36	-0.96	79.75	-3.40
	(Pouchou @ Pichoir, 1984)	67.93	-3.86	63.79	-1.63	80.45	-4.30
	(Heinrich, 1986)	66.39	-1.49	62.56	0.32	78.65	-1.97
(Sewell et al., 1985)	(Marenkov, 1982)	68.78	-5.15	64.26	-2.39	81.40	-5.53
	(Pouchou @ Pichoir, 1984)	69.35	-6.03	64.67	-3.04	82.06	-6.39
	(Heinrich, 1986)	67.85	-3.73	63.48	-1.14	80.32	-4.13
programs of JXA8200 (Philibert, 1963); (Reed, 1965); (Duncumb & Reed, 1968); (Berger & Zeltzer, 1964)	(Henke, 1966) (Henke, et al., 1982) (Heinrich, 1966) (Pouchou &P choi r, 1991).	66,02	-0,93	63,18	-0,67	78,03	-1,16
(Pouchou & Pichoir, 1991)		65,94	-0,81	62,57	0,3	77,67	-0,71
Certified concentration, wt. %		65.41		62.76		77.13	

Table 11. Comparison of methods to compute matrix effects and coefficients of absorption to measure silver by EPMA.

Particle size. μm	Calculation technique	Sample Element	Albite			Quart	Calcit	Corun	Rutile	Pyrite	
			Na	Al	Si	Si	Ca	Al	Ti	Fe	S
1	(Pouchou, & Pichoir, 1984)	C_i, wt. %	4.78	6.12	19.40	30.13	25.43	42.95	50.68	44.09	49.2
		R. er., %	45.50	40.50	39.62	35.54	36.49	18.89	15.46	5.28	7.82
	(Belozerova et al., 1998)	C_i, wt. %	5.54	8.06	25.04	35.58	36.87	47.23	57.48	46.56	52.8
		R. er., %	36.83	21.67	22.07	23.88	7.92	10.77	4.12	-0.02	1.08
	(Belozerova et al., 2003)	C_i, wt. %	6.82	9.93	30.86	44.15	42.28	51.40	58.42	46.67	52.9
		R. er., %	22.23	3.50	3.95	5.54	-5.59	2.89	2.55	-0.24	0.86
2	(Pouchou, & Pichoir, 1984)	C_i, wt. %	7.06	8.50	28.32	40.83	34.03	50.71	57.44	45.66	53.1
		R. er., %	19.50	17.39	11.86	12.64	15.00	4.19	4.19	1.91	0.56
	(Belozerova et al., 1998)	C_i, wt. %	7.12	8.87	29.59	43.62	36.86	50.85	58.05	45.67	53.3
		R. er., %	18.81	13.80	7.91	6.68	7.94	3.93	3.17	1.89	0.22
	(Belozerova et al., 2003)	C_i, wt. %	7.38	9.20	30.68	45.33	37.47	51.19	58.07	45.68	53.3
		R. er., %	15.85	10.59	4.51	3.02	6.42	3.28	3.14	1.86	0.22
3	(Pouchou, & Pichoir, 1984)	C_i, wt. %	8.28	9.93	31.54	45.50	38.80	52.26	59.15	46.09	52.9
		R. er., %	5.59	3.50	1.83	2.65	3.10	1.26	1.33	0.94	0.91
	(Belozerova et al., 1998)	C_i, wt. %	8.28	10.01	31.82	46.17	39.54	52.15	59.16	46.02	52.9
		R. er., %	5.59	2.72	0.96	1.22	1.25	1.47	1.32	1.14	1.01
	(Belozerova et al., 2003)	C_i, wt. %	8.33	10.08	32.04	46.52	39.62	52.18	59.16	46.02	52.9
		R. er., %	5.02	2.04	0.28	0.47	1.05	1.42	1.32	1.14	0.99
4	(Pouchou, & Pichoir, 1984)	C_i, wt. %	8.57	10.26	31.71	46.00	40.02	52.37	59.25	46.39	53.1
		R. er., %	2.28	0.29	1.31	1.58	0.05	1.06	1.17	0.34	0.60
	(Belozerova et al., 1998)	C_i, wt. %	8.56	10.27	31.72	46.19	40.19	52.25	59.19	46.34	53.0
		R. er., %	2.39	0.19	1.28	1.18	-0.37	1.28	1.27	0.45	0.71
	(Belozerova et al., 2003)	C_i, wt. %	8.57	10.28	31.76	46.25	40.20	52.25	59.19	46.34	53.0
		R.er., %	2.28	0.10	1.15	1.05	-0.40	1.28	1.27	0.45	0.71
Massive grain	(Pouchou, & Pichoir, 1984)	C_i, wt. %	8.71	10.27	32.06	46.69	40.57	53.00	59.60	46.59	53.4
		R. er., %	0.68	0.19	0.22	0.11	-1.32	-0.13	0.58	-0.09	0.07
	(Belozerova et al., 1998)	C_i, wt. %	8.71	10.27	32.07	46.76	40.62	52.97	59.63	46.55	53.3
		R. er., %	0.68	0.19	0.19	0.04	1.45	0.07	0.53	0.03	0.16
	(Belozerova et al., 2003)	C_i, wt..	8.71	10.27	32.07	46.76	40.62	52.97	59.63	46.55	53.3
		R. er., %	0.68	0.19	0.19	0.04	1.45	0.07	0.53	0.02	0.16
Stoichiometric composition		C_s, wt..	8.77	10.29	32.13	46.74	40.04	52.93	59.95	46.55	53.4

Table 12. Particle composition versus the method of calculation of their size. C_i is obtained concentration; R. er. $= 100(C_s - C_i)/C_s$ %.

The EPMA technique (Belozerova, 2003) for calculating composition developed for approximately spherical particles, comparable in size to the X-ray generation volume, takes into account the particle-size factor. The size factor correction significantly improves the results in spite of a simple analytical function of average atomic number and particle size.

Taking into account the particle-size factor reduces the error of composition determination from 0.5–45 to 0.2–22% relative percent, for particles sized as 1–3 μm.

The relative error increases with decreasing the element concentration from 0.02% for bulk sample to 22.2% for 1 μm particle. The size-factor introduction markedly improves the quality of determinations of particles comparable in size with the area of x-ray radiation excitation.

3. Conclusion

This chapter shows the quality dependence of EPMA on every analysis stage, beginning from the representativeness of the material, sample preparation and conditions for analytical signal excitation and registration to the availability of reference samples and the calculation methods.

We have shown how important it is to correctly select the study area and to have properly prepared samples.

It has been found that the quality of study performance is dependent on the optimum conditions of measuring and processing analytical signal.

The comparison of different methods of taking into account the processes occurring in substance in the electron-excited x-ray radiation proves the necessity to correctly select the methods of their consideration in every study.

The method of assessing the selected set of reference samples and defining their appropriateness for EPMA is described.

The urgency to develop and certify new control samples is critical for the methods of analytical chemistry and especially EPMA.

We have shown the influence of inhomogeneity of samples for comparison on the quality of EPMA results.

Thus, the example of EPMA for the case study of determination content quality suggests, that every aspect of analytical technique is responsible for the quality of element tests.

4. Acknowledgment

The author is grateful to Mrs. T. Bunaeva and Mrs. M. Khomutova for editing the English version of this.

5. References

Beckford, J. L. W. (2010). Quality: a Critical Introduction, Routledge, ISBN 978-0415996358, New York.

Belozerova, O.Y., Afonin, V.P. & Finkelshtein, A.L. (1998). Modified biexponential model and its application to the X-ray microanalysis of gold-containing alloys. *Russian Journal Analytical Chemistry*, Vol. 53, No. 10, pp.1060–1065, ISSN: 0044-4502.

Belozerova, O.Yu., Finkelshtein, A.L. & Pavlova, L.A. (2003). Electron-Probe X-Ray Microanalysis of Individual Particles of Solid Snow Sediment with Size Factor Correction. *Micron*, Vol. 34, No. 1, pp. 49-55, ISSN: 0968-4328.

Berger M.J. & Zeltzer S.M. (1964). *Tables of energy losses and ranges of electrons and positrons.* National Academy of Sciences – National Reseach Council. Publ. 1133, DC, pp. 205-267, Washington.

Borkhodoev, V. Ya. (2010). Assessment of reference samples homogeneity in electron microprobe analysis. *X-ray Spectrometry*, Vol. 39, No. 1, pp. 28–31, ISSN: 1097-4539.

Borovskii, I.B. (1953). To the 70th Anniversary of I.P. Bardin, In: *Problems of Metallurgy,* Borovskii, I.B. pp. 135-153, AN SSSR, Moscow.

Borovskii, I.B. & Il'in N.P., (1956). New method of chemical content investigation in micro volume. *Dokl. Akad. Nauk SSSR,* Vol. 106, No. 4, pp. 654-657, ISSN: 0869-5652.

Borovskii, I.B., Vodovatov, F. F., Zhukov, A.A. & Cherepin, V. T. (1973). *Lokal'ny Metody Analiza Mineralov*, Metallurgiya, Moscow.

Brizuela, H. & Riveros, J.A. (1990). Study of mean excitation energy and K-shell effect for electron probe microanalysis. *X-Ray Spectrometry*, Vol. 19, № 4, p. 173–176, ISSN: 1097-4539.

Buseck, P.R. & Goldstein, J.I. (1969). Olivine Compositions and Cooling Rates of Pallasitic Meteorites. *Geological Society of America Bulletin*, Vol. 80, No. 11, pp. 2141-2158, ISSN: 1943-2674.

Castaing, R. (1951). Application of electron probes to local chemical and crystallographic analysis (Thesis, Univ. of Paris), Translated by P. Duwez and D.B, wittry, CalTech, 1955 (*Special Technical Report* under U.S. Army).

Castaing, R. & Guinier, A. (1953). Point-by-point chemical analysis by X-ray spectroscopy: Electronic Microanalyzer, In: *Analytical Chemistry special issue "Symposium on X-rays as analytical chemical tool"*, Vol. 25, pp. 724-726.

Duncumb, P. & Reed S.J.B.(1968). The calculation of stopping power and backscatter effects in electron probe microanalysis. *Quantitative electron probe microanalysis : proceedings of a seminar held at the National Bureau of Standards*, pp. 133 -154, Gaithersburg, Maryland, June 12-13, 1967.

Farkov, P.M., Il'icheva, L.N., & Finkelshtein, A.L. (2005). X-ray Fluorescence Determination of Carbon, Nitrogen, and Oxigen in Fish Plant Samples. *Russian Journal Analytical Chemistry*, Vol. 60 , No. 5, pp. 485-489, ISSN: 0044-4502.

Finkelshtein, A.L. & Farkov, P.M., (2002). Approximation of X-ray Attenuation Coefficients in Energy Range 0,1 to 100 keV. *Analytic and control*, Vol. 6, No. 4, pp. 377-382, ISSN: 2073-1442 (Print), ISSN: 2073-1450 (Online).

Hubbell, J. H. & Seltzer, S. M. (April 2011). Tables of X-Ray Mass Attenuation Coefficients and Mass Energy-Absorption Coefficientsfrom 1 keV to 20 MeV for Elements Z=1 to 92 and 48 Additional Substances of Dosimetric Interest, In: *NIST X-ray Attenuation Datadases*, 05.04.2011, Available from: http:// NIST Home/PML/Physical Reference Data/X-Ray Mass Attenuation Coefficients (http://www.nist.gov/pml/data/xraycoef/index.cfm).

Heinrich, K.F.J. (1966). X-ray absorption uncertainty. *The electron microprobe; proceedings of the symposium sponsored by the Electrothermics and Metallurgy Division, the Electrochemical Society,* pp. 296-377, Washington, D.C., October, 1964.

Heinrich, K.F.J. (1986). Mass absorption coefficients for electron probe microanalysis. *11th International congress on X-ray optics and microanalysis. Proceedings of a conference held in London, Ontario, Canada,* pp. 67-119, 4–8 August 1986.

Henke, B.L. (1966). Application of multilayer analyzers to 15-150 angstrom fluorescence spectroscopy for chemical and valence band analysis. *Advance in X-ray Analysis,* Vol. 9, n.d., pp. 430-440, ISSN: 0376-0308

Henke, B.L., Lee, P., Tanaka, T.J., Shimabukuro, R.L. and Fujikawa, B.K. (1982). Low Energy X-ray Interaction Coefficients: Photoabsorption, Scattering and Reflection.E=10-2000 eV. Z=1-94. *Atomic Data and Nuclear Data Tables,* Vol. 27, pp. 1-144, ISSN: 0092-640X.

Henke, B.L., Gullikson, E.M., & Davis, J.C. (1993). X-ray interactions: photoabsorption, scattering, transmission, and reflection at E=50-30000 eV, Z=1-92. *Atomic Data and Nuclear Data Tables, Vol. 54,* No. 2, pp. 181-342.

Goldstein, J.I., Newbury, D.E., Echlin, P., Joy, D.C., Lyman, C.E., Fiori, C., Romig A.D. Jr. & Lifshin, E. (1992). *Scanning Electron Microscopy and X-Ray Microanalysis,* Plenum Press, NewYork.

Kanakin S.V. & Karmanov, N.S. (2006). ICA and main scope of the software package MARshell32, *Proceedings of the V All-Russian X-Ray analysis Conference,* Irkutsk, May–June, 2006.

Lavrentiev, Yu.G., Berdichevskiy, G.V., Chernyavsriy, L.I. & Kuznetsova, A.I. (1980). KARAT – is program for quantitative electron probe X-ray microanalysis, *Apparatus and Methods for X-Ray Analysis,* Mashinostroenie, Leningrad, Vol. 23, n.d., pp. 217-224.

Marenkov O.S. (1982). *Tables and Formulas X-ray analysis. Methodical recommendations.* Mashinostroenie, Leningrad.

MI. (1988). Standard sample homogeneity of monolithic materials for the spectral analysis. Technique of measurement. Standards Publish, Moscow.

NIST (1990) Standard Reference Materials Catalog 1990-1991, Ed: R.S. McKenzie, NIST Special Publication 260, U.S. Department of Commerce, National Institute of Standards and Technology.

Pavlova, L. A. Belozerova, O. Yu., Paradina, L. F. & Suvorova, L. F. (2000). *X-Ray Electron Probe Analysis of Environmental and Ecological Objects,* Nauka, ISBN 5-02-031533-8, Novosibirsk, Russia.

Pavlova, L. A., Paradina, L. F. & Belozerova, O. Yu. (2001) Electron Probe Microanalysis of Environmental Samples: Preparation of Reference Materials, Correction for Particle Size Effects and Representative Results on Sediments Recovered from Snow, Power Station Fly Ash and Bone Samples. *Geostandard Newsletter: the Journal of Geostandard and Geoanalysis,* Vol. 25, No 2-3, pp. 333-344, ISSN: 0150-5505.

Pavlova, L.A., Pavlov, S. M, Anoshko, P.N. & Tyagun, M.L. (2003). Determination of calcium and sodium in Baykal omul otoliths by Electron probe microanalysis, *Analytic and control*, Vol. 7, No. 3, pp. 242-247, ISSN: 2073-1442 (Print), ISSN: 2073-1450 (Online).

Pavlova, L. A., Suvorova, L. F, Belozerova, O. Yu. & Pavlov, S.M. (2003). Quality of determinations obtained from laboratory reference samples used in the calibration of X-ray electron probe microanalysis of silicate minerals. *Spectrochimica Acta Part B*, Vol. 58, No. 2, pp. 289-296, ISSN: 0584-8547.

Pavlova, L.A., Pavlov, S. M, Paradina, L. Ph., Karmanov, N.S., Kanakin, S.V., Anoshko, P.N. & Levina, O.V. (2004). Investigation of Baykal flora and fauna representative using the techniques of electron probe microanalysis and electron microscopy. *Ecological chemistry*, Vol. 13, No. 4, pp. 249-256, ISSN: 0869-3498.

Pavlova, L.A. & Kravtsova, R.G. (2006). Studing modes of silver occurrence, discovered in the lithochemical stream sediments of Dukat gold-silver deposit by electron probe x-ray microanalysis. *Methods and objects of chemical analysis*, Vol. 1, No. 2, pp. 132–140, ISSN: 1991-0290.

Pavlova, L. A. (2009). Quality of electron probe X-ray microanalysis determinations obtained from laboratory reference materials of the coppery alloys and basaltic glasses. *Spectrochimica Acta Part B: Atomic Spectroscopy*, Vol. 64, No 8, pp. 782-787, ISSN: 0584-8547.

Paradina, L.F. & Pavlova, L.A. (1999). Definition of vanadium, rubidium and strontium impurities in a vitreous substance of Siberia. *Russian Journal Analytical Chemistry*, Vol. 54 , No. 1, pp. 78–82, ISSN: 0044-4502.

Philibert J.A. (1963). A method for calculation of the absorption correction in electron probe microanalysis. In: *X-Ray Optics and X-Ray Microanalysis*, V.E. Cosslet & A. Engstrom, pp. 379-392, Academic Press, ISSN: 0028-0836, N.Y.

Pouchou J.L. & Pichoir F. A. (1984). New model for quantitative X-ray microanalysis, Part 1. Applications to the analysis of homogeneous samples. *La Recherche Aerospatiale*, n.d., No. 3, pp. 13–38, ISSN: 0034-1223.

Pouchou J.L. & Pichoir F. A. (1991). Quantitative analysis of homogenous or stratified microvolumes applying the model "PAP". In: *Electron Probe Quantitation*, K.F.J. Heinrich and D.E. Newbury (Eds.), 31-75, Springer, ISBN: 0306438240, Plenum Press.

Reed S.J.B. (1965). Characteristic fluorescence correction in electron probe microanalysis. *British Journal of Applied Physics*, Vol. 16, No 7, pp. 913–926, ISSN: 0508-3443.

Saloman, E.B., Hubbell, J. H. & Scofield J.H. (1988) X-ray attenuation cross srctions for energies 100 eV to 100 keV and elements Z=1 to Z=92. *Atomic Data and Nuclear Data Tables*, Vol. 38, pp. 1-197, ISSN: 0092-640X/88.

Sewell, D.A., Love, G. & Scott, V.D. (1985). Universal correction procedure for electronprobe microanalysis. I. Measurement of X-ray depth distribution in solids. *Journal of Physics D: Applied Physics*, Vol. 18, No. 7, pp. 1233–1243, ISSN: 0022-3727.

Szaloki, I., Osan, J. & Van Grieken, R. E. (2004). X-ray Spectrometry. *Analytical Chemistry*, Vol. 76, No. 12, pp. 3445-3470, ISSN: 1097-4539.

Thompson, M., Potts, P.J. & Webb, P.C. (1997). GeoPT1. International proficiency test for analytical geochemistry laboratories — report on round 1 (July 1996). *Geostandard Newsletter: the Journal of Geostandard and Geoanalysis*, Vol. 21, No. 1, pp. 51–58, ISSN: 0150-5505.

Wilson, S.A. & Taggart, J.E. (2000). Development of USGS microbeam reference materials for geochemical analysis, *Proceedings of the 4th International Conference on the Analysis of Geological and Environmental Materials*, Pont a Mousson, France, 29th Aug.-1st Sep., 2000.

16

Laser Diffuse Lighting in a Visual Inspection System for Defect Detection in Wood Laminates

David Martin, Maria C. Garcia-Alegre and Domingo Guinea

Spanish Council for Scientific Research (CSIC), Madrid,
Spain

1. Introduction

Nowadays, wood companies are ever more interested in automatic vision systems (Li & Wu, 2009), (Åstrand & Åström, 1994), for an effective surface inspection that greatly increases the quality of the end product (Smith, 2001), (Armingol et al., 2006). The inspection process, in most visual inspection systems, pursues online defects identification, to reach optimum performance (Malamas et al., 2003), (Spínola et al., 2008).

The usual wood inspection systems are visual ones, based on standard cameras and lighting (Batchelor & Whelan, 1997), (Cognex, 2011), (Parsytec, 2011), (Pham & Alcock, 2003) to operate in highly structured environments (Silvén et al., 2003). The quality control in visual surface inspection systems must be robust to cope with wood variable reflectance and high speed requirements.

The surface inspection methods proposed in the literature for visual inspection aim at ad-hoc surface inspection systems to solve each specific problem (Pham & Alcock, 1999). Usual inspection systems are based on visible lighting and few of them use diffuse components to illuminate the rough and bright surfaces. A visual wood defect detection system proposed by (Estévez et al., 2003) is composed by a colour video camera, where the standard lighting components are a mixture of two frontal halogen and ceiling fluorescent lamps. The commercial light diffusers use a light source and different components to illuminate the surface in an irregular way to eliminate shadows, but present some problems such as, short useful life, extreme sensitivity and high cost.

On the other hand, one of the major drawbacks in automated inspection systems for wood defect classification is the erroneous segmentation of defects on light wood regions, (Ruz et al., 2009). Moreover, the speed of wooden boards at the manufacturing industry is at about 1 m/s, which implies high computational costs (Hall & Aström, 1995).

Current work presents a surface inspection system that uses laser diffuse lighting to cope with different type of defects and wood laminated surfaces to improve defect detection without any previous defect information.

The work will not only highlight the specific requirements for a laser diffuse lighting in a visual inspection system but also those of unsupervised defect detection techniques to cope with the variability of wood laminated surfaces and defect types, leading to a heterogeneous and robust visual surface inspection system. The manuscript is organized as follows: section 2 displays images of different wood laminated surfaces, captured by a visual surface inspection system with standard lighting. In Section 3, an innovative surface inspection system using laser diffuse lighting is presented, as well as the acquired images. Section 4 describes an unsupervised wood defect detection and segmentation algorithm to process images acquired with the two types of lighting. Section 5 displays the results, and the conclusions are presented in Section 6.

2. Standard lighting visual systems

Most inspection systems are based on visible lighting (Guinea et al., 2000), (Montúfar-Chaveznava et al., 2001). These systems can adequately tackle defect detection, but presents some drawbacks:

1. Inspection systems composed by multiple cameras and equal standard lighting for all cameras, make difficult the inspection of the whole wood boards in the production line due to non-uniform illumination of the surface.
2. When fluorescent light is replaced by commercial light diffusers to improve lighting, the maintenance cost greatly increases.

Images acquired with a standard visual surface inspection system, composed of a CCD visual camera and fluorescent light, are displayed in Figure 1. The images show knots and splits and are manually classified by a human expert to validate the performance of the automated visual inspection system. Images show different background colour as they come from different wood laminates.

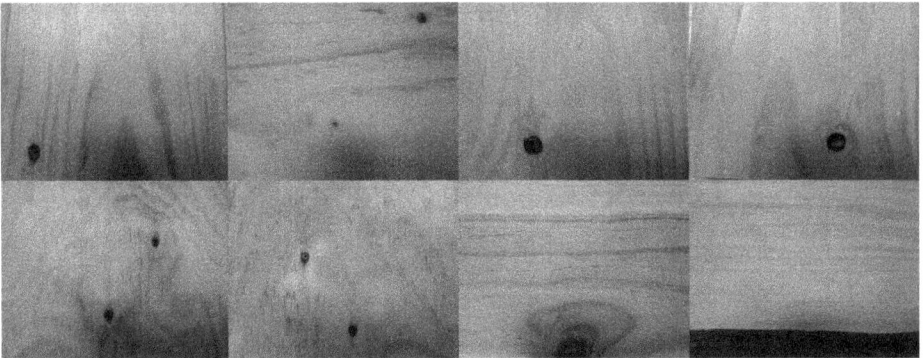

Fig. 1. Defects, knots and splits, on images acquired from a standard lighting visual surface inspection system.

3. Laser diode diffuse lighting in visual systems

Main innovation of current work is a CCD sensor-based industrial inspection system for defect detection on wood laminates, where lighting is based on a laser technique that comprises two different illumination modes: Diffuse Coaxial Lighting and Diffuse Bright-Field Lighting.

Few visual inspection systems use laser-lighting (Palviainen & Silvennoinen, 2001), (Yang et al., 2006), and even less laser diode diffuse components, to illuminate both rough and bright surfaces. The commercial light diffusers are composed of a light source and different optical components to illuminate correctly the surface in an irregular way. However, they present drawbacks such as, short useful life, extreme sensitivity and high cost, that can be overcome with the use of laser diode diffuse lighting to highlight the relevant features.

Main advantages of the proposed lighting system are:
1. Inspection of different wood surfaces without any reconfiguration of the system
2. Detection of different types of wood defects with the same diffuse lighting
3. Inspection of defects with different areas, ranging from 10 to 100 mm², and shapes.
4. Each CCD camera has its own laser diode diffuse lighting

The inspection vision system can be configured to work in two lighting modes (Martin et al., 2010):
- Diffuse Coaxial Lighting
- Diffuse Bright-Field Lighting

The proposed design permits to reuse the laser-optical system (laser-lighting and optical components) in both configurations to tackle high-speed and small-defect industrial surface inspections over the whole inspection region. The laser diffuse lighting provides a high intensity beam but only on a small area, removing the image shadows generated by the surface roughness, and facing the surface variable reflectance.

The wood samples are illuminated using a green diode-pumped solid-state (DPSS) laser, which provides highly uniform illumination in the region of interest (ROI). Coherent laser light is unusual for surface illumination in inspection vision systems due to the emergence of interference patterns. The two images displayed in Figure 2, are captured using only a laser-optical system without any wood surface sample, to exhibit the interference patterns caused by the optical components with coherent laser lighting.

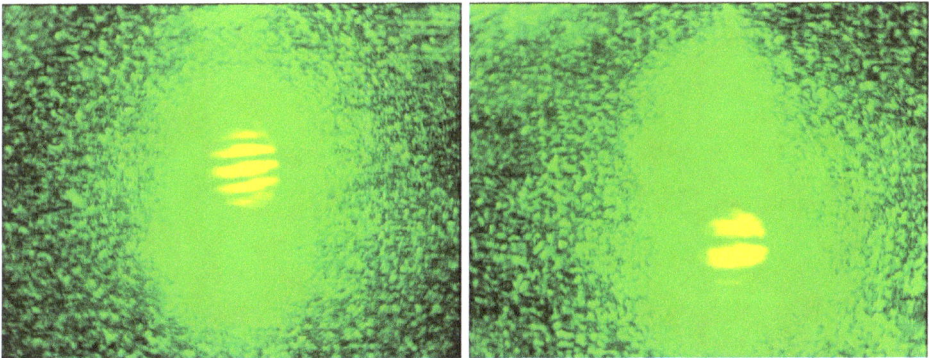

Fig. 2. Two interference patterns (yellow) caused by coherent green laser light on the surface inspection lenses.

Another interference effect appears when the laser beam is reflected on rough wood surfaces. The result of this effect in images acquired by a standard CCD camera is a bright and dark spot pattern, namely "speckle pattern", which is specific for each type of surface. The pattern is distributed randomly in the whole space and is caused by constructive and destructive interference effects.

In spite of the interference patterns, the laser lighting source is robust for integration in an industrial environment due to its low-cost and durability in comparison with commercial light diffusers. The solution to the interference problem has been achieved by means of a dispersion technique to obtain laser diffuse lighting and remove the speckle interference pattern. The components proposed for laser beam scattering are:

1. A convergent lens that increases the width of the collimated beam
2. A spinning diffuser that disperses the collimated beam

The wood defect inspection and 1 m/s wood board speed, require high intensity in the inspection region and short exposure time for suitable image acquisition. Former requirements are interrelated, as if intensity increases the exposure time, for real-time inspection, can be reduced. This high intensity allows a correct inspection as the laser diode diffuse lighting illuminates only a small inspection area, about 10-100 mm^2.

In current work, a second set of images have been acquired with diffuse bright-field lighting. The configuration is composed by the laser-optical components (laser, dispersion lens and spinning diffuser) and the imaging components (CCD sensor camera and focusing lens). The orientation of the imaging components related to the laser diffuse lighting is approximately 30°, Figure 3. A black box covers the imaging components so that the light reaching the CCD sensor only comes from the focusing lens.

Fig. 3. Laboratory benchmark: diffuse bright-field lighting with a 160 mm focal length lens and a green DPSS laser.

The second set of images obtained with this configuration is shown in Figure 4.

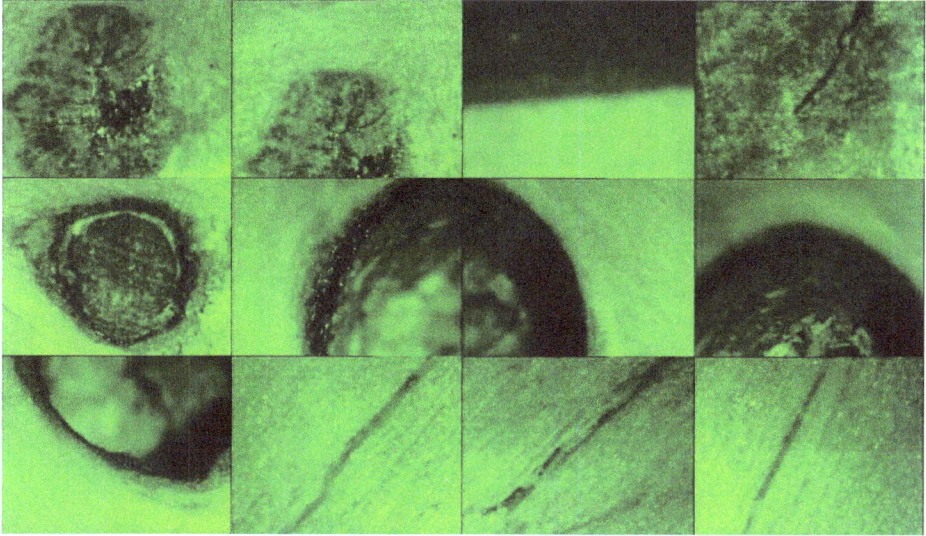

Fig. 4. Wood images acquired with a laser diffuse lighting.

The main characteristics of the proposed visual inspection system endowed of laser diffuse lighting are:
i.   High intensity in the inspection region and consequently short exposure time
ii.   High-speed and small-defect surface inspection on the whole inspection region
iii.   Elimination of the image shadows generated by the surface roughness
iv.   Deals with the variability of laminated wood such as, colour or texture
v.   Algorithms required for defect detection are simpler and thus shorter the computing time, as all images present the same background colour.

4. Wood defect detection and image segmentation

Unsupervised visual processing algorithms are proposed for both wood defect detection and image segmentation. The sooner the line inspector accurately detects the appearance of defects, the shorter the problem is fixed. The defect detection process is accomplished with an algorithm that searches for the seeds of the defects, characteristic pixels belonging to the defect, which determines the location of the defects in the image. Then the image segmentation algorithm uses a region growing method to calculate the size of the defect. The region growing method is based on Active Contours (Chan & Vese, 2001). The validation of the automatic visual inspection results is performed by a human expert.

An unsupervised defect detection algorithm has been developed and tested, to cope with variations in the laminated material, such as defect type, pose, size, shape and colour, as well as with environment variations: surface illumination and speed of the laminated material in the production line. All these variables make automatic wood defect detection a challenging task. The flow chart of Figure 5 summarizes the operation of the unsupervised defect detection algorithm.

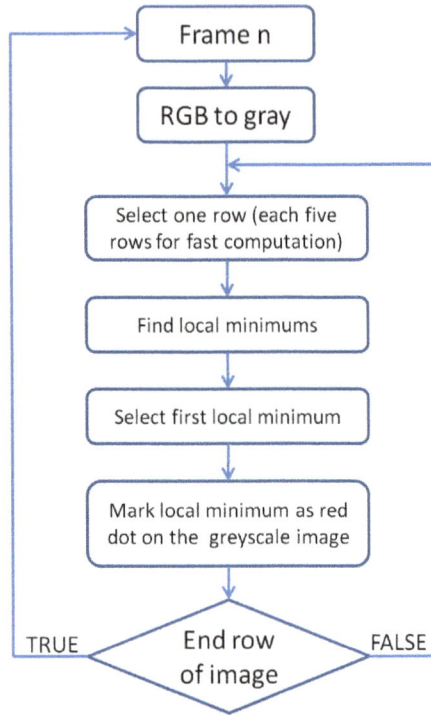

Fig. 5. Flow chart of the wood defect detection algorithm.

The wood defect detection algorithm proceeds as follows:
1. RGB colour image is converted to greyscale to reduce computational time
2. The developed algorithm searches for local minimum at each fifth row of the image and uses expert knowledge to associate defects with lower grey-level image pixels. The grey value of each pixel in an image row is compared to its neighbours and if its value is lower than the values of each neighbours, the pixel is defined as a local minimum. Two parameters are set up to configure the algorithm restrictions:
    *   "Minimum valley depth", determines the minimum depth as a greyscale value. The valley depth value here selected is 190. Then, only valleys that exceed this value are returned.
    *   "Minimum valley separation", specifies the minimum distance between valleys as a positive integer (pixels). That is, the algorithm ignores small valleys that occur in the neighbourhood of a deeper valley. The selected value is 100 pixels.
3. The position of the local minimum is visualised on the greyscale image with a red dot.
On the other hand, image segmentation allows for the calculation of the area of the defect. This is of great aid for quality control analyses, as the greater the size of the defects is, the lower is the quality of the wood laminates. The region growing algorithm groups pixels together into regions of similarity, beginning from an initial set of pixels (red dots). The method works iteratively for increasing the initial pixels set, comparing its grey value with that of its neighbours for increasing the area of the defect to reach the borders. These seeds, previously calculated by the unsupervised defect detection algorithm, allow applying the

active contours method only in the regions where defects are detected, thus reducing computing costs.

5. Results

The wood defect detection and image segmentation algorithms, here proposed, have been tested with the two set of images acquired with both standard and laser diffuse lighting visual system.

The results of the unsupervised wood defect detection algorithm applied to the first set of images, are displayed in Figure 6. The results show that the algorithm is capable of detecting the position of the defects in the images several times (red dots). This multi-detection implies an accurate detection process caused by applying the algorithm only to each fifth row, for fast defect detection. A gap of five rows is an appropriate value for wood defect detection, as defects usually intersect in more than five rows. Then, the larger the defects are the greater the number of the detected pixels is (red dots). Knots defects have been correctly detected except in the sixth image due to the fact that the background was extremely similar to the defect grey-level. The split appearing on the eighth image has been detected. Moreover, comparing the results of the automatic defect detection algorithm with those of the expert, in the first set of images, the groundwork of the algorithm succeeded in 87.5%. Longer term processing would be required to detect a higher percentage of defects in images.

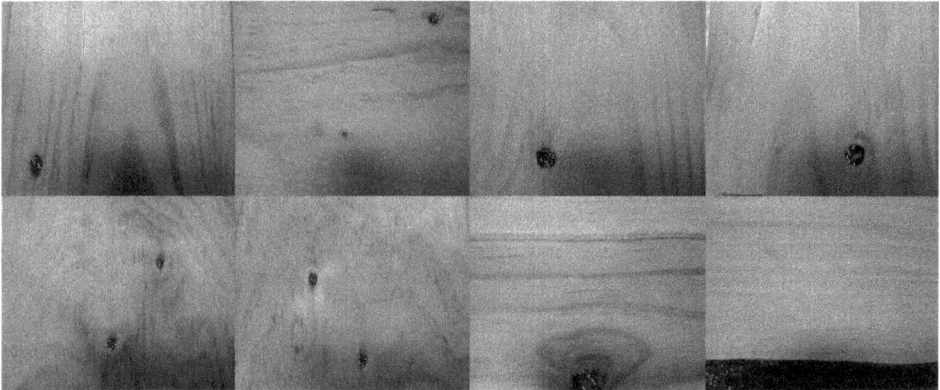

Fig. 6. Results of wood defect detection in images (first set) acquired with standard lighting visual system.

The results obtained with the unsupervised wood defect detection algorithm on the second set of images, are displayed in Figure 7. The second set of images has been captured with the laser diffuse lighting visual system, and in that case all defects were detected.

The multi-detection process enhances defect detection due to the large size of the defect in the analysed images. The different types of defects are easily perceived by the expert. In the last image, split defect was not completely detected as a consequence of the similarity between the grey-level of the defect and that of the background. The success in comparing automatic defect detection and expert visual defect detection, is 98%. This percentage is better than the one obtained in the processing of the first set of images.

Therefore, the use of a surface inspection system with laser diffuse lighting greatly improves the success of the automatic wood defect detection algorithm.

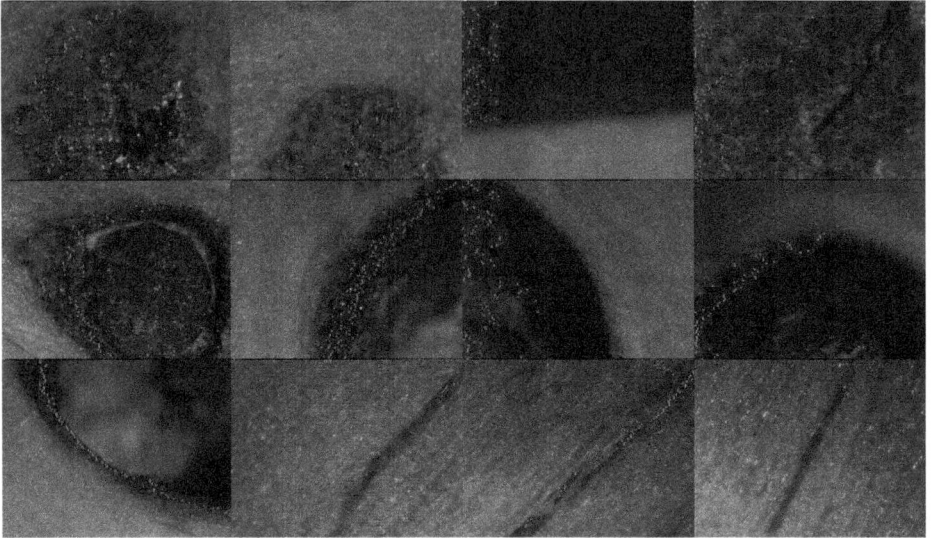

Fig. 7. Results of wood defect detection in images, second set, acquired with a laser diffuse lighting surface inspection system.

The result of the image segmentation algorithm on the first set of images is displayed in Figure 8. The image segmentation algorithm is based on a region growing method that uses Active Contours. The results are displayed in Figure 8, marking the complete border of the defect, in green colour. The defect detected pixels (red dots) grow until the border of the defect is reached, whenever the border of the defect is obtained before last iteration. Moreover, the border can be extra grown, as happens in the sixth and seventh images of Figure 8. However, the wood defects are well shaped, in spite of that present in the eighth image that would need some more iterations to complete the whole defect contour. Finally, the areas of the defects are calculated for further off-line quality control analysis.

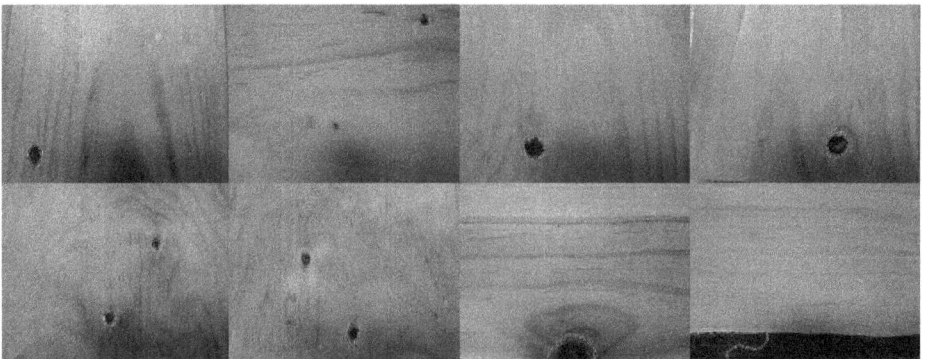

Fig. 8. Segmentation of images acquired with standard lighting visual system (first set).

The result of the image segmentation of the second set of images is displayed in Figure 9, where the partial border of the defect is marked in green colour. After 200 iterations of the algorithm, departing from the initialization pixels (red dots), the complete area of the defects present in the images is not totally segmented. Defects can be more accurately segmented by increasing the seed pixels (red dots) and the number of iterations of the region growing algorithm, but this implies a greater computing time.

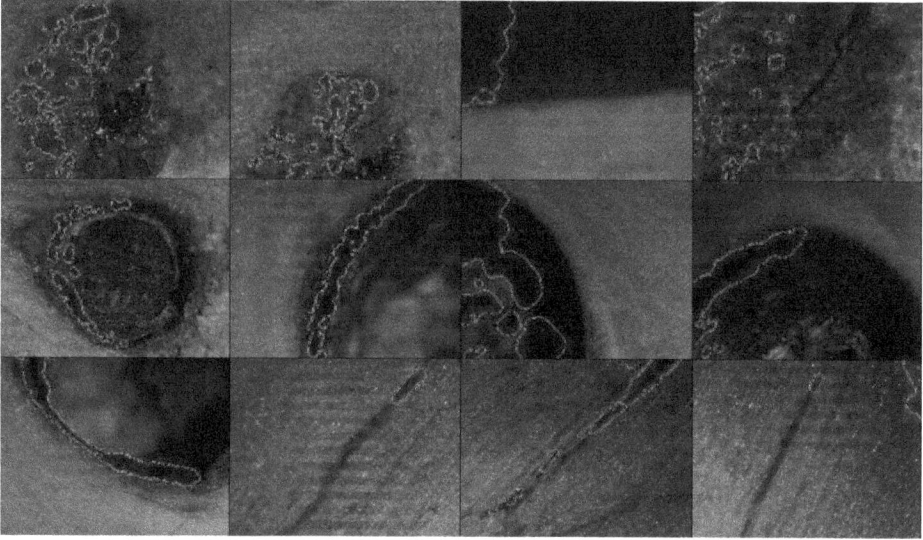

Fig. 9. Segmentation of the images acquired from the laser diffuse lighting surface inspection system (second set).

Summarizing, results displayed in Figure 9, are better than those shown in Figure 8, but total segmentation of the defects is not achieved as a greater number of seed pixels would be required.

Thus, to increase the number of seed pixels, the "Minimum valley separation" variable has been set to 10 pixels. The results obtained with the set of images acquired with laser diffuse lighting are shown in Figure 10. In first, second, fourth, tenth and twelfth images, the seeds (red dots) partially cover the wood defects, but in the third, fifth, sixth, seventh, eighth, ninth and eleventh images the defects are totally covered by the seeds. Then, the image processing would only require the wood defect detection algorithm as the calculated seeds can detect the shape of the whole defects. This point is extremely relevant as implies lower computing costs.

Next, region growing method is applied to obtain the complete segmentation of the defects (green borders), and the results are displayed in Figure 11. It can be remarked that segmentation results of the region growing method are close to those of Figure 10, obtained with the wood defect detection algorithm.

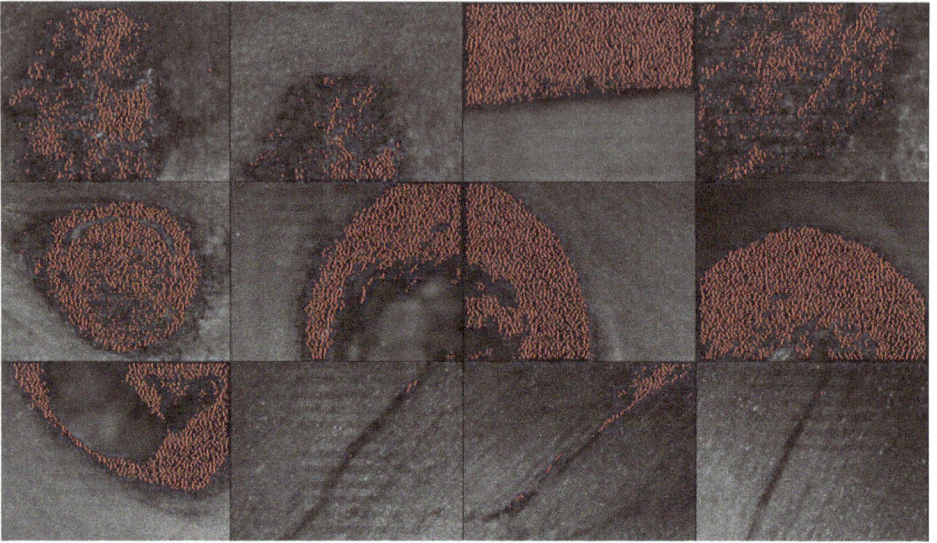

Fig. 10. Wood defect detection on images acquired with laser diffuse lighting and "Minimum valley separation" = 10 pixels.

Fig. 11. Image segmentation using a large number of seeds: "Minimum valley separation"=10 pixels.

Finally, the computation time for the defect detection and segmentation algorithms is calculated for each image in Figure 10 and Figure 11, and displayed in Table 1. The second column corresponds to the computation time of the wood defect detection algorithm increasing the number of seed pixels (red dots). The third column is the computation time of the region growing algorithm for the image segmentation (green borders).

Images acquired from laser diffuse lighting surface inspection system	Defect detection algorithm (seconds)	Region growing algorithm (seconds)
Image 1	0.51	10.03
Image 2	0.23	8.65
Image 3	0.37	9.17
Image 4	0.21	9.82
Image 5	0.21	10.34
Image 6	0.21	9.90
Image 7	0.23	9.17
Image 8	0.31	12.42
Image 9	0.21	9.21
Image 10	0.23	7.96
Image 11	0.20	8.18
Image 12	0.18	8.09

Table 1. Computation time, in seconds, for each of the processing algorithm and images acquired from laser diffuse lighting.

6. Conclusions

The global goal of current work is the defect detection on wood images for quality control with short computing time. To this aim, the work compares wood images captured with both: a standard and an innovative surface inspection system that uses laser diffuse lighting.

Images from wood laminates with knots and splits defects acquired from the innovative lighting visual system are the most suitable for defect detection. The main characteristics of the proposed system to be used in the wood manufacturing industry are:

i.   Detection of small defects ranging from one to few millimetres
ii. Independence from the variable reflectance of the wood laminates. Images present both a high background smoothness and high-contrast of the defects, independently of the colour of the wood laminates
iii. Defect detection time in less than 0.5 s
iv. Defect detection success is 98% on a sample of two hundred images
v.   The laser diffuse light source is robust enough for integration in an industrial environment, due to its low cost and durability, in comparison with commercial light diffusers

A smart combination of algorithms for wood inspection has been proposed, based on an unsupervised method for defect detection and a region growing algorithm for defect segmentation. Results indicate that processing time for wood defect detection algorithm is better than for region growing algorithm. Moreover, first algorithm reaches both objectives: wood defect detection and segmentation.

Finally, it has been demonstrated that the use of the proposed wood defect detection algorithm with images acquired from visual surface inspection system using laser diffuse lighting, appears as the best choice for real-time wood inspection.

7. References

Armingol, J. Mª; Otamendi, J.; de la Escalera, A.; Pastor, J.M. & Rodríguez, F.J. (2006). Intelligent Visual Inspection System for Statistical Quality Control of a Production Line. *Frontiers in Robotics Research*, Nova Publishers, Chapter 1, pp. 1-33

Åstrand, E. & Åström, A. (1994). A Single Chip Multi-Function Sensor System for Wood Inspection, *Proceedings of the 12th IAPR International Conference on Pattern Recognition*, Vol. 3, pp. 300-304, ISBN: 0-8186-6275-1, Jerusalem, Israel.

Batchelor, B.G. & Whelan, P.F. (1997). *Intelligent Vision Systems for Industry*, Springer-Verlag

Chan, T.F. & Vese, L.A. (2001). Active Contours Without Edges, *IEEE Transactions on Image Processing*, Vol.10, pp. 266–277

Cognex Corporation (2011). *Machine Vision Systems*. Natick, MA 01760-2059, United States of America. Available from http://www.cognex.com/

Estévez, P.A.; Perez, C.A. & Goles, E. (2003). Genetic Input Selection to a Neural Classifier for Defect Classification of Radiata Pine Boards. *Forest Products Journal*, Vol. 53, pp. 87–94.

Guinea, D.; Preciado, V.M., Vicente, J.; Ribeiro, A. & García-Alegre, M.C. (2000). CNN based visual processing for industrial inspection, *Proceedings of SPIE Machine Vision Applications in Industrial Inspection VIII*, California, Vol. 3966, pp. 315-322

Hall, M. & Aström, A. (1995). High Speed Wood Inspection Using a Parallel VLSI Architecture, *Proceedings of the International Workshop Algorithms and Parallel VLSI Architectures III*, Leuven, Belgium, pp. 215-226

Li, M.-X. & Wu, C.-D. (2009). A VPRS and NN Method for Wood Veneer Surface Inspection, *Proceedings of the 2009 International Workshop on Intelligent Systems and Applications*, ISBN: 978-1-4244-3893-8, Wuhan, China, pp. 1-4

Malamas, E.N.; Petrakis, E.G.M.; Zervakis, M.; Petit, L. & Legat, J.D. (2003). A Survey on Industrial Vision Systems, Applications and Tools. *Image Vision Comput.*, Vol. 21, pp. 171–188

Martin, D.; Guinea, D.M.; García-Alegre, M.C.; Villanueva, E. & Guinea, D. (2010). Multi-modal Defect Detection of Residual Oxide Scale on a Cold Stainless Steel Strip. *Machine Vision and Applications*, Vol.21, pp. 653-666

Montúfar-Chaveznava, R.; Guinea, D.; Garcia-Alegre, M.C. & Preciado, V.M. (2001). CNN computer for high speed visual inspection, *Proceedings of SPIE Machine Vision Applications in Industrial Inspection IX*, California, Vol.4301, pp. 236-243

Palviainen, J. & Silvennoinen, R. (2001). Inspection of Wood Density by Spectrophotometry and a Diffractive Optical Element Based Sensor. *Measurement Science and Technology*, Vol.12, pp. 345-352

Parsytec Computer GmbH (2011). *Surface Inspection Systems*. Auf der Huels 183, 52068 Aachen, Germany. Available from http://www.parsytec.de/

Pham, D.T. & Alcock, R.J. (1999). Automated visual inspection of wood boards: selection of features for defect classification by a neural network. *Proceedings of the Institution of Mechanical Engineers -- Part E -- Journal of Process Mechanical Engineering (Professional Engineering Publishing)*, Vol. 213, pp. 231-245

Pham, D.T. & Alcock, R.J. (2003). *Smart Inspection Systems. Techniques and Applications of Intelligent Vision*, Academic Press

Ruz, G.A., Estévez, P.A. & Ramírez, P.A. (2009). Automated visual inspection system for wood defect classification using computational intelligence techniques. *International Journal of Systems Science*, Vol. 40, pp. 163–172

Silvén, O.; Niskanen, M. & Kauppinen, H. (2003). Wood Inspection with Non-supervised Clustering. *Machine Vision and Applications*, Vol.13, pp. 275-285

Smith, M.L. (2001). Surface Inspection Techniques - Using the Integration of Innovative Machine Vision and Graphical Modelling Techniques, *Engineering Research Series*. London and Bury St Edmunds, UK: Professional Engineering Publishing

Spínola, C.G.; García, F.; Martin, M.J.; Vizoso, J.; Espejo, S.; Cañero, J.M.; Morillas, S.; Guinea, D.; Villanueva, E.; Martin, D. & Bonelo, J.M. (2008). Device for Detecting and Classifying Residual Oxide in Metal Sheet Production Lines, Patent P2007_00865, Patent Cooperation Treaty (PCT): Publication No.: WO/2008/119845, International Application No.: PCT/ES2007/000768, ACERINOX S.A.

Yang, D.; Jackson, M.R. & Parkin, R.M. (2006). Inspection of Wood Surface Waviness Defects Using the Light Sectioning Method. *Proceedings of the Institution of Mechanical Engineers, Part I: Journal of Systems and Control Engineering*, Professional Engineering Publishing, Vol.220, pp. 617-626

17

Quality Control Through Electronic Nose System

Juan C. Rodríguez-Gamboa, E. Susana Albarracín-Estrada
and Edilson Delgado-Trejos
MIRP, Research Center, Instituto Tecnológico
Metropolitano (ITM), Medellín
Colombia

1. Introduction

Quality control is defined as: "a process selected to guarantee a certain level of quality in a product, service or process. It may include whatever actions a business considers as essential to provide for the control and verification of certain characteristics of its activity. The basic objective of quality control is to ensure that the products, services or processes provided meet particular requirements and are secure, sufficient, and fiscally sound"1 In order to apply Quality Control through the Electronic Nose System, all the stages involved in the process must be taken into account, this case refers to the use of electronic nose systems as a tool for quality control tasks. Therefore best practices must be implemented that will lead to obtaining good quality measures, which will later become good results (Badrick, 2008; Duran, 2005)

Section 2 of this chapter presents an overview of the parts or subsystems involved in an electronic nose system and the operating principle.

Section 3 deals with the issue of food quality control using electronic nose systems. This section discusses how to use the electronic nose system for these types of applications, and also presents some issues for consideration when analyzing products such as coffee, fruits and alcoholic beverages.

Section 4 covers other applications of electronic nose systems, especially applications in the medical field for detection and diagnosis of diseases. This section focuses more on viable alternatives for the detection of diseases, rather than on quality control.

It is important to note that quality control is mainly used to find errors in processes, so the deductions presented here have gone through a series of tests and experiments to obtain the desired results and thus facilitate further research and shed light on the question of how these types of applications should be addressed.

2. A look at the electronic nose systems

Existing systems for electronic olfaction (EOS), also commonly known as electronic noses, are basically arrays of chemical sensors, connected to a computer or processing systems which apply advanced techniques of digital signal processing and statistical pattern recognition. Their main objective is to enable the qualification of odours through classification tasks, discrimination, prediction, and even quantification of products, elements or components according to their organoleptic characteristics (Duran & Baldovino, 2009; Wilson & Baietto, 2009; Zhou et al., 2006).

2.1 Elements of an Electronic Nose System (EOS)

An electronic nose system can be seen as an instrument or measuring equipment of artificial olfaction, consisting of a series of modules that work together, which analyzes gas samples, vapours and odours. An instrument or equipment of this type has at least 4 parts, each with specific functions which are detailed below (Duran & Baldovino, 2009; Tian et al., 2005).

2.1.1 Matrix or array of gas sensors

In general, the gas sensors are devices that consist of two main parts, the first is an active element which changes its physical or chemical properties in the presence of that which it detects and the second part is a transducer, which converts the changes in the properties of the active element into an electrical signal. These sensors typically have a selective membrane, preventing passage of particles or unwanted material, acting as a first noise filter. In Figure 1 shows a simplified diagram of a device of this type, in which the main parts of a gas sensor and the nature of the inputs and outputs can be seen. (Grupo E-Nose, 2011, Tian et al., 2005).

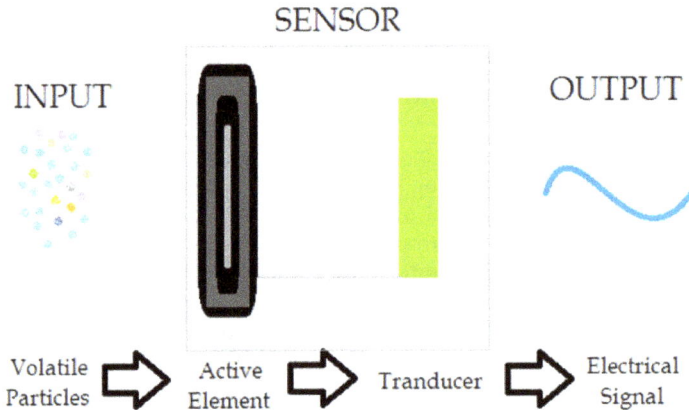

Fig. 1. Simplified schematic diagram of a gas sensor.

There are different types of gas sensors for use in EOS, the most common are: MOX (Metal Oxide Semiconductor), QCM (Quartz Crystal microbalance), SAW (Surface Acoustic Waves), MOSFET (Metal Oxide Semiconductor Field Effect Transistor), CP (Conducting Polymers), and FO (Fiber Optics). This chapter deals specifically with MOX sensors built with semiconductor materials such as Tin oxide ($SnO2$), Zinc oxide (ZnO), Titanium oxide ($TiO2$), among others. Their operating principle is based on the change of conductivity of a sensitive material when it absorbs or reacts with the gases in the environment, Figure 2 shows several commercial sensors of this type (Berna, 2010).

Fig. 2. Commercial gas sensors manufactured by Figaro and FIS, with different sizes and pin configuration.[2]

The majority of gas sensors are general purpose and usually have high sensitivity, detecting very low concentrations of volatile, but have disadvantages when trying to determine concentrations of a single component, because the output signal cannot be unambiguously assigned to the component by its generality (Duran, 2005, 2009).

Due to the fact that all EOS have a gas sensor array, it is desirable that the array be located in a special chamber or compartment in which the right conditions can be ensured for the proper operation. Mainly adequate insulation must be ensured to prevent pollutants from entering and the appropriate temperature and pressure must be maintained, these parameters are important or critical depending on the type of sensor used (Duran, 2005). Another advantage of using a chamber of sensors is that it facilitates the measurement process, because the volatiles will be in a higher concentration and they will have more contact with the active element of the sensor, which enables better and faster response from the sensors. It has also been experimentally determined that if the chamber of sensors is more hermetic, it can further exploit these advantages. Figure 3 shows a photograph of a chamber of sensors used in one of our projects with EOS (Velásquez et al., 2009).

2.1.2 Volatile delivery system
Basically it is a system that is responsible for transporting volatiles emitted by the samples or elements to be scanned into the chamber of sensor. Sometimes the sample is manually injected into the chamber of sensors, which results in error and delays; other times an automated system is responsible for transporting the volatile odorous molecules to the chamber of sensors, with the injection of a gas or air (Duran 2005, 2009).

---

[2] Images of sensors were taken from different Internet pages.

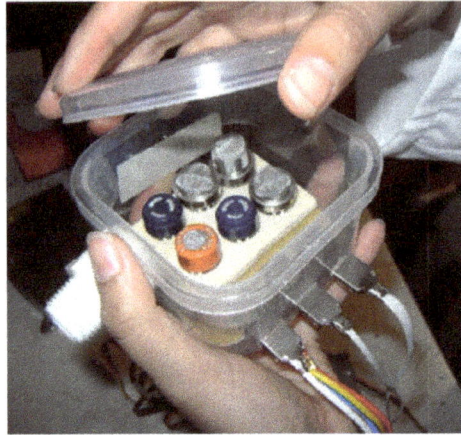

Fig. 3. The chamber of sensors provides hermetic isolation and guarantees reliable measurements.

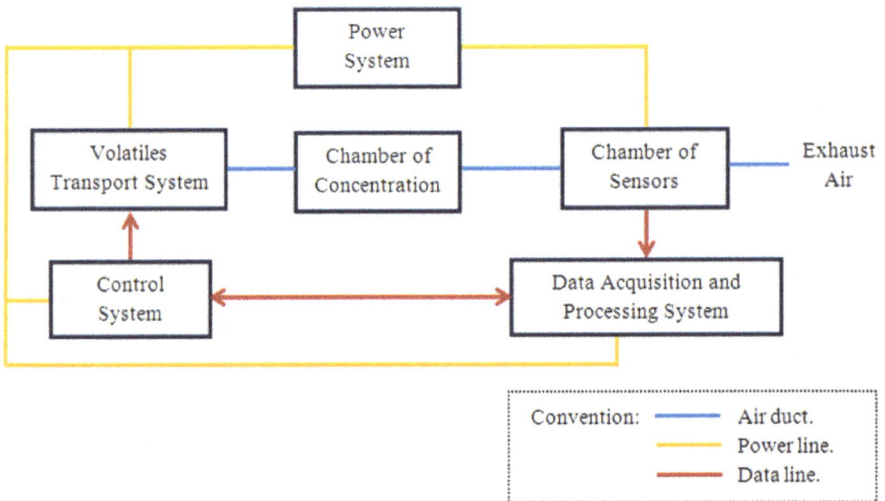

Fig. 4. Block diagram representing the electronic nose system.

Additionally most EOS have some kind of cleaning mechanism for the chamber of sensors, so that subsequent measures are made based on the same initial conditions and thus reproducibility of results is ensured. We recommend a different camera or hermetic compartment be used, called the "Chamber of Concentration", for containing the sample to be analyzed provided the environmental and physical conditions of the system allow it. Figure 4 shows the representation of an electronic nose system; note that the volatiles transport system is fundamental because it affects the operation of the EOS in the 3 different processes that can be carried out: concentration of volatiles, measurement and cleanup. (Rodriguez & Duran, 2008).

2.1.3 Control system and data acquisition

The control system takes care of proper handling of the Volatiles Transport System, for example: valves, air pump and other devices that are part of this system. It is also in charge of controlling additional subsystems or variables that the electronic nose system may have, such as temperature and humidity control, among others (Duran, 2005).

The data acquisition system is responsible for capturing the signals provided by the gas sensors and then delivering them to the process processing or computing system that has the appropriate software for processing such information (Rodriguez & Duran, 2008).

The control and data acquisition systems can be integrated into a single device, which can be a data acquisition card, a microcontroller, a DSP (Digital Signal Processor) or a computer; it must also have adequate power stage to handle the elements that consume more power and must have the proper memory settings to store large amounts of data obtained from the sensors.

We recommend working with a data acquisition card connected to a computer, to achieve good storage capacity, correct handling of information processing and graphical representation. Although in some cases when portability is required, a DSP or microcontroller can be used.

A significant part of the control system is the power source, which must be of a few amps, depending on the number of gas sensors and additional elements used; a source of 3 Amps is enough when working with an EOS that contains an array of 8 gas sensors.

2.1.4 Processing system

The processing system in most cases consists of a computer with an appropriate software for manipulating the data obtained by the sensors. Pre-processing techniques are applied to the data in order to extract the static parameters of the measures and reduce the amount of information to be analyzed. Subsequently multivariate analysis techniques and pattern recognition can be applied, such as PCA (Principal Component Analysis) and ANN (Artificial Neural Networks) to perform tasks such as: classification, discrimination, prediction, quantification of samples according to their organoleptic characteristics (Wilson & Baietto, 2009; Berna, 2010).

2.2 Operation of an electronic nose system

The operation of an electronic nose system depends on the component parts and the features of the equipment. In order to obtain measurements with a EOS the first step is to adjust the adequacy of the sample to be examined, this depends on the type of element to be analyzed, which sometimes must be heated, cut, mixed with other elements and simply placed near the sensors array or in the chamber of concentration ready to be analyzed. (Duran & Baldovino, 2009).

The concentration process begins when the sample is placed in the chamber of concentration. After this a few minutes should be given to allow the sample to release enough volatile particles, only then can the measurement process begin, for which the volatiles must be deposited or transported from then chamber of concentration to the chamber of sensors. During the measurement process, the data acquisition system records all the changes in the output signal of each of the gas sensors. When the measurement process is finished the cleaning process of chamber of sensors begins and the stored data can be processed and analyzed immediately (off-line processing), using the pre-processing software and signal processing, in order to obtain an olfactory footprint that represents the sample, to perform the tasks of classification, discrimination and other (Berna, 2010; Wilson & Baietto, 2009).

3. Quality control of food using electronic nose systems

A great part of electronic nose system applications are used in the food industry, where studies can be found with meat, milk and dairy products, eggs, different grains, fruits, oils, alcoholic and non alcoholic beverages, among others (Berna, 2010).

a)   b)

Fig. 5. a) Growing Coffee.[3]. b) Image of green coffee beans.[4]

Food quality control is one of the many applications that can benefit from the use of electronic nose (EOS). E.g., it can determine the type of product that is being analyzed, it can be classified by region, quality, time of ripening or storage, the food life span can be determined or predicted, as well as the level of deterioration or decomposition, the food life span can be determined or predicted and can determine flavors (Berna, 2010; El Barbri et al., 2008).

This chapter discusses the use of EOS in the quality control of foodstuffs such as coffee, fruits and alcoholic beverages.

3.1 Quality control of coffee with an electronic nose system

In the quality control of coffee, the organoleptic characteristics are a determinant of its quality and therefore, they are significant to locating the predominant defects of coffee beans, as they negatively affect its flavor and odor (Rodriguez et al., 2010; Pardo et al., 2000). It is important to keep in mind that coffee production (Fig. 5. a.) is such an artisan process, that its control is somewhat complex and highly dependent on the historical traditions and cultural knowledge of those involved in the process, the lack of modernization of coffee farms, the incidence of fungal and other diseases in the crops and the need for chemicals sometimes influences the product quality (Rodriguez et al., 2010).

Another important factor to take into consideration in the quality of coffee is the climatic and edaphological conditions or nature of the soil. The Colombian coffee zone is located on

---

[3] Photo owned by FNC, by Patricia Rincon Mautner. http://www.colombia.travel/es/turista-internacional/actividad/590-clima-y-ubicacion-geografica-del-cafe

[4] Taken from the website of Herbolario Esencia.
http://herboesencia.es/a-e/cafe-verde-coffea-arabica/

hillsides between 1000 m and 2000 m above sea level, with temperatures between 17°C (290°K) and 23°C (296°K) and relative humidities between 70% and 85%. The Table 1, show other data associated with optimal climatic conditions for growing coffee ([CENICAFE], 2011).

Average Solar Radiation	Solar Brightness	Temperature	Rainfall	Relative Humidity	Daily evaporation	Winds
Between 300 and 450 cal/cm ^{2} per day.	Between 4 to 5 hours daily.	Between 17 and 23 ° C or 290°K and 296°K	Between 1800 and 2800 mm annually.	Between 70 and 85%	Between 3 to 4 mm.	Below 5 km/h.

Table 1. Average Climatic conditions in coffee growing regions ([CENICAFE], 2011).

The CENICAFE web page (2011) states that: "The soils of the Colombian coffee region are relatively young, e.g. they are still under development and the nature of the material which is derived from petrographic material is grouped into the following classes: Metamorphic, igneous and sedimentary, which occur on different levels and patterns of coverage with volcanic ash. These soils are highly variable in their characteristics and their distribution throughout the coffee zone, its location in reliefs from flat or gently undulating to steep with 75% slope, and the variety of their physical (from rocky and sandy to loam and clay) and chemical conditions (low to high content of organic matter and minerals)".

3.1.1 What should be taken into account when considering a product such as coffee?
Coffee is a product that is collected manually and subjected at certain processes before obtaining the green coffee beans (Fig. 5. b.), in this condition it is more difficult to make an organoleptic analysis of coffee; therefore the best way to analyze coffee is the same way as tasters do, who perform tests on toasted and ground coffee, therefore the coffee must be subjected to a process of roasting and grinding in order to obtain a powder which is mixed with water at an average temperature of 60°C (333°K), which enables the emission of volatile particles. This mixture is introduced into the chamber of concentration (Fig. 6) in order to cluster the volatile particles which are then carried to the chamber of sensors for the measurement process. Figure 7 shows in detail the procedure used for the preparation of the mixture before the measurements (Falasconi et al., 2005).

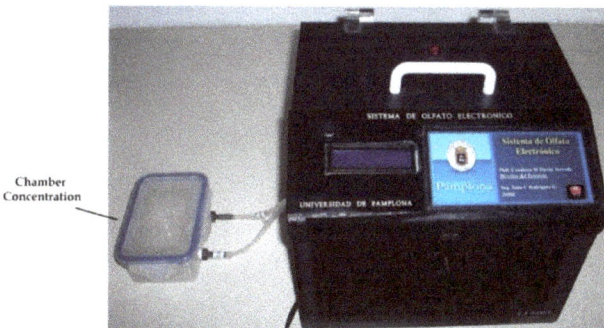

Fig. 6. Chamber of concentration, container for the different samples to be analyzed.

3.1.2 Some results obtained with the coffee

There have been several tests of different varieties of export quality "Excelso" coffee, with "regular" coffee and coffee with marked defects in the grains. Tests have been accompanied by personnel trained in coffee tasting, who issued their concept based on their personal perception of each coffee sample, helping in the designation of various patterns for facilitating subsequent classification tasks with different measures (Rodriguez et al., 2010).

In one of the tests measures were taken from samples of export quality coffee of two different varieties, Excelso Europe and Excelso UGQ-10%, which are classified as good quality cafes. These measurements were compared to those of regular coffee (for domestic consumption, commonly known as "Pasillas"), in which the experts detected some defects such as traces of fermentation, chemical contamination and signs of "Repose" (caused by prolonged storage or storage in unfavorable conditions). It is noteworthy that the defects mentioned are those most commonly found in coffee, influenced by poor product handling techniques (Rodriguez et al., 2010).

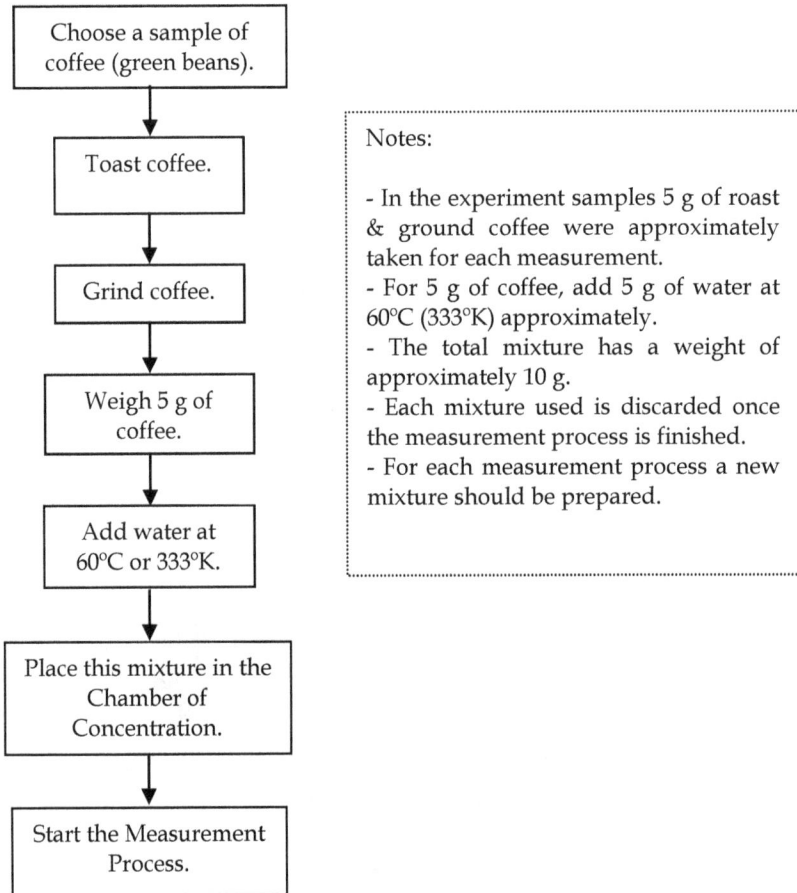

Notes:

- In the experiment samples 5 g of roast & ground coffee were approximately taken for each measurement.
- For 5 g of coffee, add 5 g of water at 60°C (333°K) approximately.
- The total mixture has a weight of approximately 10 g.
- Each mixture used is discarded once the measurement process is finished.
- For each measurement process a new mixture should be prepared.

Fig. 7. Procedure used for preparing samples of coffee before starting the measurement process.

Figure 8 shows the analysis of these measurements, using the technique PCA (Principal Component Analysis). The different measurement groups can be seen, clearly differentiated in samples of regular and export type coffee. The measurements taken from export quality coffee are highlighted in green circles, while the measurements taken from regular coffee are in red circles.

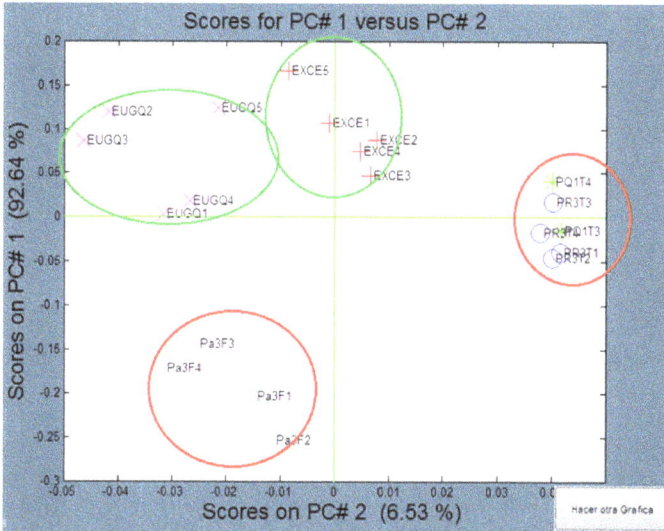

Fig. 8. Results of PCA analysis between measures of good quality coffee (green circles) and coffee with defects (red circles).

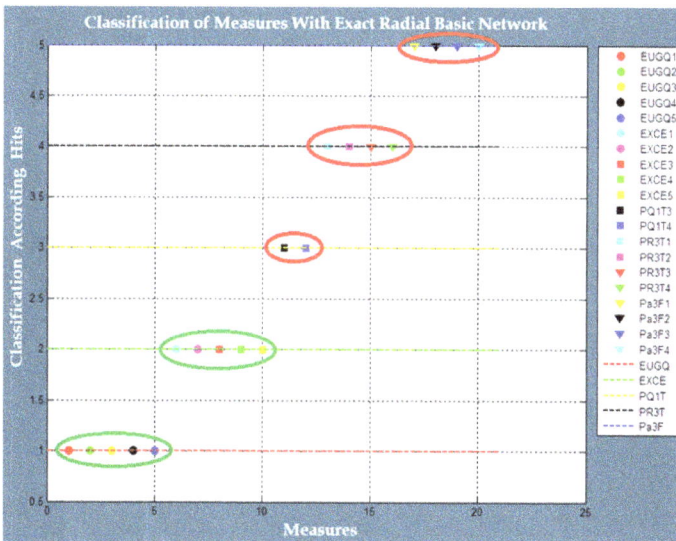

Fig. 9. Classification results of the measurements with a radial basis neural network, between good quality coffee (green circles) and coffee with defects (red circles).

This group of measurements was classified using a radial basis neural network (Figure 9). It can be seen how the various measurements of the same type are located within a horizontal axis, forming 5 different subgroups, which belong to two major groups of export type coffee (green circles) and regular coffee (red circles).

3.2 Quality control of fruits

For the analysis of fruits invasive and noninvasive techniques can be used. Invasive techniques involve damaging the fruit to take a sample, in order to perform various tests with the same fruit at the same moment and also facilitate extraction of volatile particles, as manipulation helps to release more volatile particles, which facilitates the measurement process. A drawback of this technique is that once the product is handled it can only serve in the measurement process, because handling accelerates the decomposition process. Meanwhile, the noninvasive techniques, just take the fruit for testing without inflicting damage therefore the same fruit can be used for further testing in order to analyze maturity stages and study the processes of decomposition. (Rodriguez & Duran, 2008; Duran & Baldovino, 2009; Berna, 2010).

Figure 10 shows the results of the analysis of some measurements made on samples of passion fruit, peaches and apples, using the PCA technique. The 2 measurement groups can be seen, which can be clearly differentiated in samples of passion fruit and peaches, in addition 2 measurements of apple were introduced as a test (Creole apple and Chilean apple), in order to test the classification accuracy of the system and the similarity that may exist between different varieties of a fruit. Also Figure 11 shows the validation of the measurements using an Artificial Neural Network Feed Forward Back Propagation, applying the technique "Leave one out", it can be seen how the system responds to the eventual absence of a measure in the training of neural network, the most significant result occurs with measurements of apples which are classified successfully despite having so few measures.

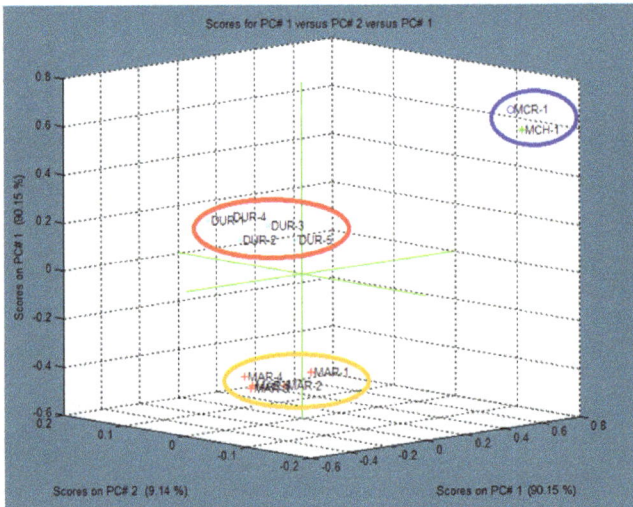

Fig. 10. Results of PCA analysis between passion fruit (yellow circle), peaches (red circle) and apples (blue circle).

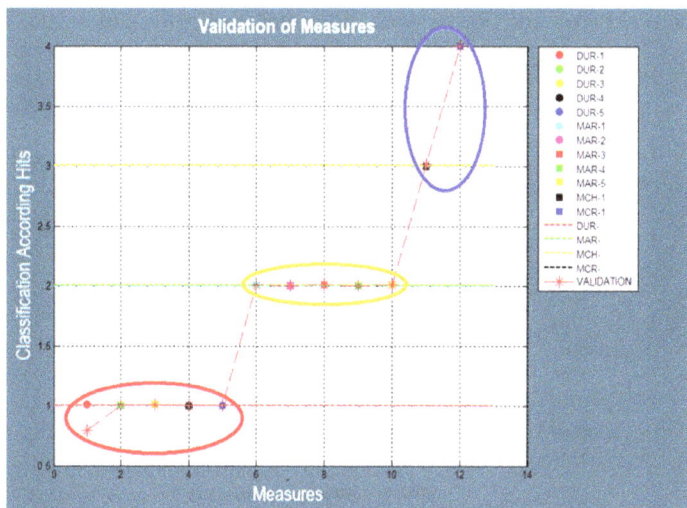

Fig. 11. Validation Results of measurements with passion fruit (yellow circle), peaches (red circle) and apples (blue circle) using a neural network Feed Forward Back Propagation.

3.3 Quality control of alcoholic beverages

Electronic nose systems have been widely used for classification, discrimination of characteristics and detection of different elements or compounds considering the organoleptic characteristics, but its application in quantification tasks has not been widely explored. In some of these studies the least square regression method is used to consider the gas concentration (Khalaf et al., 2009) and for the quantification of mixed contaminants in the air (Zhou et al., 2006), also the new technologies have been used as systems based on micro-electromechanical sensors for the quantification of components in vapor mixtures (Zhao et al, 2007).

Below is a study with an electronic nose system, where a digital signal processor DSP was adapted and artificial neural network "Feed-forward back propagation" was implemented, which was trained with the aim of identifying and quantifying levels of Ethanol and Methanol in different samples. As result the percentage of Ethanol and Methanol of the samples were obtained, and the electronic nose system was improved, called "A-NOSE" (Rodriguez et al., 2010), when the processing software was implemented in a different device from the personal computer.

The artificial neural network that was used to perform the identification and quantification of Ethanol and Methanol was Multi Layer Perceptron (MLP) Feed-forward back propagation network, which was trained and tried in R2006a Matlab software; as a result of training of the artificial neuronal network the weight matrices and bias vectors were obtained, that were used to codify the artificial neural network program in C++ language with software CodeWarrior and subsequently downloaded this program in the digital signal processor DSP56F801 of Motorola.

The initial samples were 95% Ethanol and 95% Methanol, which were diluted with distilled water to obtain 50%, 25% and 10% concentrations. Different measurements with the Ethanol and Methanol were realized in their different concentrations to realize the training of the artificial neuronal network and additionally other measures with wines (red, white, fruity, orange wine) and Aguardiente (national drink) were performed. The percentages of wine were close to 10% Ethanol and 0% Methanol and Aguardiente was close to 30% Ethanol and 0% Methanol, values in accordance to the values specified on the product labels.

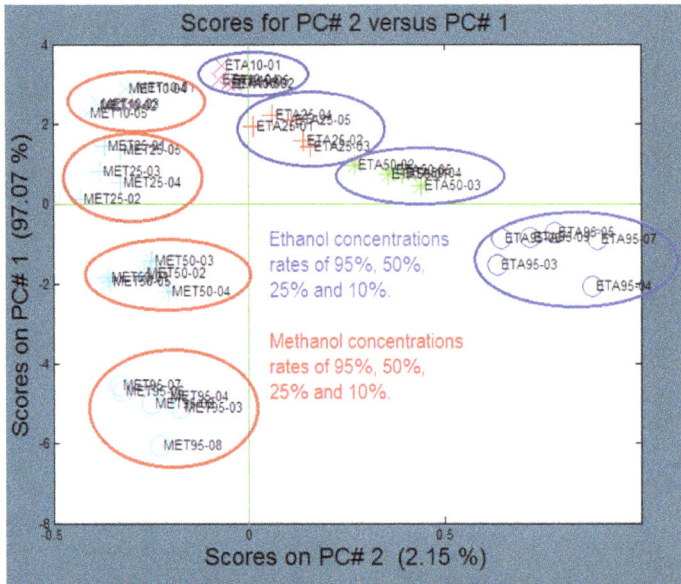

Fig. 12. Results of PCA analysis between Ethanol measurements (blue circles) and Methanol measurements (red circles).

Figure 12 the results of PCA analysis can be seen applied to measurements of different ethanol and methanol concentrations. It can be inferred that the measurements follow a trend, which yield a characteristic equation that models the behavior for different concentrations of Ethanol and Methanol. It can also be seen that as the concentration of Ethanol and Methanol is lower measurements tend to find a common point, this may be because they have are both alcohol.

Another test analyzed samples of different kinds of wines (e.g.: red wine, white wine, orange wine) and Aguardiente (national drink), the results are shown in Figures 13 and 14. It can be seen that wine measurements are close to 10% Ethanol, the results obtained by the neural network (feed-forward back propagation) that was trained for this purpose showed results very close to 10% and 0% Ethanol Methanol.   It should be clarified that the neural network was trained with data from measurements of different ethanol and methanol concentrations and were then tested with data from measurements of different drinks.

4. Other applications of electronic nose systems

The applications of electronic nose systems are very diverse. The previous sections have covered some of the possible applications in the agro-food industry, but there are many other still to be mentioned, for example: The identification and diagnosis of respiratory

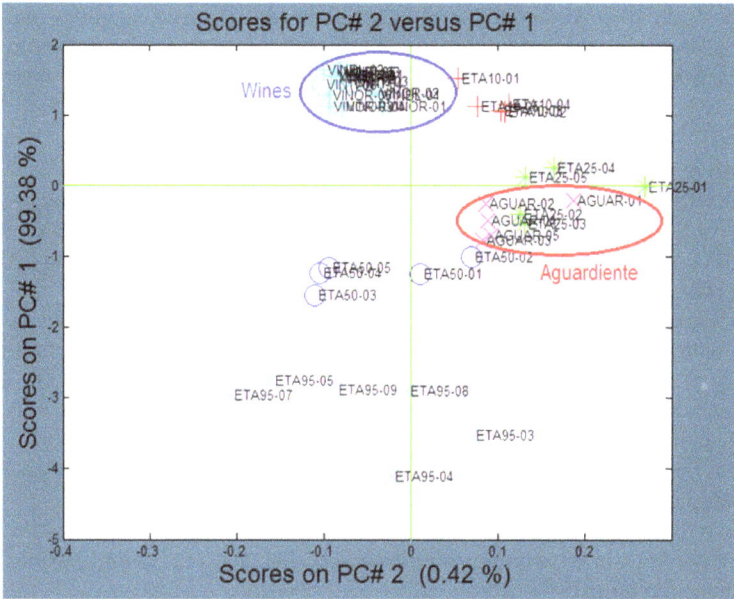

Fig. 13. Results of PCA analysis for the classification of wines and Aguardiente, according to the concentration of Ethanol.

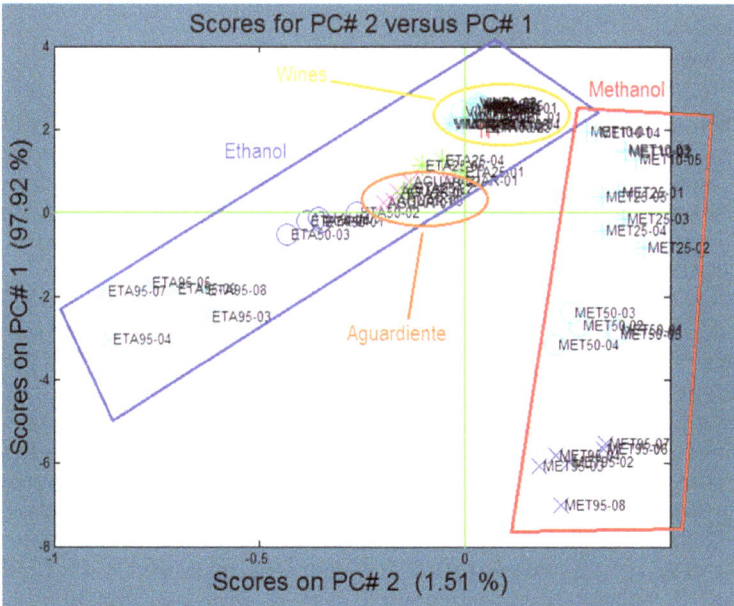

Fig. 14. Results of PCA analysis for the classification of wines and Aguardiente, according to the concentration of Ethanol and Methanol.

diseases (Xu et al., 2008, Velasquez et al ., 2009), the detection of narcotics and explosive substances (Oakes, L.; Dobrokhotov, V., 2010), determination of air quality and the environment (Zhou et al., 2006), among others, although there are still many possible applications to be explored.

4.1 Detection of diseases using electronic nose systems

There are a variety of respiratory diseases, which in some cases are caused by smoking and exposure to contaminated environments. This is the case of Chronic Obstructive Pulmonary Disease COPD, which has a mortality rate exceeding 15.9% (Velásquez et al., 2009; Velásquez, 2008).

COPD is a chronic lung disease characterized by airflow limitation that is not fully reversible, with progressive deterioration and is associated with abnormal lung inflammatory response to noxious particles or gases (Velásquez et al., 2009; Velásquez, 2008). The main cause of COPD is prolonged consumption of cigarettes, it is said that up to 20% of smokers have COPD.

This disease is more common in:
- White people.
- People over 60 years of age.
- People who work in environments polluted by chemical vapors and harmful dust that can damage the lungs.
- People who suffer from chronic asthma.
- People with a family history of emphysema.

Other COPD risk factors include:
- Passive smoking.
- Air pollution.
- Low birth weight and other lung infections (Velásquez, 2008).

4.1.1 Analysis of measurements taken from people with COPD and healthy controls

Below are some results of the analysis of measurements taken from healthy controls, nonsmokers and patients diagnosed with COPD. All patients diagnosed with COPD were long time smokers from 16 years up to even 50 years and most of them have already quit smoking, due to the fact that many receive medical treatment. (Velásquez et al., 2009; Velásquez, 2008).

Figure 15 shows the results of PCA analysis of samples of healthy controls and patients with COPD. The low dispersion of measurements of healthy controls and high dispersion of measurements of patients with COPD can be seen; due the fact that to not all patients have the disease at the same level.

Figure 16 has separated the samples from different patients with COPD, from these results it can be inferred that according to the health of the person the different patients can be classified. Future research should conduct more measurements on patients at different stages of the disease and could also be extended to other respiratory diseases and even gastric related diseases.

A study that deserves attention is (Xu et al., 2008) who developed a solid trap/thermal desorption-based odorant gas condensation system designed and implemented for measuring low concentration odorant gas. The results showed that the technique was successfully applied to a medical electronic nose system. The developed system consists of a flow control unit, a temperature control unit and a sorbent tube. The theoretical analysis and experimental results indicate that gas condensation, together with the medical electronic nose system can significantly reduce the detection limit of the nose system and increase the system's ability to distinguish low concentration gas samples.

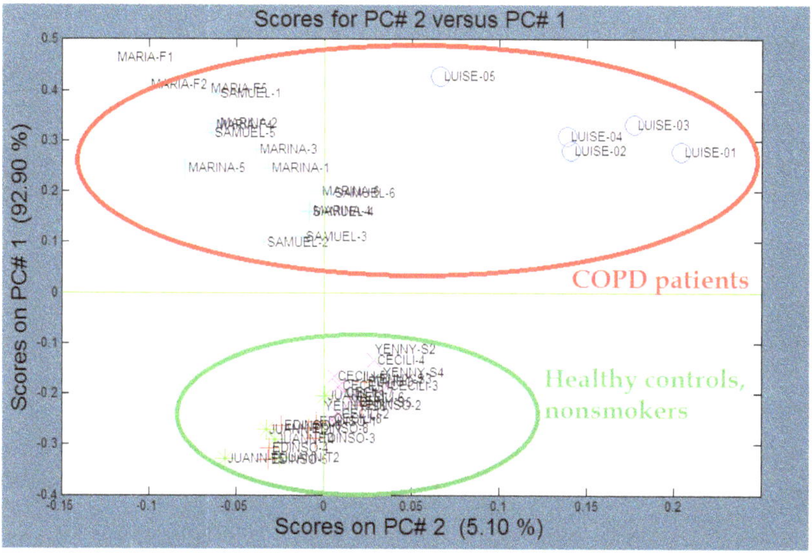

Fig. 15. Results of PCA analysis for the classification of COPD patients and non-smokers.

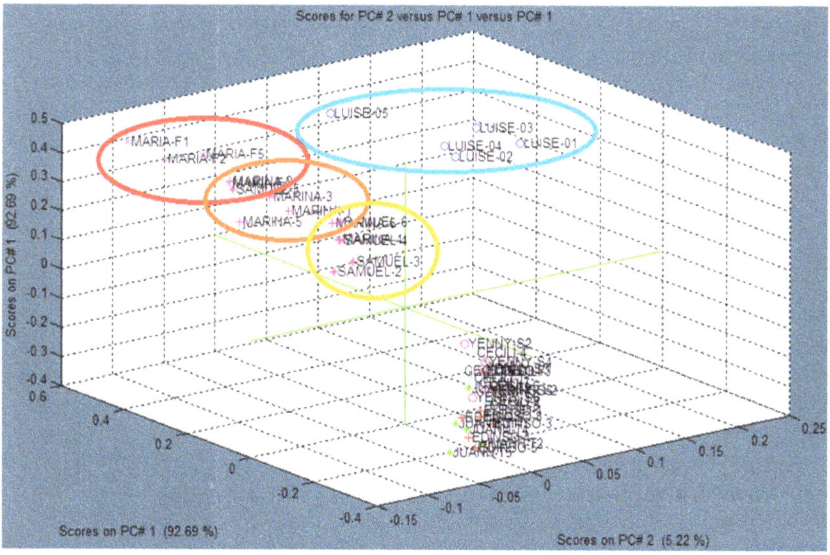

Fig. 16. Results of PCA analysis with emphasis on measurements of patients with COPD. Using DGN (Center), displaying the first 2 components.

4.2 Determination of air quality using electronic nose systems

Such applications have a bright future in the industry, because the environment is very susceptible to leakage and contamination by gases, which in many cases can be harmful and even lethal to humans.

NASA has done some work on this issue, for example (Ryan et al., 2009), Whom Developed an Electronic Nose to be used in Environmental Monitoring in the International Space Station, the Electronic Nose (Enose) is an array of 32 polymer sensors, the pattern of response may identify contaminants in the environment. An engineering test model of the ENose was used to monitor the air of the Early Human Test experiment at Johnson Space Center for 49 days. Examination of the data recorded by the ENose shows that major excursions in the resistance recorded in the sensor array may be correlated with events recorded in the Test Logs of the Test Chamber. The ability to monitor the constituents of breathing air in a closed chamber in which air is recycled is important to NASA for use in closed environments such as the space shuttle and the space station.

In the same way an electronic nose system could be developed for places such as airports or customs, in order to detect narcotics or prohibited hallucinogenic substances and in hostile or war environment to detect explosives or mines planted in the soil.

5. Conclusions

The operation of the electronic nose system depends on the component parts and the features of the equipment. Inside we find the gas sensor array, the volatile particle delivery system, control system, data acquisition and data processing system.

We recommend a different chamber or hermetic compartment be used for containing the sample to be analyzed, called "Chamber of Concentration", provided the environmental and physical conditions of the system allow it.

The volatile particle transport system is fundamental because it affects the operation of the electronic nose system in the 3 different processes: concentration of volatile particles, measurement and cleanup.

Measurements with electronic nose systems begin by ensuring the adequacy of the sample to be examined, this depends on the type of element to be analyzed, which sometimes must be heated, cut, mixed with other elements or simply placed near the sensor array or in the chamber of concentration.

During the measurement process, the data acquisition system records all the changes in the output signal of each of the gas sensors. When the measurement process is finished the cleaning of the chamber of sensors begins, which is very important to restore the initial conditions of the system and to ensure the reproducibility of the measurements.

Once the measurement process is finished the stored data is processed and analyzed using the pre-processing software which allows to extraction of static parameters from the measurements and reduces the amount of information to be analyzed. Subsequently the processing software is applied, in order to obtain an olfactory footprint that represents the sample, to perform classification, discrimination and other tasks.

Coffee is preferably analyzed in the same way as by tasters, who perform tests on toasted and ground coffee, therefore the coffee must be roasted and ground in order to obtain a powder which is mixed with hot water, to facilitate the emission of volatile particles and this mixture is introduced into the chamber of concentration for the measurement process. This procedure for the preparation of the mixture can be applied similarly to other elements before the start of the measurements.

To identify measurement patterns or to carry out the training applications using computational intelligence it is very important to have expert staff on hand, as in the case of coffee quality control which had the support of trained coffee tasters, who issued their concept based on personal perception of each coffee sample, helping in the designation of the various patterns to facilitate subsequent classification tasks with different measurements.

Electronic nose systems have been widely used for classification, discrimination of characteristics and detection of different elements or compounds considering the organoleptic characteristics, and even for quantification tasks. This can be carried out with tools like multivariate analysis techniques and pattern recognition, such as PCA (Principal Component Analysis) and ANN (Artificial Neural Networks).

The processing software of electronic nose system can be implemented on a digital signal processor DSP using an artificial neural network like the alcohol research case presented in section 3 which used a Feed-forward back propagation network, which was trained with the aim of identifying and quantifying Ethanol and Methanol of different samples. As result the percentage of Ethanol and Methanol of the samples were obtained.

The artificial neural network that was used for the identification and quantification of Ethanol and Methanol was trained and tried using R2006a Matlab software; the training results were used to codify the the artificial neural network program in C++ language and subsequently downloaded this program in the digital signal processor DSP56F801 of Motorola.

The results of PCA analysis for samples of healthy controls and patients with COPD showed differences in the low dispersion of the measurements taken of healthy controls and high dispersion of the measurements taken of patients with COPD; due to the fact that not all patients have the disease at the same level.

6. Acknowledgements

This study was supported by the P09225 grant and carried out within the MIRP Research Group, Research Center, Instituto Tecnológico Metropolitano ITM, Medellín–Colombia.

7. References

Badrick, T. (2008). The Quality Control System. The Clinical Biochemist – Reviews, (August,2008), p.p. 67-70, ISSN 0159 – 8090.

Berna, A. (2010). Metal Oxide Sensors for Electronic Noses and Their Application to Food Analysis. Sensors, Vol.10, (April 2010), pp. 3882-3910, ISSN 1424-8220.

[CENICAFE] Centro Nacional de Investigaciones del Café Colombia (March 2011). Sistemas de Producción, 25.03.2011, Available from
http://cenicafe.org/modules.php?name=Sistemas_Produccion&file=condclim

Duran, C. & Baldovino, D. (2009). Monitoring System to Detect the Maturity of Agro-industrial Products Through of an Electronic Nose. Revista Colombiana de Tecnologías de Avanzada, Vol.1, No.13, (December 2009), pp. 1-8, ISSN 1692-7257.

Duran, C. (2005). Diseño y optimización de los subsistemas de un sistema de olfato electrónico para aplicaciones agroalimentarias e industriales. Universitat Rovira i Virgili. Tarragona, España.

El Barbri, N.; Llobet, E.; El Bari, N.; Correig, X. & Bouchikhi, B. (2008). Electronic Nose Based on Metal Oxide Semiconductor Sensors as an Alternative Technique for the

Spoilage Classification of Red Meat. Sensors, Vol.8, (January 2008), pp. 142-156, ISSN 1424-8220.

Falasconi, M.; Pardo, M.; Sberveglieri, G.; Ricco, I. & Bresciani, A. (2005). The novel EOS[835] electronic nose and data analysis for evaluating coffee ripening, Sensors and Actuators B: Chemical, Volume 110, Issue 1, (September 2005), p.p 73-80, ISSN 0925-4005.

Grupo E-Nose, (March 2011) ¿Qué es una Nariz Electrónica?, 15.04.2011, Available from http://www.e-nose.com.ar/paginas/funcionamiento.htm

Khalaf, W.; Pace, C. & Gaudioso, M. (2009). Least Square Regression Method for estimating gas concentration in an Electronic Nose System. Sensors, Vol. 9, pp. 1678-1691, ISSN 1424-8220.

Oakes, L.; Dobrokhotov, V. (2010). Electronic Nose for Detection of Explosives. American Physical Society. (March 2010).

Pardo, M.; Niederjaufner, G.; Benussi, G.; Comini, E.; Faglia, G.; Sberveglieri, G.; Holmberg, M. & Lundstrom, I. (2000). Data preprocessing enhances the classification of different brands of Espresso coffee with an electronic nose, Sensors and Actuators B: Chemical, Volume 69, Issue 3, (October 2000), p.p 397-403, ISSN 0925-4005.

Ryan, M.; Homer, M.; Buehler, M.; Manatt, K. & Zee, F. (2009). Monitoring the Air Quality in a Closed Chamber Using an Electronic Nose. Jet Propulsion Laboratory, California Institute of Technology. Johnson Space Center, NASA, Houston TX 77058.

Rodríguez, J.;   Durán, C.;   Reyes, A. (2010). Electronic Nose for Quality Control of Colombian Coffee through the Detection of Defects in "Cup Tests". Sensors, Vol.10, (December 2009), pp. 36-46, ISSN 1424-8220.

Rodriguez, J. & Duran, C. (2008). Electronic odor system to detect volatile compounds. Revista Colombiana de Tecnologías de Avanzada, Vol.2, No.12, pp. 20-26, ISSN 1692-7257.

Tian, F.; Yang, S. & Dong, K. (2005). Circuit and Noise Analysis of Odorant Gas Sensors in an E-Nose. Sensors, Vol.5, (February 2005), pp. 85-96, ISSN 1424-8220.

Velásquez, A.; Durán, C.; Gualdron, O.; Rodríguez, J. & Manjarres, L. (2009). Electronic Nose to Detect Patients with COPD From Exhaled Breath. Proceedings of the 13th International Symposium on Olfaction and Electronic Nose. AIP Conference Proceedings, Volume 1137, pp. 452-454, ISBN: 978-0-7354-0674-2, Brescia, Italy, April 15-17, 2009.

Velásquez, A. (2008). Sistema Multisensorial Electrónico No Invasivo Para La Detección De La Patología Respiratoria Epoc. Universidad de Pamplona. Pamplona, Norte de Santander, Colombia.

Wilson, A. & Baietto, M. (2009). Applications and Advances in Electronic-Nose Technologies. Sensors, Vol.9, (June 2009), pp. 5099-5148, ISSN 1424-8220.

Xu, X.; Tian, F.; Yang, S.; Jia, Q. & Ma, J. (2008). A Solid Trap and Thermal Desorption System with Application to a Medical Electronic Nose. Sensors, Vol.8, (November 2008), pp. 6885-6898, ISSN 1424-8220.

Zhao, W.; Pinnaduwagel, L.; Leis, J.W.; Gehl, A.C.; Allman, S.L.; Shepp,A. & Mahmud, K.K. (2007) Quantitative analysis of ternary vapor mixtures using a microcantilever-based electronic nose. Applied Physics Letters, 91 (4). ISSN 0003-6951.

Zhou, H.; Homer, M.; Shevade, A. & Ryan, M. (2006). Nonlinear Least-Squares Based Method for Identifying and Quantifying Single and Mixed Contaminants in Air with an Electronic Nose. Sensors, Vol.6, (December 2005), pp. 1-18, ISSN 1424-8220.

18

Mammographic Quality Control Using Digital Image Processing Techniques

Mouloud Adel and Monique Rasigni
Université Paul Cézanne, Institut Fresnel, UMR-CNRS 6133
Domaine Universitaire de Saint Jérôme,
France

1. Introduction

Breast cancer is the leading cause of cancer mortality among middle aged women. Survival and recovery depend on early diagnosis. At present mammography is one of the most reliable methods for early breast cancer detection. However relevance of diagnosis is highly correlated to image quality of the mammographic system. Hence periodic controls in mammographic facilities are necessary in order to make sure they work properly. In particular global image quality is evaluated from a mammographic phantom film. A phantom is an object with the same anatomic shape and radiological response as an average dense fleshed breast and in which are embedded structures that mimic clinically relevant features such as microcalcifications, nodules and fibrils. For each category of features, the targets have progressively smaller sizes and contrast so that the largest one is the most readily visible and the next is less visible and so on. Using a phantom makes it possible to free from the variable of differences in breast tissue positioning and level of compression from patient to patient. The process is as follows: the mammographic phantom film is analysed independently by several readers and a score is obtained by each of them depending on the number of objects they see. The independent object visibility scores are then averaged and the resulting score is assigned to the phantom film.

Automating this score by using computer image processing of digitized phantom films should make the evaluation of mammographic facilities easier and less subjective. In addition image processing should enable us to take into account other parameters such as, for instance, noise, texture and shape of the targets that a reader eye cannot estimate quantitatively, and so to perform a more elaborate analysis. In collaboration with ARCADES (Association pour la Recherche et le Dépistage des Cancers du Sein et du col de l'utérus) which set, since 1989, a breast cancer screening program in South of France, a project aimed at automating phantom film evaluation is in progress. Such a project consists first in digitizing phantom films with the adequate spatial resolution and then in processing the obtained images in order to detect, segment and characterize the objects contained in the phantom.

Little work has been done to automate quality control in mammographic facilities. Fast Fourier transform is used (Brooks et al., 1997) to establish some visibility criteria for the phantom test object. (Chakraborty et al., 1997) compares phantom images with a pattern image to obtain relations between some of the image parameters and the physical conditions in which the images have been obtained. His work concerned only microcalcifications. (Dougherty, 1998) studies the most prominent microcalcification group and the most prominent mass using a manual threshold and some mathematical morphology operators. (Castellano et al., 1998) used binary masks to locate microcalcifications and they studied image resolution scales contained in the phantom. (Blot et al., 2003) used grey level co-occurrence matrices to score structures embedded in the phantom. (Mayo et al., 2004) used region growing and morphological operators to segment and characterize microcalcifications. They also studied horizontal resolution areas using morphological operators. This chapter presents a feasibility study aiming at automating phantom scoring using image processing techniques on digitized phantom films.

In the following sections, we describe the phantom used, the mammographic phantom image acquisition and digitization, the noise reduction and the contrast enhancement schemes used for processing phantom images, the segmentation step for each object (microcalcifications, masses and fibres) and the results obtained on nine phantoms films.

2. Description, acquisition and digitization of phantom films

2.1 Phantom description

The phantom used in this study is the MTM 100/R (Computerized Imaging Reference Systems, Inc., 2428 Almeda Avenue Suite 212, Norfolk, VA 23513, U.S.A., Phantom Serial Number: 2788). The MTM 100/R is used in France for Mammography Accreditation. It is made of tissue equivalent material in which are embedded objects simulating 7 pentagonal-shaped groups of microcalcifications (M1 to M7), 7 masses (N1 to N7) and 7 fibres (F1 to F7). For each category of features, the targets have progressively smaller size and contrast so that the largest one is the most readily visible, the next is less visible and so on. For convenience M1 stands for the microcalcification group containing the largest specks and M7, the smallest ones, with a similar convention for the other target structures. Inside are also present vertical (V) and horizontal (H) spatial resolution scales (line pair target : 20 lp/mm), a delimited zone (Z) for measuring a reference optical density, two different optical density contiguous areas (C1 and C2) for defining contrast, three cavities (D) for x-ray dose measurement and at last small balls (B) for x-ray alignment control. Figure 1 shows a schematic diagram of the MTM 100/R phantom.

2.2 Acquisition and digitization of phantom films

Phantom films are digitized with an ultra high resolution drum scanner (Scanmate 11000-ScanView A/S.Meterbuen 6. DK-2740 Skovlunde. Denmark) which may digitize from 50 to 11000 dpi (dots per inch) and code images on 256 (8bits / pixel) or 16384 (14bits / pixel) grey levels. Spatial scanning resolution was chosen so that it approximately corresponds to the resolving power of a viewer (with a standard resolving power $\sim 4.10^{-3}$rd) using a twice magnifying lens (such lenses are often used by radiologists for reading clinical mammograms). So phantom films were digitized with a resolution of 50 µm per pixel (or 508 dpi) and were coded on 256 grey levels.

For each category of structures (microcalcifications, masses and fibers) a sub-image was extracted from the digitized phantom image so that each sub-image contained one target and was roughly centered on it. Subimages sizes were 256×256-pixels for microcalcification groups and masses and 412×412-pixels for fibres. Fig. 2 shows an example of subimages extracted from a digitized phantom film.

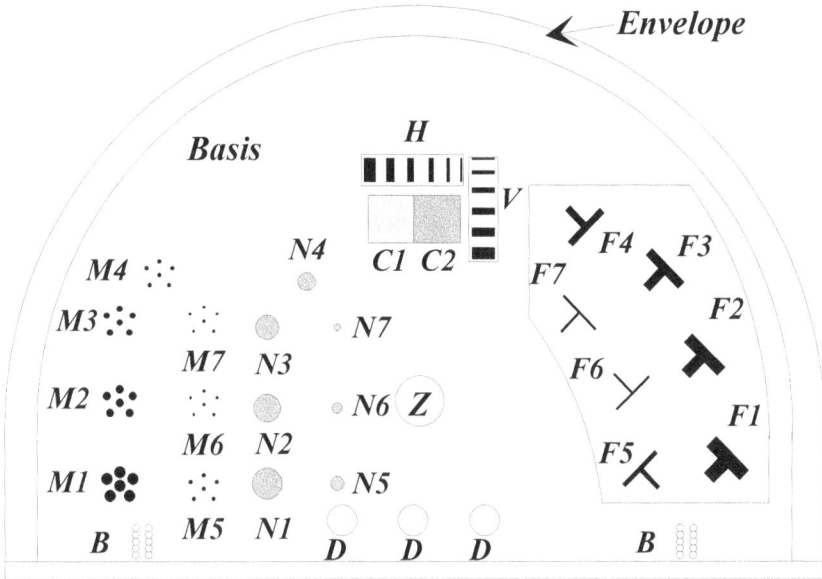

Fig. 1. Schematic diagram of the phantom MTM 100/R showing the locations and the relative size of features.

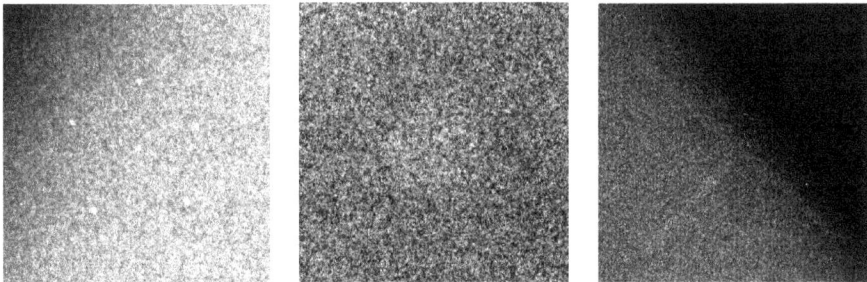

Fig. 2. Subimages extracted from a digitized phantom film. (a) : group of microcalcifications (b) : mass; (c) : fibre

3. Image processing of digitized phantom films

3.1 General description

Two pre-processing steps are applied to each extracted subimage before the segmentation step as shown in Fig. 3. Because of the noisy nature of these subimages, a noise reduction method is used as a first processing step. A contrast enhancement step is then applied. At last image segmentation is done.

3.2 Local contrast modification method description

In classical image processing techniques, a fixed shape and a fixed size window around each pixel is used in order to convolve it with a defined filter. In order to take the local features around each pixel into account, a variable shape and size neighbourhood is defined (Dhawan et al., 1986; Dhawan & Le Royer, 1988) using local statistics.

Noise reduction step consist in filtering two kinds of noise. The non uniform background considered as a "low frequency noise" and the radiographic noise (high frequency noise): film granularity and quantum mottle. Shadow correction of the background is adapted to each object (microcalcifications, masse and fibre), whereas the radiographic noise filter and the contrast enhancement steps are based on the same method, described in the next section. This method consists in computing a local contrast around each pixel using a variable neighbourhood whose size and shape depend on the statistical properties around the given pixel. The obtained image is then transformed into a new contrast image using various contrast modification functions. At last an inverse contrast transform is applied on the new contrast image to yield an enhanced version of the original image. Contrast enhancement step consists in enhancing image features while preserving details for the segmentation step. Image Segmentation is adapted to the objects to be detected and is presented in the following sections.

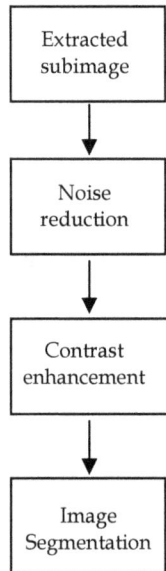

Fig. 3. Flowchart of the image processing steps applied to phantom images.

Each pixel (i,j) is assigned an upper window W_{max} centered on it, whose size is $(2N_{max}+1)\times(2N_{max}+1)$. We also define an inner area around the pixel (i,j) whose size is $(c\times c)$ and an external area whose size is $(c+2) \times (c+2)$, where c is an odd number. Let $I(i,j)$ be the grey level of pixel (i,j) in image I, and T a given threshold. Pixel (k,l) within W_{max} is assigned a binary mask value "0" if $|I(k,l) -I(i,j)| > T$, else it is assigned a binary mask value "1". Then the percentage P_0 of zeros is computed over the region between the external $(c+2)\times(c+2)$ and the inner $(c\times c)$ areas, for each c in the range $[1, 3, 5, ... ,2N_{max} - 1]$. The process stops if this percentage is greater than 60% or if the upper window W_{max} is reached. The value of 60% has been chosen because beyond this limit, we may consider too many pixels "0" are surrounding the inner area and so the notion of neighbourhood with the central pixel (i,j) in terms of grey levels is no longer satisfactory. Let c_0 be the upper c value beyond which the percentage P_0 is greater than 60%. The pixel (i,j) is assigned the window $W=(c_0+2)\times(c_0+2)$. In the window W such as $W \leq W_{max}$ we finally define the "center" as the set of pixels having the mask value "1", and the "background" as the set of pixels having both the mask value "0" and which are 8-neighbourhood connected at least to a pixel "1". Pixels "0" which do not verify the previous constraint belong neither to the "center" nor to the "background" and are not taken into account later on. Fig. 4 gives an example the way the "center" and the "background" areas are determined.

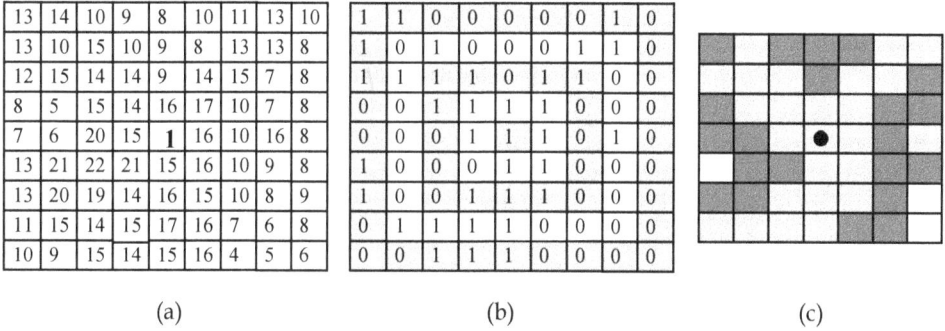

13	14	10	9	8	10	11	13	10
13	10	15	10	9	8	13	13	8
12	15	14	14	9	14	15	7	8
8	5	15	14	16	17	10	7	8
7	6	20	15	**1**	16	10	16	8
13	21	22	21	15	16	10	9	8
13	20	19	14	16	15	10	8	9
11	15	14	15	17	16	7	6	8
10	9	15	14	15	16	4	5	6

1	1	0	0	0	0	0	1	0
1	0	1	0	0	0	1	1	0
1	1	1	1	0	1	1	0	0
0	0	1	1	1	1	0	0	0
0	0	0	1	1	1	0	1	0
1	0	0	0	1	1	0	0	0
1	0	0	1	1	1	0	0	0
0	1	1	1	1	0	0	0	0
0	0	1	1	1	0	0	0	0

(a)　　　　　　　　　　(b)　　　　　　　　　　(c)

Fig. 4. Adaptive neighbourhood selection with a threshold value T=5; (a) W_{max} window around bold-faced pixel value 15, N_{max}=4; (b) Mask values associated to test pattern in (a); (c) The "center" (light grey)and the "background" (dark grey) areas around the bold-faced pixel. White cells correspond to pixels belonging neither to the "center" nor to the "background".

After determining the "center" and the "background" regions around each pixel (i,j), a local contrast image C is computed from :

$$C(i,j) = \frac{|M_c(i,j) - M_b(i,j)|}{\max(M_c(i,j), M_b(i,j))} \qquad (1)$$

where $M_c(i,j)$ and $M_b(i,j)$ are the mean values, in image I, of pixels labelled as the "center" and as the "background" regions around pixel (i,j) respectively. Note that C(i,j) is within the range [0,1].

The local contrast image C obtained in the previous step is transformed into a new image C' such as $C'(i,j)=\psi(C(i,j))$, where ψ is a contrast modification function depending on features to be detected. This function meets some requirements in the interval [0,1] :
- $\psi(0) = 0$ and $\psi(1) = 1$.
- $\psi(x) \geq 0$ for $x \in [0,1]$.
- ψ is an increasing function in the range [0,1].

In image C' each pixel value is a contrast value. In order to obtain the corresponding image in the grey level domain, an inverse contrast transform of the process used to obtain image C from I (Eq. 1) is used as follows:

$$E(i,j) = M_b(i,j)(1 - C'(i,j)) \text{ if } M_b(i,j) \geq M_c(i,j)$$

$$E(i,j) = \frac{M_b(i,j)}{(1 - C'(i,j))} \text{ if } M_b(i,j) < M_c(i,j)$$

This transform gives a new image E which is an enhanced version of image I. It is then possible to evaluate the efficiency of the method from comparison between images E and I.

3.2.1 Performance evaluation on simulated images

Several functions including square root, exponential, polynomial and trigonometric were tested (Guis et al., 2003). Actually functions which are over the line y=x increase the contrast but enhance the noise too. In the other hand, functions which are under the line y=x yield noise reduction. Because of the noisy nature of real images of phantom, the second kind of functions was chosen for enhancing the objects contained in these images. To choose suitable function ψ, computer simulated images containing objects similar to those observed in the phantom film were generated with various contrast and noise levels. The aim of this simulation was to perform a quantitative evaluation of the noise reduction method described in previous sections. For each target, 6 noise-free images were generated each of them with a different contrast level. Three noise levels were then assigned to each contrast level image. Contrast level is defined as the difference between the mean grey level of the object and the mean grey level of the background divided by the mean grey level of the background. According to studies on radiographic noise, two types of noise sources, namely film granularity and quantum mottle, are present in an X-ray image. Spatially correlated Poisson noise model has to be considered in the case of mammographic films. In our simulations, a signal-dependent spatially uncorrelated Gaussian noise is used as a first-order approximation of the Poisson noise model (Quian et al., 1994; Aghadasi et al., 1992; Kuan et al., 1985) namely: $n(i,j) = \sqrt{f(i,j)}u(i,j)$, where f is the noise free image and where u is a zero-mean Gaussian noise with standard deviation σ. The computer simulated image or noisy image g is then given by $g(i,j) = f(i,j) + n(i,j)$.

Contrast levels of noise free images were in the range [10%; 60%] with a step size of 10%. Background grey-level was set to 128. Concerning noisy images, standard deviation σ of the zero-mean Gaussian noise u was adjusted so that the signal to noise ratio (SNR) takes the values 21dB, 15dB and 9dB which simulate a low, an intermediate and a high noise level respectively. Computer simulated images consist of 256×256-pixels for microcalcification groups and nodules, and 336×336-pixels for fibres. The whole computer simulated images were coded on 256 grey levels.

Two criteria are used to test the effectiveness of the algorithm on computer simulated images. The first one, namely output to input Signal to Noise Ratio (SNR) ρ, quantifies noise suppression, and the second one, namely the mean-squared error (MSE), in addition to quantify noise removal gives also an information on structure distortion and therefore better interprets the first criterion. One can notice that parameter ρ is all the more higher as the method removes much more noise, whereas MSE parameter is all the smaller as the method denoises and preserves structures in the image.

Results obtained on simulated images show that the trigonometric function $\psi(x) = \tan\left(\frac{\pi}{4}x\right)$ gives the best balance between noise reduction and edge sharpness preservation and that $\psi(x) = \sqrt{x}$ is suitable for contrast enhancement.

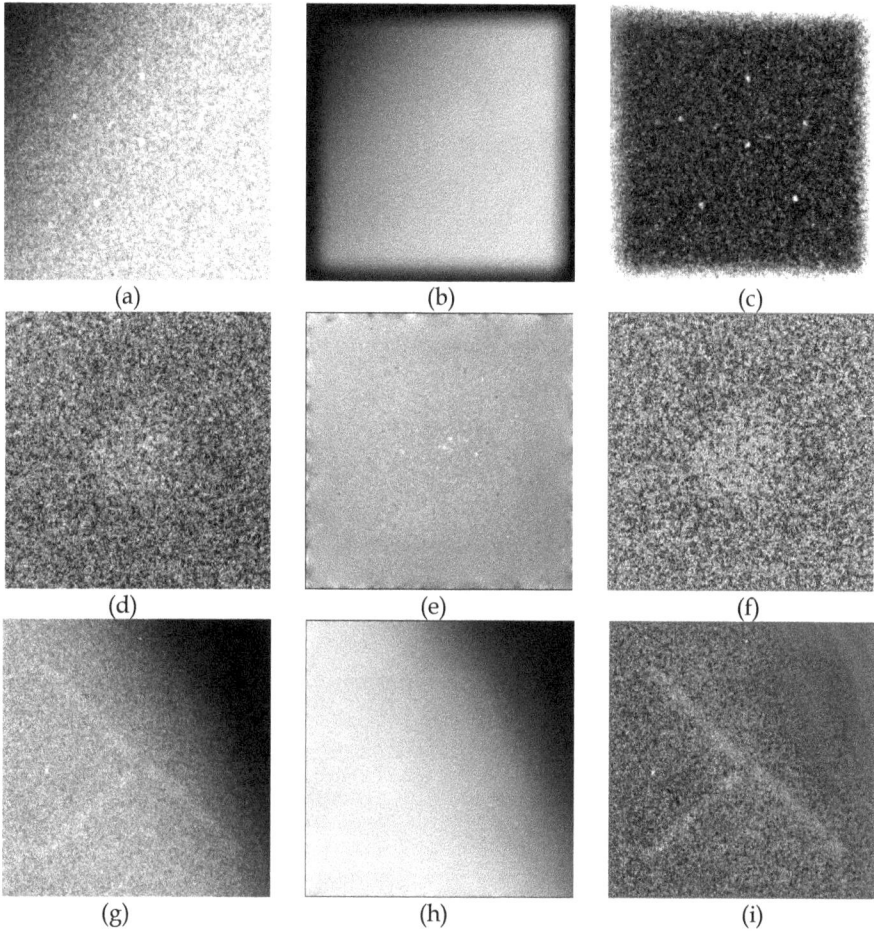

Fig. 5. Correction of Background inhomogeneity on extracted subimages. (a), (d) and (g) : original extracted subimages; (b), (e) and (h) background images; (c), (f) and (i) corrected images.

3.2.2 Preprocessing of real phantom images

Before applying noise reduction and contrast enhancement steps, inhomogeneous background of subimages containing microcalcification is extracted using the classical multiresolution Burt decomposition into level 3 (level 0 is the original image). A linear interpolation is then applied to obtain background image. The same method was not suitable to correct background for subimages containing masses and fibres, due to the object to image size ratio. Using the local contrast modification method described above with a big window size for W_{max}, enabled us to obtain background image. At last corrected image for the whole objects is obtained by subtracting the background image from the original image as shown in Fig. 5.

Applying the noise reduction and contrast enhancement steps described in the previous section, yield the images shown in Fig. 6.

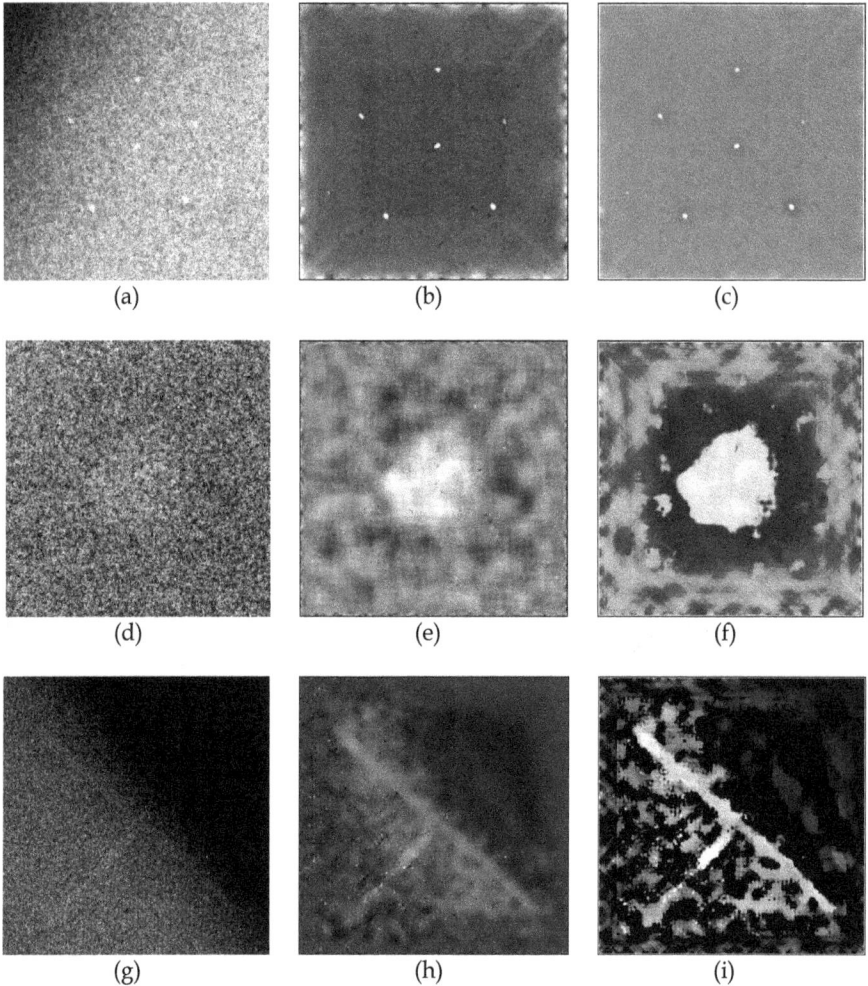

Fig. 6. Noise reduction and contrast enhancement steps on extracted subimages (a), (d) and (g): original extracted subimages; (b), (e) and (h) denoised images; (c), (f) and (i) contrast enhanced images.

4. Segmentation of extracted subimages

4.1 Microcalcification segmentation case

Microcalcifications segmentation was based on the computation of a cross-correlation between a template image M_{mic} and the preprocessed resulting image after noise reduction and contrast enhancement I_{net}. The different steps of microcalcification groups are summerized in figure 7. The template image was built after a supervised learning on real phantom images. A global thresholding was then applied on the thresholded image for extracting microcacifications. The connected component labelling step is done to determine the number of detected objects in I_{seuil} image. The microcalcification extraction step consisted in defining around each detected object, a window centered on it and using a threshold based on the computation of the mean and standard deviation of pixels within this window. Fig 8 shows an example of microcalcifications group segmentation.

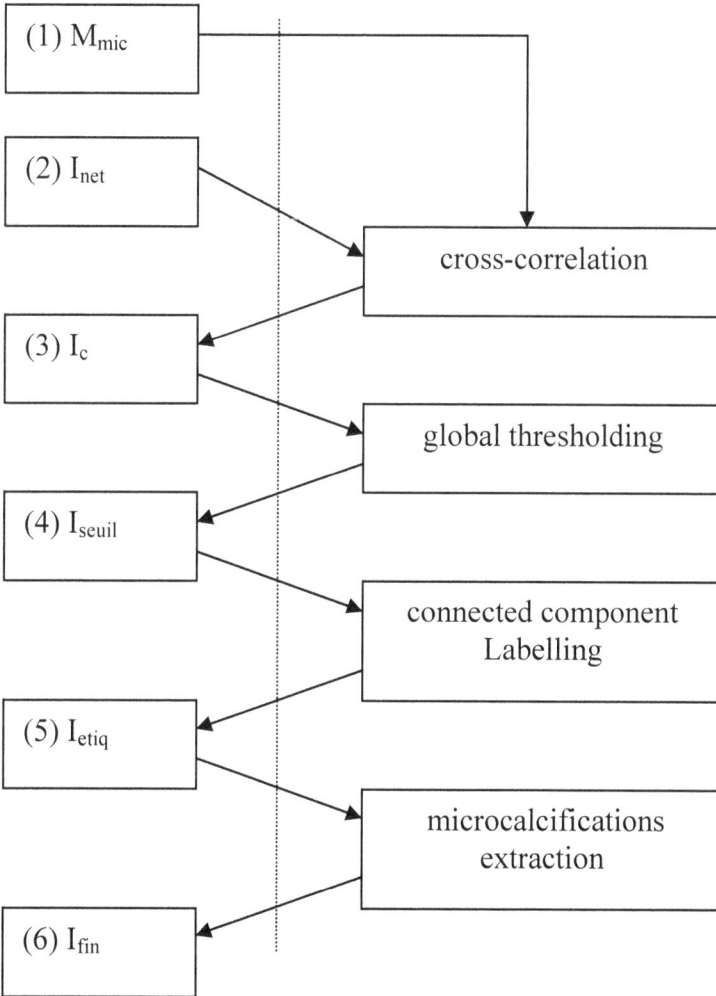

Fig. 7. General scheme of microcalcification group segmentation. (1) M_{mic}: template image, (2) I_{net} : resulting image after noise reduction and contrast enhancement, (3) I_c : resulting image after cross-correlation,(4) I_{seuil} : thresholded cross-correlated image, (5) I_{etiq} : connected component labelled image, (6) I_{fin} : resulting image with detected microcalcifications.

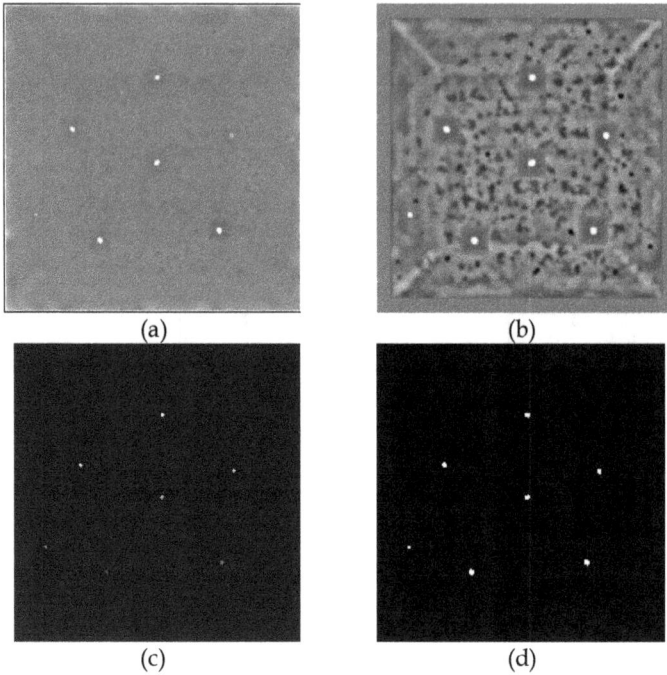

Fig. 8. Segmentation of microcalcifications. (a): extracted subimage after noise reduction and contrast enhancement I_{net} ; (b): result of template matching I_c; (c): connected component labelled image I_{etiq} ; (d): resulting image I_{fin}.

4.2 Mass segmentation case
Mass segmentation was done by using an active contour. First a square was set as an initial contour and the energy used depended only on image gradient. The algorithm used for that purpose is described as follows:
Step 1: Each point i of the active contour evolved along the normal of segment (i-1,i+1) until it met a mass edge.
Step 2: When the four initial points reached the mass edges, other points were added between each couple of points (i and i+1).
Step 3: Each added point in the previous step evolved as initials points in step 1.
The algorithm stopped when a fixed but great number of iterations was reached. Fig. 9 shows an example of a mass segmentation.

4.3 Fibre segmentation case
As for microcalcifications, fibre segmentation used a template matching between two template images (see Fig. 10) and the preprocessed resulting fibre image after applying noise reduction and contrast enhancement steps. An automatic global thresholding is then used, followed by a logical filter OR and a connected component labelling step. Figs. 11 and 12 show the general scheme of a fibre segmentation and an example of fibre segmentation respectively.

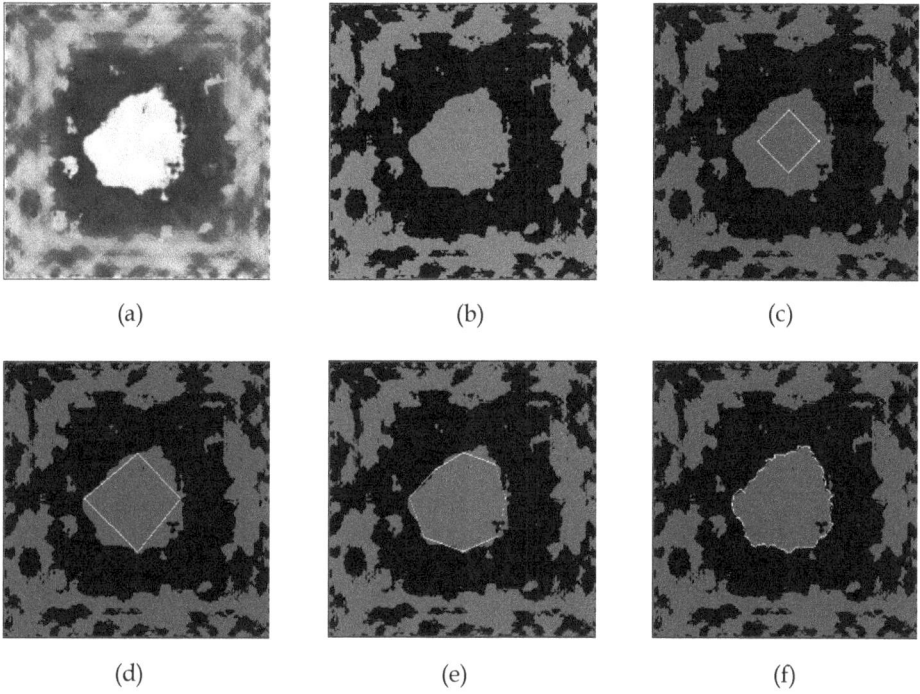

Fig. 9. Segmentation of a mass subimage (a): extracted subimage after noise reduction and contrast enhancement; (b): Thresholded image. (c): Initialization of active contour. (d) First iteration of the active contour. (e): Second iteration of active contour. (f): final segmentation.

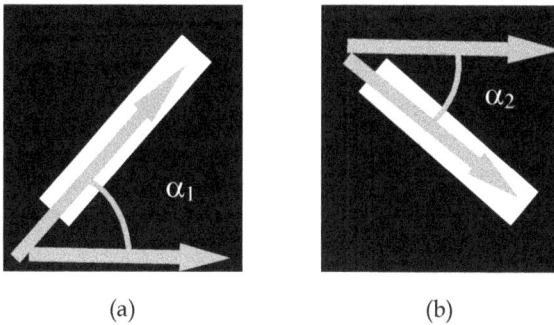

Fig. 10. Template images for fibre segmentation.

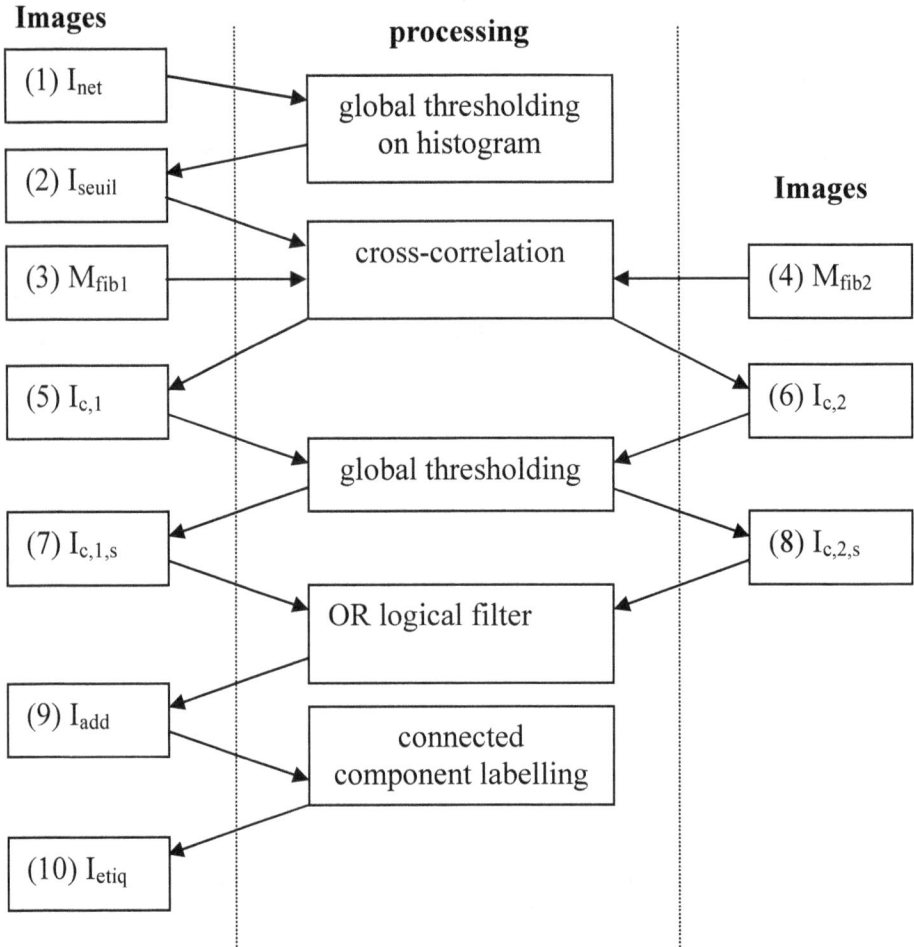

Fig. 11. General scheme of fibre segmentation. (1) I_{net} : resulting image after noise reduction and contrast enhancement , (2) I_{seuil} : image I_{net} thresholded , (3) M_{fib1} : template image 1, (4) M_{fib2} : template image 2, (5) $I_{c,1}$: obtained image after cross correlation between M_{fib1} and I_{seuil} , (6) $I_{c,2}$: obtained image after cross correlation between M_{fib2} and I_{seuil} (7) $I_{c,1,s}$: image $I_{c,1}$ thresholded , (8) $I_{c,2,s}$: image $I_{c,2}$ thresholded , (9) I_{add} : resulting image after logical filter OR between $I_{c,1,s}$ and $I_{c,2,s}$ (10) I_{etiq} : connected component labelled image.

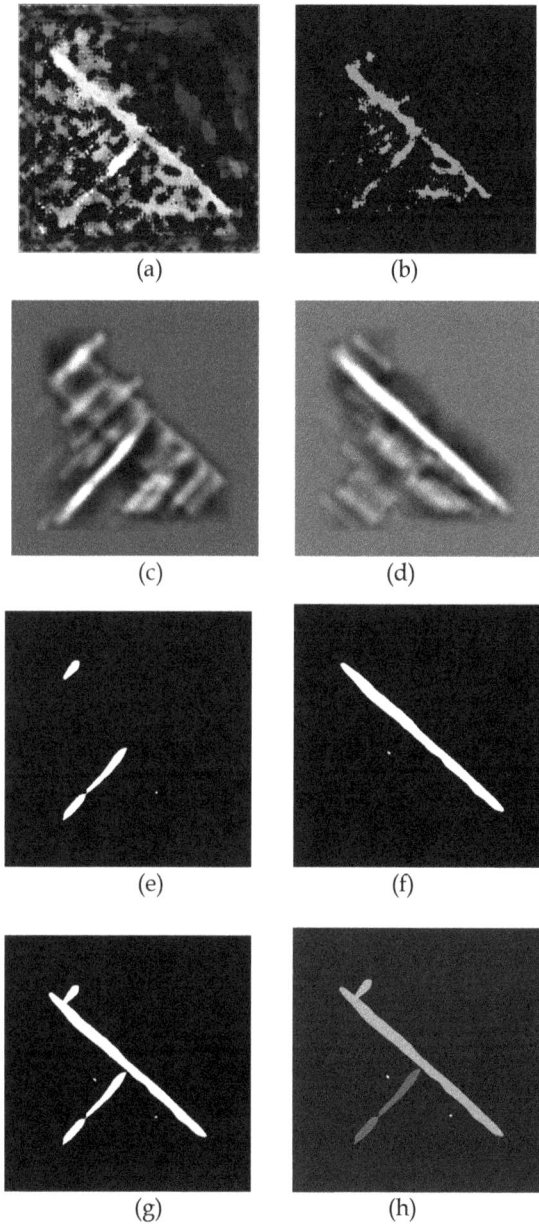

Fig. 12. Segmentation of a fibre subimage. (a): extracted subimage after noise reduction and contrast enhancement; (b): Thresholded image (c): Result of template matching with template image in Fig. 10(a); (d): Result of template matching with template image in Fig. 10(b); (e): Resulting image after thresholding image (c); (f): Resulting image after thresholding image (d); (g): Resulting image after applying OR logic filter on images (e) and (f); (h): Connect component labelled image.

5. Results and discussion

Nine phantom images from different mammographic facilities were tested. For each phantom image, only 4 subimages of each target were extracted. Two main reasons leaded us to do this choice: first, readers could not detect more than 4 objects on the phantoms used in our study, and second, a mammographic facility is considered to have good quality phantom films if at least 4 objects are detected from each embedded target.

216 microcalcifications (36 microcalcification groups) were studied. The most prominent microcalcification group M1 and M2 were almost all detected. Microcalcifications that were not detected were those with poor contrast. False detections were due to film emulsion tearing and appeared on M3 and M4 groups. Table 1 summarizes results obtained on these nine phantom films.

Thirty six masses and 36 fibres were studied. Three masses among 36 were not detected because of the non convergence of the iterative active contour algorithm. This appeared on masses containing holes when being preprocessed. When better initialising the active contour, it was possible to detect the whole masses.

The whole fibres were detected but as seen in Fig 12 some other objects appeared on the final segmentation image. These small objects will be removed in a further processing.

			1	2	3	4	5	6	7	8	9
M1 group	Detected microcalcifications number		6	6	6	6	6	6	6	6	6
	False detection										
M2 group	Detected microcalcifications number		6	6	6	6	6	6	6	6	6
	False detection		1					1		3	
M3 group	Detected microcalcifications number		6	6	6	6	5	6	5	6	5
	False detection		5	4	1	3	2				
M4 group	Detected microcalcifications number		6	6	6	6	3	6	5	4	6
	False detection		3	3	2			1			1

Table 1. Detection results on microcalcification groups.

6. Conclusion

This chapter presents a feasibility study of automating breast phantom scoring using image processing techniques. The main contribution in this project is noise reduction and contrast enhancement of noisy images extracted from digitized phantom films. The segmentation step which uses known methods shows that quality control in mammographic facilities could be done using image processing techniques. Next step in this project is to adapt image processing techniques used for digitized film to digital phantom images acquired directly from Full-Filed Digital Mammograms. In this case it will be possible to control the quality of digital mammographic systems using software similar to the one described in this study.

7. References

K. W. Brooks, J. H. Trueblood, K. J. Kearfott, D. T. Lawton "Automated analysis of the American College of Radiology mammographic accreditation phantom images". Medical Physics 24 : 907-923, 1997.

D. P. Chakraborty "Computer analysis of mammography phantom images (CAMPI). Application to the measurement of microcalcification image quality of directly acquired digital images". Medical Physics 24 : 1269-1277, 1997.

G. Dougherty "Computerised evaluation of mammographic image quality using phantom images". Computerized Medical Imaging and Graphics 22: 365-373, 1998.

A. D. Castellano Smith, I. A. Castellano Smith, D. R. Dance. "Objective assessment of phantom image quality in mammography: a feasibility study". The British Journal of Radiology 71 : 48-58, 1998

L. Blot, A. Davis, M. Holubinka ,R. Marti, R. Zwiggelaar "Automated quality assurance applied to mammographic imaging". EURASIP Journal on Applied Signal Processing 7 : 736-745, 2003

P. Mayo, F. Rodenas, G. Verdu, J. I. Villaescusa, J. M. Campayo. "Automatic evaluation of the image quality of a mammographic phantom". Computers Methods and Programs in Biomedicine 73, 115-128, 2004

A. P. Dhawan, G. Buelloni R. Gordon, "Enhancement of mammographic features by optimal adaptative neighborhood image processing," IEEE Trans. Medical Imaging 5(1), 8-15,1986.

A. P. Dhawan E. Le Royer "Mammographic feature enhancement by computarized image processing". Computer Methods and Programs in Biomedecine 27, 23-25, 1988.

V. Guis, M. Adel, M. Rasigni, G. Rasigni, B. Seradour, P. Heid "Adaptive neighborhood contrast enhancement in mammographic phantom images". Optical Engineering 42: 357-366, 2003.

W. Qian, L.P. Clarke, M. Kallergi, R. A. Clark, "Tree-Structured Nonlinear Filters in Digital Mammography". IEEE Trans. Medical Imaging 13(1), 25-36, 1994.

F. Aghadasi, R. K. Ward, B. Palcic, "Noise filtering for mammographic images", *14th Annual International Conference of the I.E.E.E. Engineering in Medecine and Biology Society* 1877-1878, 1992.

D. T. Kuan, A. A. Shawchuk, T. C. Strand, P. Chavel, "Adaptive noise smoothing filter for images with signal-dependent noise". IEEE Trans. Patt. Anal. Mach. Intell. 7(2), 165-177, 1985.

Permissions

The contributors of this book come from diverse backgrounds, making this book a truly international effort. This book will bring forth new frontiers with its revolutionizing research information and detailed analysis of the nascent developments around the world.

We would like to thank Dr. Ahmed Badr Eldin, for lending his expertise to make the book truly unique. He has played a crucial role in the development of this book. Without his invaluable contribution this book wouldn't have been possible. He has made vital efforts to compile up to date information on the varied aspects of this subject to make this book a valuable addition to the collection of many professionals and students.

This book was conceptualized with the vision of imparting up-to-date information and advanced data in this field. To ensure the same, a matchless editorial board was set up. Every individual on the board went through rigorous rounds of assessment to prove their worth. After which they invested a large part of their time researching and compiling the most relevant data for our readers. Conferences and sessions were held from time to time between the editorial board and the contributing authors to present the data in the most comprehensible form. The editorial team has worked tirelessly to provide valuable and valid information to help people across the globe.

Every chapter published in this book has been scrutinized by our experts. Their significance has been extensively debated. The topics covered herein carry significant findings which will fuel the growth of the discipline. They may even be implemented as practical applications or may be referred to as a beginning point for another development. Chapters in this book were first published by InTech; hereby published with permission under the Creative Commons Attribution License or equivalent.

The editorial board has been involved in producing this book since its inception. They have spent rigorous hours researching and exploring the diverse topics which have resulted in the successful publishing of this book. They have passed on their knowledge of decades through this book. To expedite this challenging task, the publisher supported the team at every step. A small team of assistant editors was also appointed to further simplify the editing procedure and attain best results for the readers.

Our editorial team has been hand-picked from every corner of the world. Their multi-ethnicity adds dynamic inputs to the discussions which result in innovative outcomes. These outcomes are then further discussed with the researchers and contributors who give their valuable feedback and opinion regarding the same. The feedback is then collaborated with the researches and they are edited in a comprehensive manner to aid the understanding of the subject.

Apart from the editorial board, the designing team has also invested a significant amount of their time in understanding the subject and creating the most relevant covers. They scrutinized every image to scout for the most suitable representation of the subject and create an appropriate cover for the book.

The publishing team has been involved in this book since its early stages. They were actively engaged in every process, be it collecting the data, connecting with the contributors or procuring relevant information. The team has been an ardent support to the editorial, designing and production team. Their endless efforts to recruit the best for this project, has resulted in the accomplishment of this book. They are a veteran in the field of academics and their pool of knowledge is as vast as their experience in printing. Their expertise and guidance has proved useful at every step. Their uncompromising quality standards have made this book an exceptional effort. Their encouragement from time to time has been an inspiration for everyone.

The publisher and the editorial board hope that this book will prove to be a valuable piece of knowledge for researchers, students, practitioners and scholars across the globe.

List of Contributors

Stahl, Edmundo
LatAmScience, LLC, U.S.A.

Javier Arrieta
Hospital Universitario de Basurto – Bilbao, Spain

Silvia Izquierdo Álvarez and Francisco A. Bernabeu Andreu
Servicio de Bioquímica Clínica, Hospital Universitario Miguel Servet, Zaragoza, Hospital Universitario Príncipe de Asturias, Alcalá de Henares, Madrid, Spain

Avinoam Pirogovsky
Quality Assurance and Control Committee, Medical Corps, IDF, Israel
Head of Standards and Regulation Department in the Division of Community Medicine, Ministry of Health, Tel Aviv, Israel

Amir Navon, Avi Yona, Avi Cohen and Yoram Chaiter
Quality Assurance and Control Committee, Medical Corps, IDF, Israel

Orna Tal
Israeli Center for Technology Assessment in Health Care; The Gertner Institute for Epidemiology and Health Policy Research, Head of Emerging Technologies Unit, Tel Aviv, Israel

Nachman Ash
Chief Medical Officer, Medical Corps, IDF, Israel

Yossy Machluf
Quality Assurance and Control Committee, Medical Corps, IDF, Israel
Weizmann Institute of Science, Rehovot, Israel

Elio Palma
Quality Assurance and Control Committee, Medical Corps, IDF, Israel
Head of Department of Occupational Medicine, Clalit Health Services, Afula, Israel

Mana Sezdi
Istanbul University, Turkey

Yosio Masuda
Breast cancer screening committee of Funabashi Municipal Medical Association, Japan
Masuda Clinic of breast and thyroid diseases, Japan

Taku Kato
Laboratory Section of Cytology, Funabashi Municipal Medical Center, Japan

Satoru Ishii
Radiotechnical Department, Funabashi Municipal Medical Center, Japan

Kanae Iwata
Department of Pharmacy, Funabashi Municipal Medical Center, Funabashi, Chiba, Japan

Noriyuki Tohnosu
Department of Surgery,Funabashi Municipal Medical Center, Japan
Breast cancer screening committee of Funabashi Municipal Medical Association, Japan

Jun Hasegawa
Breast cancer screening committee of Funabashi Municipal Medical Association, Japan
Funabashi Futawa Hospital, Japan

Shahid Pervez
Professor, Section of Histopathology, Department of Pathology & Microbiology, Aga Khan University Hospital Karachi, Pakistan

Sezer Saglam
Istanbul University, Oncology Institute, Department of Medical Oncology, Fatih, Istanbul, Turkey

Seyfettin Kuter
Istanbul University, Oncology Institute, Department of Medical Pyhsics, Fatih, Istanbul, Turkey

Aydin Cakir
Istanbul University, Oncology Institute, Department of Medical Pyhsics, Fatih, Istanbul, Turkey

Giuseppe Vermiglio, Giuseppe Acri, Barbara Testagrossa, Federica Causa and Maria Giulia Tripepi
Environmental, Health, Social and Industrial Department – University of Messina, Italy

Sachiko Shimizu, Rie Tomizawa, Maya Iwasa, Satoko Kasahara, Tamami Suzuki, Fumiko Wako, Ichiroh Kanaya, Kazuo Kawasaki, Atsue Ishii, Kenji Yamada and Yuko Ohno
Osaka University, Japan

Petr Košin, Jan Šavel and Adam Brož
Budweiser Budvar, N.C., Czech Republic

Sunday J. Ameh, Obiageri O. Obodozie and Karnius S. Gamaniel
Department of Medicinal Chemistry and Quality Control, National Institute for Pharmaceutical Research and Development (NIPRD), Garki, Abuja, Nigeria

Mujitaba S. Abubakar
Department of Pharmacognosy and Drug Development, Ahmadu Bello University, Zaria, Nigeria

Mag̣ạji Garba
Department of Pharmaceutical and Medicinal Chemistry, Ahmadu Bello University, Zaria, Nigeria

Francesc Centrich and Teresa Subirana
Laboratori de l'Agència de Salut Pública de Barcelona, Spain

Mercè Granados and Ramon Companyó
Universitat de Barcelona, Spain

Yefim Haim Michlin
Technion - Israel Institute of Technology, Israel

Genady Grabarnik
St' Johns University, USA

Liudmila Pavlova
Vinogradov Institute of Geochemistry, Siberian Branch of Russian Academy of Sciences, Irkutsk, Russia

David Martin, Maria C. Garcia-Alegre and Domingo Guinea
Spanish Council for Scientific Research (CSIC), Madrid, Spain

Juan C. Rodríguez-Gamboa, E. Susana Albarracín-Estrada and Edilson Delgado-Trejos
MIRP, Research Center, Instituto Tecnológico, Metropolitano (ITM), Medellín, Colombia

Mouloud Adel and Monique Rasigni
Université Paul Cézanne, Institut Fresnel, UMR-CNRS 6133, Domaine Universitaire de Saint Jérôme, France

www.ingramcontent.com/pod-product-compliance
Lightning Source LLC
Chambersburg PA
CBHW070725190326
41458CB00004B/1044